# The Design, Synthesis and Biological Evaluation of Compounds with Medicinal Value

# The Design, Synthesis and Biological Evaluation of Compounds with Medicinal Value

Editor

**Maria Stefania Sinicropi**

MDPI • Basel • Beijing • Wuhan • Barcelona • Belgrade • Manchester • Tokyo • Cluj • Tianjin

*Editor*
Maria Stefania Sinicropi
Department of Pharmacy,
Health and Nutritional Sciences,
University of Calabria
Italy

*Editorial Office*
MDPI
St. Alban-Anlage 66
4052 Basel, Switzerland

This is a reprint of articles from the Special Issue published online in the open access journal *Applied Sciences* (ISSN 2076-3417) (available at: https://www.mdpi.com/journal/applsci/special_issues/compound_medicinal_valve).

For citation purposes, cite each article independently as indicated on the article page online and as indicated below:

| |
|---|
| LastName, A.A.; LastName, B.B.; LastName, C.C. Article Title. *Journal Name* **Year**, *Volume Number*, Page Range. |

**ISBN 978-3-0365-1299-0 (Hbk)**
**ISBN 978-3-0365-1300-3 (PDF)**

Cover image courtesy of Prof. Dr. Maria Stefania Sinicropi

# Contents

# About the Editor

**Maria Stefania Sinicropi** is associate Professor of Medicinal Chemistry at the Department of Pharmacy, Health and Nutritional Science, "Dept. of Excellence", University of Calabria, since December 2006.

The main focus of her research activity concerns the following: the study of new molecules as anticancer drugs; the design, synthesis and biological evaluation of compounds of pharmaceutical interest; the study and biological evaluation of nutraceuticals. Currently, she is the scientific coordinator of the 2nd level Master degree in "Nutrition and Nutraceutical Supplements" at the University of Calabria. She has coordinated many research projects and is the teacher of some courses in the pharmaceutical field. She has trained several PhD students and postdoctoral fellows and collaborates with numerous national and international research groups in the field of medicinal chemistry and nutraceutics.

applied sciences

MDPI

Editorial

# Special Issue on "The Design, Synthesis and Biological Evaluation of Compounds with Medicinal Value"

**Maria Stefania Sinicropi**

Department of Pharmacy, Health and Nutritional Sciences, University of Calabria, 87036 Arcavacata di Rende (CS), Italy; s.sinicropi@unical.it

check for updates

**Citation:** Sinicropi, M.S. Special Issue on "The Design, Synthesis and Biological Evaluation of Compounds with Medicinal Value". *Appl. Sci.* **2021**, *11*, 1904. https://doi.org/10.3390/app11041904

Received: 10 February 2021
Accepted: 17 February 2021
Published: 22 February 2021



## 1. Introduction

During the last few decades, in industrialized countries a significant increase in infectious, cardiovascular, inflammatory and neurodegenerative diseases was registered, as well as different forms of cancer, diabetes, and so on [1–4]. Among them, microbial infections and cancer are still the major causes of death among the world population due to the increased drug resistance phenomena. For these reasons, there is an urgent need to design and synthesize new antimicrobial agents, particularly active against Gram (-) pathogens, that could be used to fight drug resistance. Along with them, there is the need of new antineoplastic drugs with higher selectivity on tumoral cells, which are able to overcome cancer cell resistance with moderate side effects.

Recently, some delivery systems have been proved particularly effective as antimicrobial and anticancer carriers due to the targeted drug delivery at the action sites, reduced drug-resistance and side effects, and the increased therapeutic index.

In this scenario, the Special Issue "*The design, synthesis and biological evaluation of compounds with medicinal value*", which includes the 13 latest original papers, will allow us to deepen and develop innovative research regarding the above mentioned relevant topics.

## 2. Promising Scaffolds in Medicinal Chemistry

The study of privileged scaffolds in medicinal chemistry supplies scientists with a solid start in the search for new and improved therapeutic molecules. In this Special Issue, in which 39 papers were submitted and 13 of them were published, some interesting and promising scaffolds with antimicrobial, antitumor, and antioxidant properties and with the ability to act on various cellular targets are reported.

In particular, three papers focus on the antimicrobial properties of new synthesized compounds. In the first one, Rastija V. et al. [5] reported a series of 4,5–dihydro-1*H*-pyrazole derivatives that have been tested for their antibacterial properties towards four bacterial strains: *Escherichia coli*, *Pseudomonas aeruginosa*, *Bacillus subtilis* and *Staphylococcus aureus*. Among them, 5-(2-chlorophenyl)-3-(4-fluorophenyl)-4,5–dihydro-1*H*-pyrazole-1-carbaldehyde has been shown to be a strong inhibitor of the monophenolase activity of mushroom tyrosinases. For the prediction of the activity for phosphodiesterase inhibition, docking studies have also been performed.

The second paper, authored by Spina C. et al. [6], concerns the metal complexes that have recently been considered with success for many applications in medicinal chemistry [7–11].

Particularly, silver complexes were employed as antimicrobial agents [12], and, recently, novel silver compounds in higher oxidation states, Ag(II) and Ag(III), have emerged as desirable alternatives to existing forms of antimicrobial agents. In this context, Spina C. et al. present a facile and one-pot strategy for the preparation of a higher oxidation state silver-silica gel, $Ag_7NO_{11}$:$SiO_2$, framework based on the direct oxidation of silver nitrate from an oxidizing alkali silicate aqueous solution. The corresponding characterization, thermal stability, aqueous degradation, and antimicrobial efficacy of the $Ag_7NO_{11}:SiO_2$ framework have also been evaluated [6].

Urease is a therapeutic target associated with several important diseases and health problems [13,14].

Ali M. et al. in the paper "Enamine Barbiturates and Thiobarbiturates as a New Class of Bacterial Urease Inhibitors" [15], exploiting the privileged structure assigned to the (thio)barbiturate (pyrimidine) scaffold, tested the capacity of two (thio)barbiturate-based compound collections to inhibit bacterial urease.

Various compounds have, in fact, been reported to be potential candidates for the treatment of certain clinical conditions caused by bacterial ureases [16].

Therefore, in this third paper [15], several compounds endowed with higher activity than acetohydroxamic acid, used as a standard tested compound, have been reported. Thanks to the conformational and docking studies, energetically low-lying conformers and the binding mechanism of these new pyrimidine derivatives have been identified as urease inhibitors.

Cancer is a high-incidence and life-threatening disease. Nowadays, high-income countries (HIC) continue to have the highest incidence for lung, colorectal, breast, and prostate cancer, although some low- and middle-income countries (LMIC) now count among those with the highest rates [17,18].

One of the major challenges for relieving its burden is to develop highly effective drugs with few/no side effects on normal mammalian cells [19–21].

In this area, three other papers focus on the anticancer activity of heterocyclic molecules.

In particular, in the first article [22], a series of novel N-benzylisatin-aryl-hydrazones was designed, synthesized and evaluated for antimicrobial and antiproliferative activities with SAR and ADME studies, aiming to develop anticancer drugs with no antimicrobial, yet high antiproliferative activities. These synthesized hydrazones, indeed, revealed no effects on any of the strains of bacteria and fungi up to 100-μg/disc concentration. However, four compounds showed two-to-four fold antiproliferative activity over Gefitinib [22].

The imidazole ring is an important scaffold in medicinal chemistry, as several derivatives have shown a wide array of biological activities [23–25]. A second paper [26] reports a series of 2,4,5-triphenyl imidazole derivatives for their activities as antiproliferative, antioxidant, AChE, and XO inhibitor compounds, designed in order to find new leads with these biological profiles. To confirm the in vitro evaluations, molecular docking and in silico analysis of their ADME properties were made. These studies allowed the selection of the best candidates and set the path for studies on new drug families [26].

Another paper, authored by Islam M. S. et al., [27] reported the synthesis of highly functionalized spirooxindoles analogues, via a single step, multicomponent, one-pot reaction, which is a very versatile method in medicinal chemistry [28].

These compounds were then tested in vitro for their antiproliferative effects against three cancer cell lines, namely, HepG2 (liver cancer), MCF-7 (breast cancer), and HCT-116 (colon cancer). A spirooxindole of the series exhibited broad activity against this panel of cell lines when compared to *cis*platin. Modeling studies, including shape similarity, lipophilicity scores, and physicochemical parameters were calculated [27].

Another paper [29], in the area of the drug delivery systems, concerns the development of a sol–gel-based coating used as an entrapping polymeric cross-linked network for an N-palmitoyl-ethanolamine (PEA) derivative, 2-methyl-pentadecanoic acid (4-nitro-phenyl)-amide or N-palmitoyl-(4-nitro-phenyl)-amine (PNPA), with anti-inflammatory and antioxidant properties.

Finally, Iacopetta D. et al. [30], in the context of supercritical fluid technology and as an innovative method to extract nutraceuticals from natural matrices, reported the extraction of nutraceuticals as polyphenolic compounds from plant matrices, such as the cactus pear, able to prevent and treat several chronic-degenerative diseases.

## 3. Nutraceuticals: A New Challenge for Medicinal Chemistry

In recent years, there is a growing interest in nutraceuticals for their health promoting or disease-preventing effects. Various nutrient, herbal and dietary supplements able to act against various disease conditions and thus promote quality of life are studied and reported in the literature [31–35].

In the Special Issue "*The Design, Synthesis, and Biological Evaluation of Compounds with Medicinal Value*", three reviews have been published in this area: one concerning the health-promoting properties of pomegranate and its bioactive compounds against principal human pathologies [36], another one that deepened the beneficial effects of β-Caryophyllene, a natural bicyclic sesquiterpene [37], and the last one that systematically analyzed the effects of Tinospora cordifolia-derived phytocomponents on cancer [38].

## 4. Others

In the Special Issue, other interesting topics have been developed.

Surgical site infection (SSI) is a frequent complication of surgical procedures. Ceresoli M. et al. [39] have critically analyzed the role of triclosan-coated sutures (TCS) on SSI prevention in abdominal surgery. A cost analysis to provide a more comprehensive representation of the value of this technology in clinical practice has also been reported.

Sender-Janeczek et al. [40] have reported new local drug delivery with antibiotics in the nonsurgical treatment of periodontitis. The effectiveness of the activity of the piperacillin and tazobactam combination in the form of an intrapocket-administered chemotherapy has been assessed.

**Funding:** This research received no external funding.

**Institutional Review Board Statement:** Not applicable.

**Informed Consent Statement:** Not applicable.

**Data Availability Statement:** Not applicable.

**Acknowledgments:** I would like to thank all the talented authors and the careful and skilled peer reviewers for their valuable contributions to this Special Issue. It was a pleasure to work with all of you. Thanks to all components of the Editorial team of Applied Sciences. This issue would not be possible without their help, experience and expertise. A special thanks to Tamia Qing, Assistant Editor from MDPI, who always provided me with valid and profitable support with kindness and devotion. Congratulations to all the authors!

**Conflicts of Interest:** The authors declare no conflict of interest.

## References

1. Quante, A.S.; Ming, C.; Rottmann, M.; Engel, J.; Boeck, S.; Heinemann, V.; Westphalen, C.B.; Strauch, K. Projections of cancer incidence and cancer-related deaths in Germany by 2020 and 2030. *Cancer Med.* **2016**, *5*, 2649–2656. [CrossRef] [PubMed]
2. Ghoncheh, M.; Pournamdar, Z.; Salehiniya, H. Incidence and Mortality and Epidemiology of Breast Cancer in the World. *Asian Pac. J. Cancer Prev.* **2016**, *17*, 43–46. [CrossRef] [PubMed]
3. Checkoway, H.; Lundin, J.I.; Kelada, S.N. Neurodegenerative diseases. *Iarc. Sci. Publ.* **2011**, 407–419.
4. Maffi, P.; Secchi, A. The Burden of Diabetes: Emerging Data. *Dev. Ophthalmol.* **2017**, *60*, 1–5.
5. Rastija, V.; Brahmbhatt, H.; Molnar, M.; Loncaric, M.; Strelec, I.; Komar, M.; Pavic, V. Synthesis, Tyrosinase Inhibiting Activity and Molecular Docking of Fluorinated Pyrazole Aldehydes as Phosphodiesterase Inhibitors. *Appl. Sci.* **2019**, *9*, 1704. [CrossRef]
6. Spina, C.J.; Ladhani, R.; Goodall, C.; Hay, M.; Precht, R. Directed Silica Co-Deposition by Highly Oxidized Silver: Enhanced Stability and Versatility of Silver Oxynitrate. *Appl. Sci.* **2019**, *9*, 5236. [CrossRef]
7. Ceramella, J.; Mariconda, A.; Iacopetta, D.; Saturnino, C.; Barbarossa, A.; Caruso, A.; Rosano, C.; Sinicropi, M.S.; Longo, P. From coins to cancer therapy: Gold, silver and copper complexes targeting human topoisomerases. *Bioorg. Med. Chem. Lett.* **2020**, *30*, 126905. [CrossRef]
8. Iacopetta, D.; Mariconda, A.; Saturnino, C.; Caruso, A.; Palma, G.; Ceramella, J.; Muia, N.; Perri, M.; Sinicropi, M.S.; Caroleo, M.C.; et al. Novel Gold and Silver Carbene Complexes Exert Antitumor Effects Triggering the Reactive Oxygen Species Dependent Intrinsic Apoptotic Pathway. *Chem. Med. Chem.* **2017**, *12*, 2054–2065. [CrossRef] [PubMed]

9. Chimento, A.; Saturnino, C.; Iacopetta, D.; Mazzotta, R.; Caruso, A.; Plutino, M.R.; Mariconda, A.; Ramunno, A.; Sinicropi, M.S.; Pezzi, V.; et al. Inhibition of human topoisomerase I and II and anti-proliferative effects on MCF-7 cells by new titanocene complexes. *Bioorg. Med. Chem.* **2015**, *23*, 7302–7312. [CrossRef]
10. Saturnino, C.; Barone, I.; Iacopetta, D.; Mariconda, A.; Sinicropi, M.S.; Rosano, C.; Campana, A.; Catalano, S.; Longo, P.; Ando, S. N-heterocyclic carbene complexes of silver and gold as novel tools against breast cancer progression. *Future Med. Chem.* **2016**, *8*, 2213–2229. [CrossRef]
11. Sirignano, E.; Saturnino, C.; Botta, A.; Sinicropi, M.S.; Caruso, A.; Pisano, A.; Lappano, R.; Maggiolini, M.; Longo, P. Synthesis, characterization and cytotoxic activity on breast cancer cells of new half-titanocene derivatives. *Bioorg. Med. Chem. Lett.* **2013**, *23*, 3458–3462. [CrossRef]
12. Dhanyalayam, D.; Scrivano, L.; Parisi, O.I.; Sinicropi, M.S.; Fazio, A.; Saturnino, C.; Plutino, M.R.; Cristo, F.D.; Puoci, F.; Cappello, A.R.; et al. Biopolymeric self-assembled nanoparticles for enhanced antibacterial activity of Ag-based compounds. *Int. J. Pharm.* **2017**, *517*, 395–402. [CrossRef] [PubMed]
13. Marien, T.; Miller, N.L. Treatment of the Infected Stone. *Urol. Clin. North Am.* **2015**, *42*, 459–472. [CrossRef]
14. Li, W.Y.; Ni, W.W.; Ye, Y.X.; Fang, H.L.; Pan, X.M.; He, J.L.; Zhou, T.L.; Yi, J.; Liu, S.S.; Zhou, M.; et al. N-monoarylacetothioureas as potent urease inhibitors: Synthesis, SAR, and biological evaluation. *J. Enzym. Inhib. Med. Chem.* **2020**, *35*, 404–413. [CrossRef] [PubMed]
15. Ali, M.; Barakat, A.; El-Faham, A.; Al-Majid, A.M.; Yousuf, S.; Ashraf, S.; Ul-Haq, Z.; Choudhary, M.I.; de la Torre, B.G.; Albericio, F. Enamine Barbiturates and Thiobarbiturates as a New Class of Bacterial Urease Inhibitors. *Appl. Sci.* **2020**, *10*, 3523. [CrossRef]
16. Hameed, A.; Al-Rashida, M.; Uroos, M.; Qazi, S.U.; Naz, S.; Ishtiaq, M.; Khan, K.M. A patent update on therapeutic applications of urease inhibitors (2012-2018). *Expert. Opin. Pat.* **2019**, *29*, 181–189. [CrossRef]
17. Torre, L.A.; Siegel, R.L.; Ward, E.M.; Jemal, A. Global Cancer Incidence and Mortality Rates and Trends–An Update. *Cancer Epidemio.l Biomark. Prev.* **2016**, *25*, 16–27. [CrossRef]
18. Ferlay, J.; Colombet, M.; Soerjomataram, I.; Dyba, T.; Randi, G.; Bettio, M.; Gavin, A.; Visser, O.; Bray, F. Cancer incidence and mortality patterns in Europe: Estimates for 40 countries and 25 major cancers in 2018. *Eur. J. Cancer* **2018**, *103*, 356–387. [CrossRef] [PubMed]
19. Sinicropi, M.S.; Iacopetta, D.; Rosano, C.; Randino, R.; Caruso, A.; Saturnino, C.; Muia, N.; Ceramella, J.; Puoci, F.; Rodriquez, M.; et al. N-thioalkylcarbazoles derivatives as new anti-proliferative agents: Synthesis, characterisation and molecular mechanism evaluation. *J. Enzym. Inhib. Med. Chem.* **2018**, *33*, 434–444. [CrossRef]
20. Iacopetta, D.; Carocci, A.; Sinicropi, M.S.; Catalano, A.; Lentini, G.; Ceramella, J.; Curcio, R.; Caroleo, M.C. Old Drug Scaffold, New Activity: Thalidomide-Correlated Compounds Exert Different Effects on Breast Cancer Cell Growth and Progression. *Chem. Med. Chem.* **2017**, *12*, 381–389. [CrossRef] [PubMed]
21. Rizza, P.; Pellegrino, M.; Caruso, A.; Iacopetta, D.; Sinicropi, M.S.; Rault, S.; Lancelot, J.C.; El-Kashef, H.; Lesnard, A.; Rochais, C.; et al. 3-(Dipropylamino)-5-hydroxybenzofuro[2,3-f]quinazolin-1(2H)-one (DPA-HBFQ-1) plays an inhibitory role on breast cancer cell growth and progression. *Eur. J. Med. Chem.* **2016**, *107*, 275–287. [CrossRef] [PubMed]
22. Al-Salem, H.S.; Abuelizz, H.A.; Issa, I.S.; Mahmoud, A.Z.; AlHoshani, A.; Arifuzzaman, M.; Rahman, A.F.M.M. Synthesis of Novel Potent Biologically Active N-Benzylisatin-Aryl Hydrazones in Comparison with Lung Cancer Drug 'Gefitinib'. *Appl. Sci.* **2020**, *10*, 3669. [CrossRef]
23. De Luca, L. Naturally occurring and synthetic imidazoles: Their chemistry and their biological activities. *Curr Med Chem* **2006**, *13*, 1–23. [PubMed]
24. Heravi, M.M.; Daraie, M.; Zadsirjan, V. Current advances in the synthesis and biological potencies of tri- and tetra-substituted 1H-imidazoles. *Mol. Divers.* **2015**, *19*, 577–623. [CrossRef]
25. Achar, K.C.; Hosamani, K.M.; Seetharamareddy, H.R. In-vivo analgesic and anti-inflammatory activities of newly synthesized benzimidazole derivatives. *Eur. J. Med. Chem.* **2010**, *45*, 2048–2054. [CrossRef]
26. Noriega-Iribe, E.; Díaz-Rubio, L.; Estolano-Cobián, A.; Barajas-Carrillo, V.W.; Padrón, J.M.; Salazar-Aranda, R.; Díaz-Molina, R.; García-González, V.; Chávez-Santoscoy, R.A.; Chávez, D.; et al. In Vitro and In Silico Screening of 2,4,5-Trisubstituted Imidazole Derivatives as Potential Xanthine Oxidase and Acetylcholinesterase Inhibitors, Antioxidant, and Antiproliferative Agents. *Appl. Sci.* **2020**, *10*, 2889. [CrossRef]
27. Islam, M.S.; Al-Majid, A.M.; El-Senduny, B.F.; Badria, F.A.; Rahman, M.A.F.M.; Barakat, A.; Elshaier, Y.A.M.M. Synthesis, Anticancer Activity, and Molecular Modeling of New Halogenated Spiro[pyrrolidine-thiazolo-oxindoles] Derivatives. *Appl. Sci.* **2020**, *10*, 2170. [CrossRef]
28. Caruso, A.; Lancelot, J.C.; El-Kashef, H.; Sinicropi, M.S.; Legay, R.; Lesnard, A.; Rault, S. A rapid and versatile synthesis of novel pyrimido[5,4-b]carbazoles. *Tetrahedron* **2009**, *65*, 10400. [CrossRef]
29. Puoci, F.; Saturnino, C.; Trovato, V.; Iacopetta, D.; Piperopoulos, E.; Triolo, C.; Bonomo, M.G.; Drommi, D.; Parisi, O.I.; Milone, C.; et al. Sol–Gel Treatment of Textiles for the Entrapping of an Antioxidant/Anti-Inflammatory Molecule: Functional Coating Morphological Characterization and Drug Release Evaluation. *Appl. Sci.* **2020**, *10*, 2287. [CrossRef]
30. Iacopetta, D.; Baldino, N.; Caruso, A.; Perri, V.; Lupi, F.R.; de Cindio, B.; Gabriele, D.; Sinicropi, M.S. Nutraceuticals Obtained by SFE-CO2 from Cladodes of Two Opuntia ficus-indica (L.) Mill Wild in Calabria. *Appl. Sci.* **2021**, *11*, 477. [CrossRef]

31. Ceramella, J.; Loizzo, M.R.; Iacopetta, D.; Bonesi, M.; Sicari, V.; Pellicano, T.M.; Saturnino, C.; Malzert-Freon, A.; Tundis, R.; Sinicropi, M.S. Anchusa azurea Mill. (Boraginaceae) aerial parts methanol extract interfering with cytoskeleton organization induces programmed cancer cells death. *Food Funct.* **2019**, *10*, 4280–4290. [CrossRef] [PubMed]
32. Tundis, R.; Iacopetta, D.; Sinicropi, M.S.; Bonesi, M.; Leporini, M.; Passalacqua, N.G.; Ceramella, J.; Menichini, F.; Loizzo, M.R. Assessment of antioxidant, antitumor and pro-apoptotic effects of Salvia fruticosa Mill. subsp. thomasii (Lacaita) Brullo, Guglielmo, Pavone & Terrasi (Lamiaceae). *Food Chem. Toxicol.* **2017**, *106*, 155–164. [PubMed]
33. Iacopetta, D.; Grande, F.; Caruso, A.; Mordocco, R.A.; Plutino, M.R.; Scrivano, L.; Ceramella, J.; Muia, N.; Saturnino, C.; Puoci, F.; et al. New insights for the use of quercetin analogs in cancer treatment. *Future Med. Chem.* **2017**, *9*, 2011–2028. [CrossRef] [PubMed]
34. Grande, F.; Parisi, O.I.; Mordocco, R.A.; Rocca, C.; Puoci, F.; Scrivano, L.; Quintieri, A.M.; Cantafio, P.; Ferla, S.; Brancale, A.; et al. Quercetin derivatives as novel antihypertensive agents: Synthesis and physiological characterization. *Eur. J. Pharm. Sci.* **2016**, *82*, 161–170. [CrossRef]
35. Cappello, A.R.; Dolce, V.; Iacopetta, D.; Martello, M.; Fiorillo, M.; Curcio, R.; Muto, L.; Dhanyalayam, D. Bergamot (Citrus bergamia Risso) Flavonoids and Their Potential Benefits in Human Hyperlipidemia and Atherosclerosis: An Overview. *Mini. Rev. Med. Chem.* **2016**, *16*, 619–629. [CrossRef]
36. Caruso, A.; Barbarossa, A.; Tassone, A.; Ceramella, J.; Carocci, A.; Catalano, A.; Basile, G.; Fazio, A.; Iacopetta, D.; Franchini, C.; et al. Pomegranate: Nutraceutical with Promising Benefits on Human Health. *Appl. Sci.* **2020**, *10*, 6915. [CrossRef]
37. Francomano, F.; Caruso, A.; Barbarossa, A.; Fazio, A.; La Torre, C.; Ceramella, J.; Mallamaci, R.; Saturnino, C.; Iacopetta, D.; Sinicropi, M.S. β-Caryophyllene: A Sesquiterpene with Countless Biological Properties. *Appl. Sci.* **2019**, *9*, 5420. [CrossRef]
38. Deepa, B.; Babaji, H.V.; Hosmani, J.V.; Alamir, A.W.H.; Mushtaq, S.; Raj, A.T.; Patil, S. Effect of Tinospora cordifolia-Derived Phytocomponents on Cancer: A Systematic Review. *Appl. Sci.* **2019**, *9*, 5147. [CrossRef]
39. Ceresoli, M.; Carissimi, F.; Piemontese, A.; Paragò, V.; Galvain, G.; Tommaselli, G.A.; Gianotti, L. The Clinical and Economic Value of Triclosan-Coated Surgical Sutures in Abdominal Surgery. *Appl. Sci.* **2020**, *10*, 1090. [CrossRef]
40. Sender-Janeczek, A.; Zborowski, J.; Szulc, M.; Konopka, T. New Local Drug Delivery with Antibiotic in the Nonsurgical Treatment of Periodontitis—Pilot Study. *Appl. Sci.* **2019**, *9*, 5077. [CrossRef]

*applied sciences*

MDPI

*Article*

# Synthesis, Tyrosinase Inhibiting Activity and Molecular Docking of Fluorinated Pyrazole Aldehydes as Phosphodiesterase Inhibitors

**Vesna Rastija [1], Harshad Brahmbhatt [2], Maja Molnar [2], Melita Lončarić [2], Ivica Strelec [2], Mario Komar [2] and Valentina Pavić [3,*]**

1. Faculty of Agrobiotechnical Sciences Osijek, Josip Juraj Strossmayer University of Osijek, Vladimira Preloga 1, 31000 Osijek, Croatia; vrastija@fazos.hr
2. Faculty of Food Technology Osijek, Josip Juraj Strossmayer University of Osijek, Franje Kuhača 20, 31000 Osijek, Croatia; brahmbhattharshad@hotmail.com (H.B.); maja.molnar@ptfos.hr (M.M.); melita.loncaric@ptfos.hr (M.L.); ivica.strelec@ptfos.hr (I.S.); mario.komar@ptfos.hr (M.K.)
3. Department of Biology, Josip Juraj Strossmayer University of Osijek, Cara Hadrijana 8/A, 31000 Osijek, Croatia

* Correspondence: vpavic@biologija.unios.hr; Tel.: +385-31-399-933

Received: 6 March 2019; Accepted: 19 April 2019; Published: 25 April 2019

**Abstract:** A series of fluorinated 4,5-dihydro-1*H*-pyrazole derivatives were synthesized in the reaction of corresponding acetophenone and different aldehydes followed by the second step synthesis of desired compounds from synthesized chalcone, hydrazine hydrate, and formic acid. Structures of all compounds were confirmed by both $^{1}$H and $^{13}$C NMR and mass spectrometry. Antibacterial properties of compounds were tested on four bacterial strains, *Escherichia coli, Pseudomonas aeruginosa, Bacillus subtilis,* and *Staphylococcus aureus.* Among synthesized compounds, the strongest inhibitor of monophenolase activity of mushroom tyrosinase (32.07 ± 3.39%) was found to be *5-(2-chlorophenyl)-3-(4-fluorophenyl)-4,5-dihydro-1H-pyrazole-1-carbaldehyde.* The PASS program has predicted the highest probable activity for the phosphodiesterase inhibition. To shed light on molecular interactions between the synthesized compounds and phosphodiesterase, all compounds were docked into the active binding site. The obtained results showed that the compound with the dimethoxyphenyl ring could be potent as an inhibitor of phosphodiesterase, which interacts in PDE5 catalytic domain of the enzyme. Key interactions are bidentate hydrogen bond (H-bond) with the side-chain of Gln817 and van der Waals interactions of the dimethoxyphenyl ring and pyrazole ring with hydrophobic clamp, which contains residuals, Val782, Phe820, and Tyr612. Interactions are similar to the binding mode of the inhibitor sildenafil, the first oral medicine for the treatment of male erectile dysfunction.

**Keywords:** fluorinated pyrazole aldehydes; tyrosinase inhibition; phosphodiesterase inhibition; antibacterial activity; molecular docking

## 1. Introduction

Pyrazoles are nitrogen-containing aromatic heterocycles possessing a five-membered ring in their structure, with two nitrogen atoms in an adjacent position [1]. Catalytic hydrogenation of pyrazoles yields 4,5-dihydro-1*H*-pyrazoles or 2-pyrazoline (Figure 1) [2].

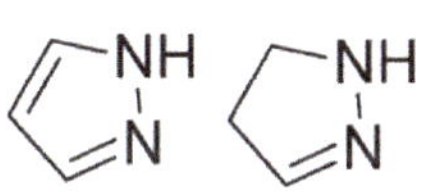


**Figure 1.** Structure of pyrazole and 4,5-dihydro-1*H*-pyrazole.

4,5-Dihydro-1*H*-pyrazoles are often synthesized in cyclocondensation of chalcones with hydrazine hydrate, while chalcones are formed in the reaction between aldehydes and acetophenones in the presence of NaOH [3], KOH [4,5]. Zhou et al. [6] synthesized chalcones in the presence of neutral $Al_2O_3$ and KOH under microwave irradiation, which in reaction with the hydrazine hydrate yielded *N*-4,5-dihydro-1*H*-pyrazoles with tyrosinase inhibiting activity.

In general, pyrazoline derivatives exhibit a variety of biological activities, depending on the substituents, which can be placed at a different position of the ring [2]. The accumulation of the fluorine on the carbon atom causes increased oxidative and thermal stability of the drugs as well as increased lipid solubility, which accelerates the drug absorption and transport in vivo [7]. Introduction of fluorine to the drug molecules can reduce their in vivo metabolic turnover by blocking potential reactive positions with fluorine and improving the stability of the molecule toward acid hydrolysis [8]. The biological activity of pyrazoline derivatives includes anticancer activity, especially fluorinated derivatives [9], anti-inflammatory activity [10,11], and anticonvulsant activity [12]. Antifertility, antibacterial, and antifungal agents are frequently fluorinated pyrazolines and pyrazoles [13]. They have also been investigated in inhibitory nNOS activity in rat brains and proven very effective [14]. Fused 4,5-dihydro-1*H*-pyrazoles were found to be an excellent antibacterial agent against *Staphylococcus aureus* and *Corynebacterium diphtheriae* [15], while 5-aryl-1-carboxamidino-3-styryl-4,5-dihydro-1*H*-pyrazoles were found to be potent antioxidants and antimicrobial agents against *Salmonella typhi, Staphylococcus aureus,* and *Streptococcus pneumoniae* [16]. An excellent antibacterial activity of 4,5-dihydro-1*H*-pyrazole derivatives was also proven against *Pseudomonas aeruginosa, Escherichia coli, Bacillus subtilis,* and *Staphylococcus aureus*, showing a potent DNA gyrase inhibitory activity as well [17]. All the described above indicates that 4,5-dihydro-1*H*-pyrazoles are very potent bioactive compounds and their structural modifications, especially the introduction of a fluorine atom on a phenyl ring, could lead to their increased bioactivity.

The growing necessity for new and potent antibiotics, due to the microorganism resistance to the existing antibiotics, has led us to synthesize pyrazole derivatives as potential antimicrobials. Thus, the aim of this work was to synthesize fluorinated 4,5-dihydro-1*H*-pyrazole derivatives in order to examine their antibacterial activity. Since 4,5-dihydro-1*H*-pyrazoles have already proven to inhibit tyrosinase activity [6], and this enzyme is responsible for human hyperpigmentation as well as browning reactions, we also investigated the compound's tyrosinase inhibiting activity. In order to indicate the other biological activity of synthesized compounds, in silico prediction according to the structural formulas was performed, as well as molecular docking to evaluated interactions of the ligand with the compatible enzyme.

## 2. Materials and Methods

### *2.1. General*

All chemicals used within this research were acquired from commercial suppliers. Aluminum TLC plates coated with fluorescent indicator F254 were used for thin-layer chromatography, with benzene:acetic acid:acetone (8:1:1) as an eluent, and checked under a UV lamp (HP-UVIS® UV-analysis lamp, Biostep GmbH –Desaga, Burkhardtsdorf, Germany) on 365 nm and 254 nm. NMR spectra were recorded on Bruker Avance 600 MHz NMR Spectrometer (Bruker Biospin GmbH, Rheinstetten, Germany) at 293 K with DMSO-$d_6$ used as a solvent and tetramethylsilane (TMS) as an internal standard. The mass spectra were recorded by LC/MS/MS API 2000 (Applied Biosystems /MDS SCIEX, Redwood City, CA, USA). Melting points were determined with electrothermal melting point apparatus (Electrothermal Engineering Ltd., Rochford, UK). Bacteria strains were isolates from various clinical specimens obtained from Microbiology Service of the Public Health Institute of Osijek-Baranja County, Osijek, Croatia.

## 2.2. General Procedure for Synthesis of Fluorinated Pyrazoles (3a–j)

All compounds were synthesized according to Rostom et al. [18]. Briefly, equimolar amounts of desired acetophenone and aldehyde were mixed together in cold methanol and 50% aqueous NaOH solution was added. The mixture was stirred for 4 h and poured over crushed ice, the product was filtered and dried.

The second step included synthesis of desired compounds from synthesized chalcone, hydrazine hydrate and formic acid under reflux for 10 h. The mixture was cooled, the obtained precipitate filtered and dried.

*3-(4-fluorophenyl)-5-(4-methoxyphenyl)-4,5-dihydro-1H-pyrazole-1-carbaldehyde* (**3a**)

Mp = 131–132 °C; Rf = 0.81; $^1$H NMR (300 MHz, ppm, DMSO-$d_6$): 8.87 (s, 1H, –CHO); 7.86–7.84 (m, 2H, arom.), 7.33–7.32 (m, 2H, arom.), 7.17–7.15 (d, 2H, arom.), 6.90–6.89 (d, 2H, arom.), 5.50–5.47 (m, 1H, CH-pyr.), 3.91–3.86 (m, 1H, –$CH_2$-pyr.), 3.73 (s, 3H, $OCH_3$), 3.22–3.18 (m, 1H, $CH_2$-pyr.); $^{13}$C NMR (DMSO-$d_6$) δ (ppm): 164.1, 162.5, 159.5, 158.6, 155.2, 133.3, 129.2, 129.1, 127.0, 115.9, 115.8, 114.0, 58.1, 55.1, 42.3; MS: *m/z* calcd for [$C_{17}H_{15}FN_2O_2$]+ ([M+H]+/M++H+Na): 298.31, found 299.0/321.20.

*3-(4-fluorophenyl)-5-(2-methoxyphenyl)-4,5-dihydro-1H-pyrazole-1-carbaldehyde* (**3b**)

Mp = 110–112 °C; Rf = 0.84; $^1$H NMR (300 MHz, ppm, DMSO-$d_6$): 8.91 (s, 1H, –CHO); 7.85–7.80 (m, 2H, arom.), 7.32–7.25 (m, 3H, arom.), 7.06–6.99 (d, 2H, arom.), 6.92–6.87 (d, 1H, arom.), 5.67–5.61 (m, 1H, CH-pyr.), 3.91–3.86 (m, 1H, –$CH_2$-pyr.), 3.80 (s, 3H, $OCH_3$), 3.09–3.02 (m, 1H, $CH_2$-pyr.); $^{13}$C NMR (DMSO-$d_6$) δ (ppm): 165.4, 162.1, 160.0, 156.4, 156.1, 129.6 (d, $J$ = 8.78 Hz), 129.3, 128.7, 127.9, 126.4, 120.8, 116.5, 116.2, 111.9, 56.3, 55.1, 41.8; *m/z* calcd for [$C_{17}H_{15}FN_2O_2$]+ ([M+H]+/M++H+Na): 298.31, found 299.20/321.10.

*5-(3-fluorophenyl)-3-(4-fluorophenyl)-4,5-dihydro-1H-pyrazole-1-carbaldehyde* (**3c**)

Mp = 157 °C; Rf = 0.85; $^1$H NMR (300 MHz, ppm, DMSO-$d_6$): 8.91 (s, 1H, –CHO); 7.86–7.84 (m, 2H, arom.), 7.42–7.39 (m, 1H, arom.), 7.33–7.31 (m, 2H, arom.), 7.13–7.08 (d, 3H, arom.), 5.58–5.56 (m, 1H, CH-pyr.), 3.95–3.90 (m, 1H, –$CH_2$-pyr.), 3.28–3.24 (m, 1H, $CH_2$-pyr.); $^{13}$C NMR (DMSO-$d_6$) δ (ppm): 164.2, 163.1, 162.5, 161.5, 159.7, 155.2, 143.9, 130.8 (d, $J$ = 8.41 Hz), 129.1 (d, $J$ = 8.41 Hz), 127.3, 121.7 (d, $J$ = 2.40 Hz), 115.9, 114.4, 114.2, 112.8, 112.7, 58.1, 42.2; MS: *m/z* calcd for [$C_{16}H_{12}F_2N_2O$]+ ([M+H]+/M++H+Na): 286.28, found 287.10/309.20.

*(E)-3-(4-fluorophenyl)-5-(4-styrylphenyl)-4,5-dihydro-1H-pyrazole-1-carbaldehyde* (**3d**)

Mp = 157 °C; Rf = 0.86; $^1$H NMR (300 MHz, ppm, DMSO-$d_6$): 8.86 (s, 1H, –CHO); 7.85–7.80 (m, 2H, arom.), 7.44–7.42 (m, 2H, arom.), 7.34–7.28 (m, 5H, arom.), 6.60–6.56 (d, 1H, styryl), 6.36–6.28 (q, 1H, styryl), 5.20–5.12 (m, 1H, CH-pyr.), 3.73–3.64 (m, 1H, –$CH_2$–pyr.), 3.29–3.22 (m, 1H, $CH_2$-pyr.); $^{13}$C NMR (DMSO-$d_6$) δ (ppm): 165.4, 162.2, 160.3, 156.1, 136.2, 135.4, 134.4, 134.1, 132.9, 131.3, 129.9, 129.6, 129.4, 129.1, 128.3, 128.1, 127.7, 127.2, 126.9, 116.5, 116.2, 57.7; *m/z* calcd for [$C_{24}H_{19}FN_2O$]+ ([M+H]+/M++H+Na): 370.42, found 371.3/393.20.

*3-(4-fluorophenyl)-5-(3-methoxyphenyl)-4,5-dihydro-1H-pyrazole-1-carbaldehyde* (**3e**)

Mp = 143–145 °C; Rf = 0.83; $^1$H NMR (300 MHz, ppm, DMSO-$d_6$): 8.91 (s, 1H, –CHO); 7.86–7.83 (m, 2H, arom.), 7.33–7.25 (m, 3H, arom.), 7.33–7.31 (m, 2H, arom.), 6.86–6.84 (m, 1H, arom.), 6.80–6.78 (m, 2H, arom.), 5.53–5.50 (m, 1H, CH-pyr.), 3.93–3.88 (m, 1H, –$CH_2$-pyr.), 3.74 (s, 3H, $OCH_3$), 3.23–3.19 (m, 1H, $CH_2$-pyr.); $^{13}$C NMR (DMSO-$d_6$) δ (ppm): 164.2, 162.5, 159.6, 159.5, 155.2, 142.8, 129.9, 129.1 (d, $J$ = 2.40 Hz), 127.3, 117.5, 115.9, 115.8, 112.7, 111.6, 58.5, 52.0, 42.4; *m/z* calcd for [$C_{17}H_{15}FN_2O_2$]+ ([M+H]+/M++H+Na): 298.31, found 299.0/321.20.

*3-(4-fluorophenyl)-5-(2-hydroxyphenyl)-4,5-dihydro-1H-pyrazole-1-carbaldehyde* (**3f**)

Mp = 204–206 °C; Rf = 0.74; $^{1}$H NMR (300 MHz, ppm, DMSO-$d_6$): 9.74 (s, 1H, OH), 8.91 (s, 1H, –CHO); 7.85–7.80 (m, 2H, arom.), 7.32–7.26 (m, 2H, arom.), 7.33–7.31 (m, 2H, arom.), 7.12–7.07 (t, 1H, arom.), 6.95–6.92 (d, 1H, arom.), 6.86–6.83 (d, 1H, arom.), 6.76–6.71 (t, 1H, arom.), 5.64–5.58 (m, 1H, CH-pyr.), 3.90–3.80 (m, 1H, –$CH_2$-pyr.), 3.12–3.05 (m, 1H, $CH_2$-pyr.); $^{13}$C NMR (DMSO-$d_6$) δ (ppm): 165.4, 162.1, 160.0, 156.1, 154.6, 129.5 (d, *J* = 8.78 Hz), 128.9, 128.1, 128.0, 127.0, 126.8, 119.3, 116.5, 116.2, 115.9, 53.4; *m/z* calcd for [$C_{16}H_{13}FN_2O_2$]+ (M++H+Na): 284.29, found 307.22.

*5-(2,5-dimethoxyphenyl)-3-(4-fluorophenyl)-4,5-dihydro-1H-pyrazole-1-carbaldehyde* (**3g**)

Mp = 153 °C; Rf = 0.82; $^{1}$H NMR (300 MHz, ppm, DMSO-$d_6$): 8.91 (s, 1H, –CHO); 7.84–7.82 (m, 2H, arom.), 7.31–7.28 (m, 2H, arom.), 6.99–6.97 (d, 1H, arom.), 6.85–6.83 (d, 1H, arom.), 6.56–6.55 (d, 1H, arom.), 5.61–5.58 (m, 1H, CH-pyr.), 3.88–3.83 (m, 1H, –$CH_2$-pyr.), 3.75 (s, 3H, $OCH_3$), 3.66 (s, 3H, $OCH_3$), 3.08–3.05 (m, 1H, $CH_2$-pyr.); $^{13}$C NMR (DMSO-$d_6$) δ (ppm): 164.1, 162.5, 159.6, 155.6, 153.1, 150.0, 130.0, 129.4, 129.0 (d, *J* = 9.61 Hz), 127.5, 115.9, 115.8, 112.7, 112.5 (d, *J* = 13.22 Hz), 56.0, 55.3, 54.6, 41.3; *m/z* calcd for [$C_{18}H_{17}FN_2O_3$]+ (M++H+Na): 328.34, found 351.20.

*5-(4-bromophenyl)-3-(4-fluorophenyl)-4,5-dihydro-1H-pyrazole-1-carbaldehyde* (**3h**)

Mp = 144 °C; Rf = 0.85; $^{1}$H NMR (300 MHz, ppm, DMSO-$d_6$): 8.89 (s, 1H, –CHO); 7.85–7.83 (m, 2H, arom.), 7.55–7.53 (d, 2H, arom.), 7.33–7.30 (m, 2H, arom.), 7.22–7.20 (d, 2H, arom.), 5.55–5.52 (m, 1H, CH-pyr.), 3.94–3.89 (m, 1H, –$CH_2$-pyr.), 3.25–3.21 (m, 1H, $CH_2$-pyr.); $^{13}$C NMR (DMSO-$d_6$) δ (ppm): 164.2, 162.5, 159.7, 155.2, 140.6, 131.6, 129.2, 129.1, 128.1, 127.3, 120.5, 115.9, 115.8, 58.1, 42.1; *m/z* calcd for [$C_{16}H_{12}BrFN_2O$]+ (M++H+Na): 347.18, found 369.00.

*5-(2-chlorophenyl)-3-(4-fluorophenyl)-4,5-dihydro-1H-pyrazole-1-carbaldehyde* (**3i**)

Mp = 125–127 °C; Rf = 0.86; $^{1}$H NMR (300 MHz, ppm, DMSO-$d_6$): 8.96 (s, 1H, –CHO); 7.87–7.82 (m, 2H, arom.), 7.53–7.50 (m, 1H, arom.), 7.34–7.28 (m, 4H, arom.), 7.17–7.14 (m, 1H, arom.), 5.79–5.73 (m, 1H, CH-pyr.), 4.06–3.96 (m, 1H, –$CH_2$-pyr.), 3.19–3.12 (m, 1H, $CH_2$-pyr.); $^{13}$C NMR (DMSO-$d_6$) δ (ppm): 165.5, 162,2, 160,1, 155.8, 138.3, 131.4, 130.2, 129.7, 129.6, 129.5, 128.2, 127.2, 116.5, 116.2, 56.9, 41.8; *m/z* calcd for [$C_{16}H_{12}ClFN_2O$]+ ([M+H]+/M++H+Na): 302.73, found 303.20/325.10.

*5-(4-(dimethylamino)phenyl)-3-(4-fluorophenyl)-4,5-dihydro-1H-pyrazole-1-carbaldehyde* (**3j**)

Mp = 168–171 °C; Rf = 0.13; $^{1}$H NMR (300 MHz, ppm, DMSO-$d_6$): 8.86 (s, 1H, –CHO); 7.86–7.83 (m, 2H, arom.), 7.33–7.30 (m, 2H, arom.), 7.05–7.04 (d, 2H, arom.), 6.68–6.66 (d, 2H, arom.), 5.43–5.40 (m, 1H, CH-pyr.), 3.87–3.82 (m, 1H, –$CH_2$-pyr.), 3.19–3.16 (m, 1H, $CH_2$-pyr.), 2.86 (s, 6H, $2CH_3$); $^{13}$C NMR (DMSO-$d_6$) δ (ppm): 164.2, 162.5, 159.4, 155.1, 149.9, 129.0, 128.9, 128.7, 127.5, 126.5, 115.9, 115.8, 112.4, 58.3, 42.2, 40.1; *m/z* calcd for [$C_{18}H_{18}FN_3O$]+ ([M+H]+/M++H+Na): 311.35, found 312.30/334.

### 2.3. Antibacterial Susceptibility Testing

Antibacterial properties against *Bacillus subtilis* and *Staphylococcus aureus* as two Gram-positive, and *Escherichia coli* and *Pseudomonas aeruginosa* as Gram-negative bacterial strains were tested for all synthesized compounds. Working cultures were grown overnight in Mueller-Hinton broth (MHB) (Fluka, BioChemica, Germany) under optimal conditions (37 °C with 50% humidity). Modified broth microdilution method [17] was used for MIC values determination as described in our previous work [18]. One hundred microliters of bacterial cultures in MHB was added to 100 µL of a serially diluted compound (250 to 0.122 µg $mL^{-1}$) in sterile TPP 96-well plates (TPP Techno Plastic Products AG Trasadingen, Switzerland). Growth control and background control were included in each plate in amounts corresponding to the highest amount in the test solution and subtracted from the results. The antibacterial standard amikacin sulphate was co-assayed under the same conditions in concentration range of 0.122–250 µg $mL^{-1}$. After incubation at 37 °C for 24 h in an atmospheric incubator with 5% $CO_2$ and 50% humidity, an additional incubation for three hours at 37 °C was performed with triphenyl tetrazolium chloride as a reducing agent indicator for microbial growth. The lowest concentrations of compound at which there was no color change or visual turbidity due to microbial growth was defined as the MIC value, derived from triplicate analyses and expressed as micrograms per milliliter.

*2.4. Tyrosinase Inhibiting Activity*

All synthesized compounds were tested for monophenolase and diphenolase inhibitory activity of mushroom tyrosinase according to a slightly modified procedure of Molnar et al. [19]. In brief, reaction mixture (1 mL) for determination of monophenolase activity contained 100 U of mushroom tyrosinase, 1 mM L-tyrosine and 100 μM inhibitor, while for diphenolase activity 0.5 mM L-DOPA instead L-tyrosine. In both cases, final phosphate buffer (pH 6.5) concentration in the reaction mixture was 100 mM, and the amount of DMSO was 1%.

IC50 value was determined for kojic acid by nonlinear regression using a dose-response inhibition model using GraphPad Prism version 7.00 for Windows (GraphPad Software, La Jolla, CA, USA).

*2.5. Docking Studies*

Selection of the biological activity for docking study was done with the help of computer program PASS [20]. Software estimates predicted activity spectrum of a compound according to the structural formulas of synthesized fluorinated pyrazole aldehydes as probable activity (Pa) and probable inactivity (Pi). Activity with Pa > Pi is considered as possible for a particular compound. Phosphodiesterase inhibition has been selected for docking study since the PASS program calculated Pa > Pi of all compounds for that activity.

The molecular docking of compounds (**3a–j**) was performed using iGEMDOCK (BioXGEM, Taiwan). Crystal coordinates of the catalytic domain of phosphodiesterase type 5 (PDE5) (PDB ID: 4OEW) in the complex with monocyclic pyrimidinones (PDB ID: 5IO) were downloaded from Protein Data Bank (PDB, https://www.rcsb.org/). The PDE5 structure was prepared, including the removal of water molecules and optimized protein structure using BIOVIA Discovery Studio 4.5 (Dassault Systèmes, San Diego, CA, USA). Avogadro 1.2.0 (University of Pittsburgh, Pittsburgh, PA, USA) was applied for optimizing the 3D structures of 28 molecules using the molecular mechanic's force field (MM+) [21]. In addition, the semiempirical PM3 method was used for geometry optimization of all structures [22].

The protein binding site was outlined according to the bounded ligand (PDB ID: 5IO) [23]. Genetic parameters were set (population size 200, generations 70, the number of a solution or poses: 2) after the preparation of the protein target and set of optimized structures of 10 fluorinated pyrazoles as ligands. Docking into the binding site and generation of protein-compound interaction profiles of electrostatic (*Elec*), hydrogen-bonding (*Hbond*), and van der Waals (*vdW*) interactions was performed for each compound in the library. Finally, by combining pharmacological interactions and energy-based scoring function, the compounds were ranked. Energy-based scoring function or total energy (*E*) is:

$$E = vdW + Hbond + Elec. \tag{1}$$

## 3. Results and Discussion

Desired 4,5-dihydro-1*H*-pyrazole derivatives were synthesized from corresponding chalcones. Chalcones were obtained in the typical aldol condensation reaction of aldehydes and acetophenones, while pyrazoles were synthesized from corresponding chalcones in the presence of hydrazine hydrate and formic acid (Figure 2). Their structures were confirmed by $^{1}$H NMR, $^{13}$C NMR, and mass spectra. All compounds show characteristic peaks for –CHO proton around 8.67 ppm, pyrazole C-5 proton peak around 5.58 ppm and pyrazole C-4 proton peaks around 3.24 ppm and 3.90 ppm. Other peaks correspond to aromatic protons (6.90–7.90 ppm) and specific substituents on the phenyl ring. Mass spectra for each compound corresponds to its molar mass. All compounds were further characterized by their melting points and $R_f$ values as indicated in the Materials and Methods Section.

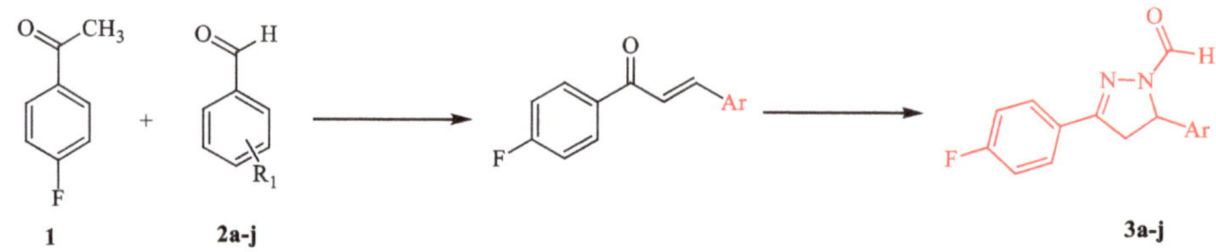


a - p-$OCH_3$-phenyl
b - 2-$OCH_3$-phenyl
c - 3-fluorophenyl
d - 4-styrylphenyl
e - 3-$OCH_3$-phenyl
f - 2-OH-phenyl
g - 2,5-$(OCH_3)_2$-phenyl
h - 4-bromophenyl
i - 2-chlorophenyl
j - 4-dimethylaminophenyl

**Figure 2.** Synthetic pathway for fluorinated pyrazole aldehydes.

After synthesis, purification, and full characterization, all compounds were investigated for their antibacterial activity against two Gram-positive and two Gram-negative bacteria (Table 1) and tyrosinase inhibiting activity (Table 2).

**Table 1.** Antibacterial activity of synthesized compounds in terms of minimum inhibitory concentration (MIC) against *Escherichia coli*, *Pseudomonas aeruginosa*, *Bacillus subtilis* and *Staphylococcus aureus* (μg $mL^{-1}$).

| Compound | Minimum Inhibitory Concentration (μg $mL^{-1}$) | | | |
|---|---|---|---|---|
| | *E. coli* | *P. aeruginosa* | *B. subtilis* | *S. aureus* |
| 3a | 62.5 | 62.5 | 62.5 | 62.5 |
| 3b | 62.5 | 62.5 | 62.5 | 62.5 |
| 3c | 62.5 | 62.5 | 62.5 | 62.5 |
| 3d | 62.5 | 62.5 | 62.5 | 62.5 |
| 3e | 62.5 | 62.5 | 62.5 | 62.5 |
| 3f | 62.5 | 62.5 | 62.5 | 62.5 |
| 3g | 62.5 | 62.5 | 125 | 125 |
| 3h | 62.5 | 62.5 | 250 | 250 |
| 3i | 62.5 | 62.5 | 250 | 250 |
| 3j | 62.5 | 62.5 | 250 | 250 |
| amikacin | 1.95 | 0.49 | 0.24 | 1.95 |

**Table 2.** Tyrosinase inhibiting activity of synthesized compounds *.

| Compound | Monophenolase Inhibition Rate (%) | Diphenolase Inhibition Rate (%) |
|---|---|---|
| 3a | 10.84 ± 1.20 | 21.08 ± 1.86 |
| 3b | 5.22 ± 1.84 | 15.98 ± 3.25 |
| 3c | 0.21 ± 1.39 | 10.11 ± 3.92 |
| 3d | 28.80 ± 4.10 | 22.81 ± 0.33 |
| 3e | 21.74 ± 2.82 | 16.55 ± 0.29 |
| 3f | 20.11 ± 1.63 | 13.67 ± 0.50 |
| 3g | 25.54 ± 2.49 | 14.34 ± 0.88 |
| 3h | 24.46 ± 3.77 | 20.79 ± 0.17 |
| 3i | 32.07 ± 3.39 | 15.11 ± 0.50 |
| 3j | 25.54 ± 4.10 | 18.38 ± 1.01 |
| Kojic acid | 100 ± 0.00 | 88.45 ± 0.83 |

* concentration of tested compound in the reaction mixture was 100 μM. Results present mean value ± standard deviation of triplicate measurements.

The antibacterial assay revealed that Gram-negative bacteria were more susceptible to the tested compounds than Gram-positive ones. The cell wall of Gram-positive bacteria ranges from 20 to 80 nm, while for Gram-negative bacteria it ranges from 1.5 to 10 nm [24,25]. Of the four tested bacteria, the Gram-positive had higher resistance against tested compounds, with up to threefold higher MIC values (62.5–250 μg $mL^{-1}$ for *B. subtilis* and *S. aureus*) than for Gram-negative bacteria (62.5 μg $mL^{-1}$ for *E. coli* and *P. aeruginosa* for all tested compounds). Electrostatic interactions can regulate the interaction of the bacterial surface with acidic and basic functional groups and various agents, which can lead to changed cell surface permeability and thus lead to the death of the cell. The target protein often tolerates the replacement of a hydrogen atom with fluorine, since it has a small atom volume. In order to increase the half-life of the drug and human exposure, many drugs on the market contain introduced fluorine atoms [26]. The introduction of a fluorine atom into a molecule can change the distribution of electrons and thus affect pKa, dipole moment, and even chemical reactivity and stability of adjacent functional groups since it is the most electronegative element. The bioavailability of the compounds can also be improved by higher membrane permeability for the compound as a result of reduced compound basicity due to the introduced fluorine [27]. Threefold higher MIC was found with the compounds **3h–j** that had various halogen (bromo/chloro) or dimethylamino group containing substituents implying that electrostatic distribution affects membrane permeation of the compound.

Determination of tyrosinase inhibitory potential revealed that none of the compounds significantly inhibited mushroom tyrosinase at 100 μM concentration, in comparison with the kojic acid as a standard inhibitor, which exhibited $IC_{50}$ of 16.96 ± 1.05 μM for monophenolase, and 13.10 ± 1.02 μM for diphenolase activity. Nevertheless, in most cases, greater tyrosinase inhibition could be observed for monophenolase than diphenolase activity. Among synthesized compounds, compound **3i** was found as the strongest inhibitor of monophenolase activity of mushroom tyrosinase (32.07 ± 3.39%), but its inhibiting activity of diphenolase activity was found twofold lower. The most probable reason for the lack of significant tyrosinase inhibitory activity of synthesized compounds is pyrazole ring *N*-substitution with aldehyde group. Zhou et al. (2013) have described that *N*-acetylation at pyrazole ring causes diminished inhibitory activity when compared to non-substituted compounds [6]. In addition, based on the report of Zhou et al. [6] it seems obvious that presence of hydroxyl groups on phenyl rings might be the prerequisite for tyrosinase inhibiting activity, which was not the case in the present study where phenyl ring A was fluorinated, and ring B had a various non-hydroxyl group containing substituents.

Experimentally proven inactivity of synthesized compounds towards mushroom tyrosinase was the motive to find another potential biological activity for these compounds. PASS online program (http://www.pharmaexpert.ru/passonline/) provides the prediction of several hundred biological activities based on structural formulas. For almost all synthesized compounds PASS program has predicted the highest *Pa* for the phosphodiesterase inhibition. All the compounds showed greater *Pa* than *Pi* (Table 3).

**Table 3.** Results of PASS program for the phosphodiesterase inhibition.

| Compound | *Pa* * | *Pi* |
|---|---|---|
| **3a** | 0.651 | 0.004 |
| **3b** | 0.584 | 0.004 |
| **3c** | 0.616 | 0.004 |
| **3d** | 0.516 | 0.004 |
| **3e** | 0.637 | 0.004 |
| **3f** | 0.396 | 0.005 |
| **3g** | 0.611 | 0.004 |
| **3h** | 0.469 | 0.004 |
| **3i** | 0.508 | 0.004 |
| **3j** | 0.460 | 0.005 |

* probable activity (*Pa*) and probable inactivity (*Pi*).

Phosphodiesterase type 5 (PDE5) is a cyclic guanosine monophosphate (cGMP-specific) enzyme and mostly expressed in smooth muscle tissue of corpus cavernosum. PDE5 inhibitors have vasodilative effects, therefore, are used for treating erectile dysfunction, pulmonary hypertension and cardiovascular diseases [28]. In order to provide virtual screening of synthesized compounds as potential inhibitors of PDE5 molecular docking study was performed. Docking score and energy of interactions between protein residue and ligand are tabulated in Table 4.

**Table 4.** Docking scores for fluorinated pyrazoles in interaction with PDE5.

| Comp. | Pose | Total Energy/kcal mol$^{-1}$ | Van Der Waals Interaction | *H Bond* | *Elec* |
|---|---|---|---|---|---|
| **3g** | 1 | −105.62 | −95.34 | −10.27 | 0 |
| **3f** | 1 | −98.66 | −75.26 | −23.40 | 0 |
| **3b** | 0 | −98.44 | −90.79 | −7.66 | 0 |
| **3d** | 0 | −96.70 | −82.39 | −14.31 | 0 |
| **3i** | 0 | −93.07 | −84.42 | −8.65 | 0 |
| **3e** | 1 | −89.18 | −84.61 | −4.58 | 0 |
| **3a** | 0 | −88.90 | −66.30 | −22.60 | 0 |
| **3c** | 0 | −88.78 | −81.78 | −7.00 | 0 |
| **3j** | 0 | −88.49 | −86.64 | −1.85 | 0 |
| **3h** | 0 | −88.09 | −70.57 | −17.52 | 0 |

According to the docking scores, compound **3g** showed the lowest total energy, which indicates it best fits into the active site of PDE5. The energy of the interactions between protein residue and ligand **3g** are tabulated in Table 5.

**Table 5.** The energy of the main interactions between protein PDE5 residue and ligand **3g.**

| H Bond | Energy | Van Der Waals Interaction | Energy |
|---|---|---|---|
| S-Gln817 | −6.81 | S-Phe820 | −18.51 |
| S-His613 | −3.41 | S-Tyr612 | −10.33 |
| | | S-Val782 | −8.70 |
| | | S-Phe786 | −7.90 |
| | | M-Leu765 | −4.20 |
| | | S-Leu765 | −4.04 |
| | | S-Met816 | −4.01 |

(M = main chain; S = side chain).

Potential surface representation of PDE5 binding site with docked compound **3g** is presented in Figure 3, while Figure 4 illustrates the interactions of ligand **3g** with receptor PDE5 in the binding site.

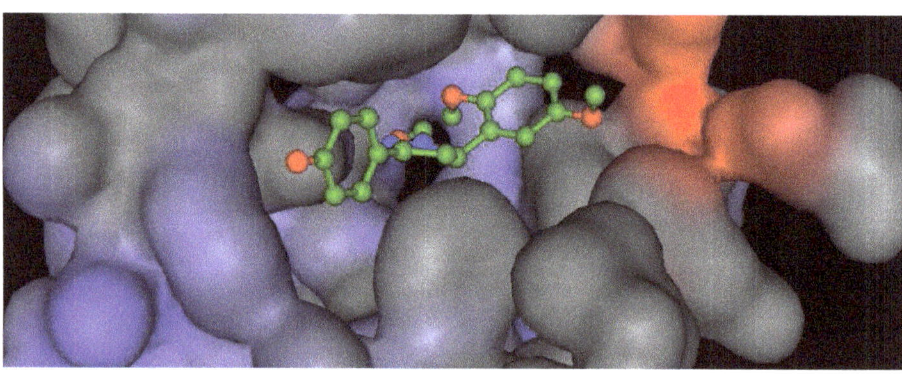

**Figure 3.** Potential surface representation catalytic domain of PDE5 with docked compound **3g**. (Range of potential: from min. 1.77 mV (blue) to max. 0.541 mV (red)).

The binding site of PDE5 was defined according to the inhibitor- - halogen derivate of monocyclic pyrimidinones (PDB ID: 5IO). Molecular docking confirmed the previous findings of characteristic binding interactions of inhibitors with the PDE5 catalytic domain [29]. Key interactions of compound **3g** are bidentate hydrogen bond (H-bond) with the side-chain of Gln817. One H-bond is formed with an oxygen atom of the carbaldehyde, and the second one with the nitrogen atom of pyrazole ring (Figure 4). Based on van der Waals interactions, pyrazole ring interacts with the side chain of Phe820. Side chain of Val782 forms $\pi$-$\pi$ interactions with pyrazole ring and dimethoxyphenyl ring. The same ring is bonded to the Tyr612 by the $\pi$-donor hydrogen bond. Interactions of fluorophenyl ring are mediated through the $\pi$-$\sigma$ interactions with Phe786 and $\pi$- sulfur interactions with Met816.

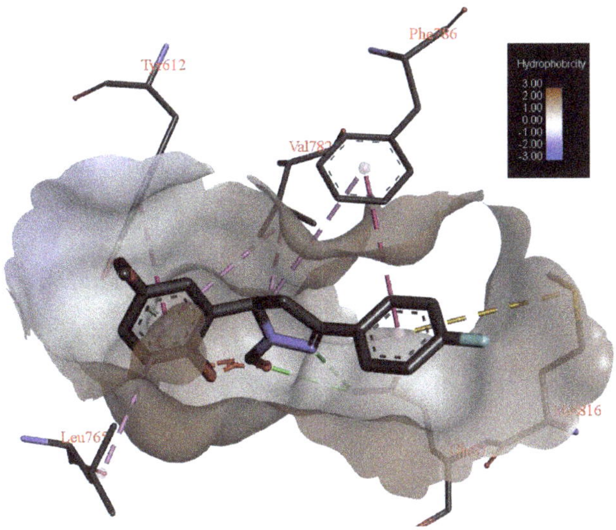


(**A**)

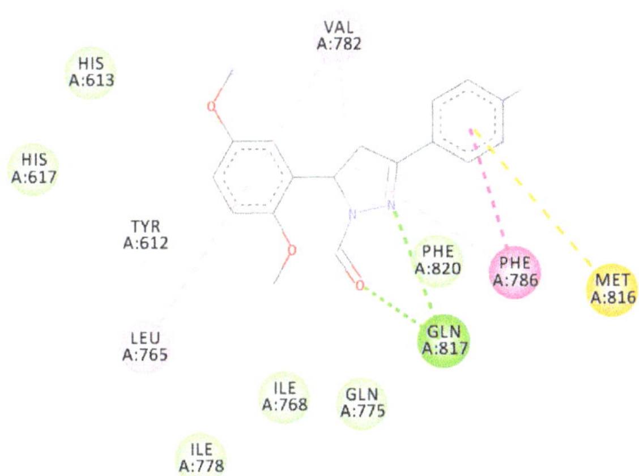


(**B**)

**Figure 4.** The main interactions of compound **3g** with residues in catalytic domain of PDE5: (**A**) 3D representation and hydrophobic surface of the binding site, (**B**) 2D representation (green = conventional hydrogen bond, light green = van der Waals, very light green = $\pi$-donor hydrogen bond, purple = $\pi$-$\sigma$ interactions, light purple = $\pi$- $\pi$ interactions, pink = alkyl and $\pi$-alkyl interactions, brown = $\pi$-sulphur bond).

Docking results of this study are in accordance with the solved crystal structure of PDE5 catalytic domain in complex with different inhibitors. The catalytic domain of PDE5 includes three subdomains: N-terminal cyclin-fold region, a linker region and a C-terminal helical bundle in which the center is an active site of PDE5 core pocket (Q pocket) of the binding site contains Gln817, Phe820, Val782, Tyr612 [30]. Typical interactions include mono or bidentate hydrogen bonds of inhibitors with Gln817 and mainly $\pi$-$\pi$ interactions of aromatic rings with hydrophobic clamp, which contains residuals, Val782, and Phe820. In the crystal structure of complex PDE5/sildenafil (PDB ID: 1UDT and 2H42) was confirmed the bidentate H-bonds are formed between the amide moiety of the pyrazolopyrimidinone of sildenafil and the side-chain of Gln817. Sildenafil (Viagra®) is a PDE5 inhibitor, which is approved as the first oral medicine for the treatments of male erectile dysfunction and for treatment of pulmonary arterial hypertension [29]. Previously mentioned ligand, 5IO [23], also formed classical bidentate H-bonds with residue Gln817, $\pi$-$\pi$ interactions of phenyl ring with Phe820 and hydrophobic interactions with residues Leu765, Val782, Ala783, and Phe786. Moreover, halogen bonding interactions, between Tyr612 and I atom have been recognized.

## 4. Conclusions

In general, Gram-positive bacteria had higher resistance against tested compounds than Gram-negative bacteria. Fluorine, as the most electronegative element, modifies electron distribution which can diversify cell surface permeability leading to cell death. Since the experiment has proven that synthesized compounds are inactive as inhibitors of tyrosinase, a docking study has been performed and has indicated that the compound with the dimethoxyphenyl ring could be potent as an inhibitor of phosphodiesterase.

**Author Contributions:** Conceptualization, M.M.; methodology, H.B., M.M., V.R., I.S., M.L., M.K. and V.P.; software, V.R.; validation, M.M., V.P. and V.R.; formal analysis, V.R. and V.P.; investigation, H.B., I.S., M.L. and M.K.; resources, M.M.; data curation, M.M., V.P., and V.R.; writing—original draft preparation, V.P.; writing—review and editing, V.P., V.R.; visualization, M.M.; supervision, V.P.; project administration, V.R.; funding acquisition, V.P.

**Funding:** This research received no external funding.

**Conflicts of Interest:** The authors declare no conflict of interest.

## References

1. Alam, J.; Alam, O.; Alam, P.; Naim, M.J. A Review on pyrazole chemical entity and biological activity. *Int. J. Pharm. Sci. Res.* **2015**, *6*, 1433–1442.
2. Alex, J.M.; Kumar, R. 4,5-Dihydro-1H-pyrazole: An indispensable scaffold. *J. Enzyme Inhib. Med. Chem.* **2014**, *29*, 427–442. [CrossRef] [PubMed]
3. Liu, P.; Hao, J.-W.; Mo, L.-P.; Zhang, Z.-H. Recent advances in the application of deep eutectic solvents as sustainable media as well as catalysts in organic reactions. *RSC Adv.* **2015**, *5*, 48675–48704. [CrossRef]
4. Shingare, R.M.; Patil, Y.S.; Gadekar, S.; Sangshetti, J.N.; Madje, B.R. Synthesis and antibacterial screening of novel 1,3,5-triaryl-4,5-dihydro-1H-pyrazole derivatives. *Morocc. J. Chem.* **2017**, *5*, 177–185.
5. Zhao, M.-Y.; Yin, Y.; Yu, X.-W.; Sangani, C.B.; Wang, S.-F.; Lu, A.-M.; Yang, L.-F.; Lv, P.-C.; Jiang, M.-G.; Zhu, H.-L. Synthesis, biological evaluation and 3D-QSAR study of novel 4,5-dihydro-1H-pyrazole thiazole derivatives as BRAFV600E inhibitors. *Bioorg. Med. Chem.* **2015**, *23*, 46–54. [PubMed]
6. Zhou, Z.; Zhuo, J.; Yan, S.; Ma, L. Design and synthesis of 3,5-diaryl-4,5-dihydro-1H-pyrazoles as new tyrosinase inhibitors. *Bioorg. Med. Chem.* **2013**, *21*, 2156–2162. [CrossRef] [PubMed]
7. Strunecká, A.; Patočka, J.; Connett, P. Fluorine in medicine. *J. Appl. Biomed.* **2004**, *2*, 141–150. [CrossRef]
8. Park, B.K.; Kitteringham, N.R.; O'Neill, P.M. Metabolism of fluorine-containing drugs. *Annu. Rev. Pharmacol. Toxicol.* **2001**, *41*, 443–470. [CrossRef]
9. Banday, A.H.; Mir, B.P.; Lone, I.H.; Suri, K.A.; Kumar, H.M.S. Studies on novel D-ring substituted steroidal pyrazolines as potential anticancer agents. *Steroids* **2010**, *75*, 805–809. [CrossRef] [PubMed]
10. Barsoum, F.F.; Girgis, A.S. Facile synthesis of bis(4,5-dihydro-1H-pyrazole-1-carboxamides) and their thio-analogues of potential PGE(2) inhibitory properties. *Eur. J. Med. Chem.* **2009**, *44*, 2172–2177. [CrossRef]

11. Bandgar, B.P.; Adsul, L.K.; Chavan, H.V.; Jalde, S.S.; Shringare, S.N.; Shaikh, R.; Meshram, R.J.; Gacche, R.N.; Masand, V. Synthesis, biological evaluation, and docking studies of 3-(substituted)-aryl-5-(9-methyl-3-carbazole)-1H-2-pyrazolines as potent anti-inflammatory and antioxidant agents. *Bioorg. Med. Chem. Lett.* **2012**, *22*, 5839–5844. [CrossRef] [PubMed]
12. Ozdemir, Z.; Kandilci, H.B.; Gümüşel, B.; Caliş, U.; Bilgin, A.A. Synthesis and studies on antidepressant and anticonvulsant activities of some 3-(2-furyl)-pyrazoline derivatives. *Eur. J. Med. Chem.* **2007**, *42*, 373–379. [CrossRef] [PubMed]
13. Sachchar, S.P.; Singh, A.K. Synthesis of some new fluorinated heteroaryl pyrazolines and isooxazolines as potential biocidal agents. *J. Indian Chem. Soc.* **1986**, *62*, 142–146. [CrossRef]
14. Camacho, M.E.; León, J.; Entrena, A.; Velasco, G.; Carrión, M.D.; Escames, G.; Vivó, A.; Acuña-Castroviejo, D.; Gallo, M.A.; Espinosa, A. 4,5-Dihydro-1H-pyrazole derivatives with inhibitory nNOS activity in rat brain: Synthesis and structure−Activity relationships. *J. Med. Chem.* **2004**, *47*, 5641–5650. [CrossRef] [PubMed]
15. Dabholkar, V.; Ansari, F. Synthesis and characterization of selected fused isoxazole and pyrazole derivatives and their antimicrobial activity. *J. Serb. Chem. Soc.* **2009**, *74*, 1219–1228. [CrossRef]
16. Gressler, V.; Moura, S.; Flores, A.F.C.; Flores, D.C.; Colepicolo, P.; Pinto, E. Antioxidant and antimicrobial properties of 2-(4,5-dihydro-1H-pyrazol-1-yl)-pyrimidine and 1-carboxamidino-1H-pyrazole derivatives. *J. Braz. Chem. Soc.* **2010**, *21*, 1477–1483. [CrossRef]
17. Liu, J.-J.; Sun, J.; Fang, Y.-B.; Yang, Y.-A.; Jiao, R.-H.; Zhu, H.-L. Synthesis, and antibacterial activity of novel 4,5-dihydro-1H-pyrazole derivatives as DNA gyrase inhibitors. *Org. Biomol. Chem.* **2014**, *12*, 998–1008. [CrossRef]
18. Rostom, S.A.F.; Badr, M.H.; Abd El Razik, H.A.; Ashour, H.M.A.; Abdel Wahab, A.E. Synthesis of some pyrazolines and pyrimidines derived from polymethoxy chalcones as anticancer and antimicrobial agents. *Arch. Pharm.* **2011**, *344*, 572–587. [CrossRef]
19. Molnar, M.; Kovač, T.; Strelec, I. Umbelliferone-thiazolidinedione hybrids as potent mushroom tyrosinase inhibitors. *Int. J. Pharm. Res. Allied Sci.* **2016**, *5*, 305–310.
20. Poroikov, V.V.; Filimonov, D.A.; Ihlenfeldt, W.-D.; Gloriozova, T.A.; Lagunin, A.A.; Borodina, Y.V.; Stepanchikova, A.V.; Nicklaus, M.C. PASS biological activity spectrum predictions in the enhanced open NCI database browser. *J. Chem. Inf. Comput. Sci.* **2003**, *43*, 228–236. [CrossRef]
21. Hocquet, A.; Langgård, M. An evaluation of the MM+ force field. *J. Mol. Med.* **1998**, *4*, 94–112. [CrossRef]
22. Stewart, J.J.P. Optimization of parameters for semiempirical methods V: Modification of NDDO approximations and application to 70 elements. *J. Mol. Model.* **2007**, *13*, 1173–1213. [CrossRef]
23. Ren, J.; He, Y.; Chen, W.; Chen, T.; Wang, G.; Wang, Z.; Xu, Z.; Luo, X.; Zhu, W.; Jiang, H.; et al. Thermodynamic and structural characterization of halogen bonding in protein–ligand interactions: A case study of PDE5 and its inhibitors. *J. Med. Chem.* **2014**, *57*, 3588–3593. [CrossRef] [PubMed]
24. Vollmer, W.; Blanot, D.; de Pedro, M.A. Peptidoglycan structure and architecture. *FEMS Microbiol. Rev.* **2008**, *32*, 149–167. [CrossRef] [PubMed]
25. Perkins, H.R. *Microbial Cell Walls and Membranes*; Springer: Dordrecht, The Netherlands, 1980; ISBN 978-94-011-6016-2.
26. Böhm, H.-J.; Banner, D.; Bendels, S.; Kansy, M.; Kuhn, B.; Müller, K.; Obst-Sander, U.; Stahl, M. Fluorine in medicinal chemistry. *ChemBioChem* **2004**, *5*, 637–643. [CrossRef] [PubMed]
27. Shah, P.; Westwell, A.D. The role of fluorine in medicinal chemistry. *J. Enzyme Inhib. Med. Chem.* **2007**, *22*, 527–540. [CrossRef]
28. Schellack, N.; Agoro, A. A review of phosphodiesterase type 5 inhibitors. *S. Afr. Fam. Pract.* **2014**, *56*, 96–101. [CrossRef]
29. Wang, X.-H.; Wang, X.-K.; Liang, Y.-J.; Shi, Z.; Zhang, J.-Y.; Chen, L.-M.; Fu, L.-W. A cell-based screen for anticancer activity of 13 pyrazolone derivatives. *Chin. J. Cancer* **2010**, *29*, 980–987. [CrossRef] [PubMed]
30. Sung, B.-J.; Hwang, K.Y.; Jeon, Y.H.; Lee, J.I.; Heo, Y.-S.; Kim, J.H.; Moon, J.; Yoon, J.M.; Hyun, Y.-L.; Kim, E.; et al. Structure of the catalytic domain of human phosphodiesterase 5 with bound drug molecules. *Nature* **2003**, *425*, 98–102. [CrossRef] [PubMed]

*applied sciences*

MDPI

*Article*

# Directed Silica Co-Deposition by Highly Oxidized Silver: Enhanced Stability and Versatility of Silver Oxynitrate

**Carla J. Spina [1,*], Roohee Ladhani [2], Carlie Goodall [2], Michelle Hay [2] and Rod Precht [2]**

1 Exciton Pharma Corp, Toronto, ON M5G 1L7, Canada
2 Exciton Technologies Incorporation, Edmonton, AB T5J 4P6, Canada; rthaver@excitontech.com (R.L.); cgoodall@excitontech.com (C.G.); mwoodward@excitontech.com (M.H.); rprecht@excitontech.com (R.P.)
* Correspondence: cspina@excitontech.com

Received: 31 October 2019; Accepted: 21 November 2019; Published: 2 December 2019

**Abstract:** Novel silver compounds in higher oxidation states, Ag (II) and Ag (III), have emerged as desirable alternatives to existing forms of antimicrobial silver compounds. Offering enhanced efficacy without sacrificing biocompatibility. Unique physiochemical characteristics associated with higher oxidation state silver confer desirable therapeutic traits. However, these same characteristics create challenges in terms of long-term stability and chemical compatibility with conventional biomedical materials. Core-shell methodologies, utilizing silica as a mesoporous or amorphous shell, have been adopted to enhance the stability of reactive active ingredients or cores. These methodologies commonly utilize controlled condensation of silicic acids in non-aqueous media by way of hydrolyzing alkyl silicates: the Stöber process or modified processes thereof. However, these strategies are not conducive to cores of higher oxidation state silver wherein hydroxyl organic precursors and by-products are incompatible with strong oxidizing agents. Addressing these challenges, we present a strategy herein for the preparation of a self-directed silver oxynitrate-silica, $Ag_7NO_{11}$:$SiO_2$, framework. The method described utilizes pH gradients generated from the oxidation reaction of soluble silver, Ag (I), with a strong oxidizing agent/alkaline silicate media to facilitate spatial control over the protonation and subsequent condensation of silicic acid from aqueous solution. The resulting $Ag_7NO_{11}$:$SiO_2$ framework confers enhanced long term and thermal stability to silver oxynitrate without impairing aqueous degradation profiles or subsequent antimicrobial and antibiofilm activities.

**Keywords:** silica; silver oxynitrate; stability; core-shell; antibiofilm; antimicrobial

## 1. Introduction

Silicon dioxide is a ubiquitous material in electronics, food, agriculture, and approved for use in pharmaceutical and drug delivery [1–3]. As a solid-state powder, silicon dioxide is used in solid or semi-solid formulations as a tabletting agent, carrier for active ingredients, thickener, or moisture scavenger. In core-shell structures silicon dioxide ($SiO_2$) may provide a protective shell or barrier, regulating release profiles and conferring protection or stabilization of active core materials [4,5]. Controlled growth of spherical silica particles of uniform size by means of (1) hydrolysis of alkyl silicates and (2) subsequent condensation of silicic acid in alcoholic solutions, or the Stöber process, is commonly employed in a variety of fields [6–8]. Based on this process, a number of methods for silica dioxide core-shell have been prepared from alkyl silicates, moderating the stability and release of active ingredients. This strategy has been applied to the preparation of metal and weakly oxidizing metal oxide core silica dioxide-shell materials [4,7,9]. However, the utility of this process is limited for strong oxidizing agents due to precursor and by-product compatibly. Silver has been known for centuries to be an effective antimicrobial agent [10–12]. Throughout these centuries the foremost state or form of this antimicrobial agent has been comprised of metallic and, or singly ionic silver: Ag (0) and Ag (I) [13–15]. As a weak oxidizing agent, with a standard electrode potential of +0.7996 V, strategies for fabrication of monodisperse Ag-$SiO_2$ core-shell from Ag (I) materials have been identified [7,9]. These strategies utilize a seeded polymerization technique, based on the Stöber method, wherein sol-gel reactions of alkyl silanes generate silver nanomaterials with amorphous silica shell coatings. Protecting the silver nanoparticle cores from oxidation without hindering antimicrobial function. Recently, higher oxidation states of silver, Ag (II) and Ag (III), have come into light as a viable alternative for these predecessor lower oxidation silvers. Higher oxidation state silver compounds demonstrate superior broad-spectrum antimicrobial efficacy against microbes in planktonic and biofilm states [16–18]. Exhibiting marked efficacy against multi-drug resistant bacterium, while remaining safe and not impairing host tissue repair or function [18,19]. Ag (II) and Ag (III) are strong oxidizing agents with standard electrode potentials of +1.980 V and +1.9 V respectively [20]. Not unlike other transition metal compounds of higher oxidation states, the same properties of higher oxidation state silver compounds that elicit unique biological activity also make them susceptible to chemical reduction and thermal degradation; limiting their utilization in a wider variety of medical applications [21–23]. Furthermore, restricting their compatibility with silica core-shell processes such as the Stöber method or modified methods thereof.

Circumventing the first step in the Stöber process, the direct use of silicic acid may also be used to generate silicon dioxide as shown in Equation (1) [24,25]. However, silicic acid is known to be unstable at concentrations above 100 ppm, resulting in uncontrolled polymerization of silica gel or silicon dioxide. Silicic acid is therefore impractical as a starting material for controlled production of silicon dioxide. Higher concentration silicon solutions may be generated through hydroxide condensation of $SiO_2$, forming stable alkaline solutions of silicate salts [26,27]. In these alkali silicate solutions, $H_2SiO_4{}^{2-}$ and $H_3SiO_4{}^{-}$ salts are believed to be the dominant ions and stable in alkali solution. Decreasing the pH of the solution may result in the formation of $H_3SiO_4{}^{-}$ and $H_4SiO_4$ and subsequent uncontrolled formation of amorphous silica solids, as expressed in Equation (2). Approaches for controlled polymerization of silica in solution include reducing silica concentration, maintaining high pH, addition of polymerization inhibitors, and eliminating nucleation sites [5,24,27,28]. These strategies provide insight into the mechanisms for controlled silica polymerization from alkali silicate solutions, nevertheless they are also unconducive to highly oxidized silver due to the reactivity of silver with organic reagents and reaction by-products. Accordingly, methods for the preparation of silica-encapsulated highly oxidized silver compounds to enhance stability while facilitating antimicrobial efficacy are needed.

$$SiO_2 + 2H_2O \Leftrightarrow H_4SiO_4 \tag{1}$$

$$H_3SiO_4{}^{-}{}_{(aq)} + H_4SiO_{4(aq)} \rightarrow (OH)_3Si\text{-}O\text{-}Si(OH)_{3(s)} + OH^{-}{}_{(aq)} \tag{2}$$

In this paper, we present a facile, one-pot method for the preparation of a higher oxidation state silver-silica gel, $Ag_7NO_{11}$:$SiO_2$, framework based on the direct oxidation of silver nitrate from an oxidizing alkali silicate aqueous solution. The corresponding characterization, thermal stability, aqueous degradation, and antimicrobial efficacy of the $Ag_7NO_{11}$:$SiO_2$ framework are evaluated over a range of relative silica concentrations based upon the one-pot reaction described herein.

## 2. Materials and Methods

### 2.1. Materials

Silver nitrate ($AgNO_3$, ACS reagent grade, ≥99.0%), nitric acid ($HNO_3$, ACS reagent grade, 68–70%), and acetone ($(CH_3)_2CO$, ACS reagent grade, 99.5%) were obtained from Sigma-Aldrich, St. Louis, Missouri, United States; potassium persulfate ($K_2S_2O_8$, ACS reagent grade, ≥99.0%) and sodium chloride (NaCl, ACS reagent grade, ≥99.0%) were obtained from VWR, Mississauga, Ontario, Canada; and silicic acid potassium salt solution ($K_2SiO_3$, 39.2 wt/wt%) was obtained from

PQ Corporation, Malvern, Pennsylvania, United States. All reagents were used without further purification. Unless otherwise mentioned, reverse osmosis (RO) water was used for all experimental procedures. *Staphylococcus aureus* (ATCC 6538) and *Pseudomonas aeruginosa* (ATCC 9027) strains were obtained from the American Type Culture Collection (ATCC), Manassas, Virginia, United States. Cultures were stored at ≤−70 °C and propagated on Mueller-Hinton agar (MHA) (at 37 °C for 24 h) immediately prior to experimentation.

## 2.2. Equipment

Scanning electron microscopy (SEM) was performed on a FEI Quanta FEG 250 ESEM (Thermo Scientific, Waltham, Massachusetts, United States) variable pressure and environmental scanning instrument, housed in the Centre for Nanostructured Imaging at the University of Toronto. SEM was performed in a low vacuum 70–130 Pa, imaging at 5–10 eV. Energy-dispersive X-ray spectroscopy (EDX) was performed on the same FEI Quanta FEG 250 ESEM instrument under equivalent conditions. Post-processing EDX analysis was performed using TEAM: Texture & Elemental Analytical Microscopy software. Transmission Electron Microscopy (TEM) was performed on a H-7000 TEM (Hitachi, Chiyoda, Tokyo, Japan), housed in the Microscopy Imaging Laboratories at the University of Toronto. TEM was performed under high vacuum at 75.0 kV. No coating methods were employed in performing TEM or SEM imaging. X-ray diffraction (XRD) was performed on a D2 Phaser (Bruker, Billerica, Massachusetts, United States) powder X-ray diffractometer, housed at Exciton Technologies Inc. in Edmonton Alberta. XRD was performed using Cu K$\alpha$ 1.54060 A, divergence slit 0.6 mm, air scatter shield 3 mm, air scatter slit 8 mm, step size 0.010°, step time 42 sec. XRD data analysis, including peak area and full width half max (FWHM) calculations was performed using DIFFRAC EVA V4.1.1 software, post-processing including stripping Cu K$\alpha_2$.

## 2.3. Preparation of the $Ag_7NO_{11}$:$SiO_2$ Framework

A series of silver oxynitrate-silica co-deposition products, wherein the molar ratio of Ag:$SiO_2$ were varied from 1:0.0 to 1:0.5, were prepared as follows. Aqueous stock solutions of silver nitrate ($AgNO_3$, 59.3 wt/wt%) were prepared immediately prior to synthesis. To an aqueous solution of $K_2S_2O_8$ (7.52 mmol), 0.0 to 5.21 mmol of $K_2SiO_3$ was added dropwise at 25 °C. The clear, colourless aqueous solution was stirred for five minutes. To this stirring solution, 1.14:1 molar equivalents of silver ($AgNO_3$:$K_2S_2O_8$), from the stock silver nitrate solution, was added dropwise into the vortex of the stirring solution, completing the addition of silver nitrate over the course of 60 s. A 50 mL total volume for the reaction solution was conserved. Over the course of silver addition, the solution transitioned rapidly from red-brown to black turbid solution in the absence of alkali silicates and from a clear bright yellow color through a translucent orange-red to a turbid brown-black suspension over the course of three to five minutes in the presence of alkali silicates. Following complete silver nitrate addition, mechanical stirring continued for a total time of 40 min. The appearance of the solution did not change during this mixing time, where the final solution appeared black and turbid. Subsequent to the 40 min of reaction time, the black suspension was filtered through a Whatman no. 40 ashless filter paper under 22 mmHg vacuum filtration. The resulting grey-black powder was washed three times with water and was subsequently rinsed three times with acetone. The product was dried under air until a steady mass of the grey-black dull powder products was obtained. Powder $Ag_7NO_{11}$:$SiO_2$ products were characterized by XRD, SEM, TEM, EDX, and potentiometric titration for silver content determination.

## 2.4. Silver Quantification

Quantitative determination of the percent silver content by mass for silver oxynitrate and the $Ag_7NO_{11}$:$SiO_2$ was completed in triplicate by potentiometric titration against sodium chloride (NaCl). In brief, samples (0.25 g) of the higher oxidation state silver compounds were digested in approximately 25 mL of dilute nitric acid (1:4, $HNO_3$:$H_2O$) overnight to liberate all silver into free Ag (I) ions in solution. This digested solution was then quantitatively transferred to a 250 mL volumetric flask and diluted with RO water. A 5.0 mL aliquot of this solution was quantitatively transferred to a 50 mL sample vial containing 10 mL of dilute nitric acid (1:4, $HNO_3$:$H_2O$) and 20 mL water. The potential is measured across the analyte solution over the duration of titration with a 0.1 M NaCl titrant. The silver content of the sample volume is determined from the second derivative of the potential voltage plotted versus titrant volume, at an equimolar ratio of silver to chloride. In this manner, the mass percent silver content for silver oxynitrate and the grey $Ag_7NO_{11}$:$SiO_2$ powders was determined.

*2.5. X-Ray Diffractometry*

The crystalline structures of silver oxynitrate and $Ag_7NO_{11}$:$SiO_2$ were determined by X-ray diffractometry (XRD). In brief, approximately 0.5 g of either the silver oxynitrate or $Ag_7NO_{11}$:$SiO_2$ powder was spread evenly into the depression in the powder XRD sample holder and placed into the X-ray diffractometer and measured from 10 to 90 °2Θ recorded over 30 min. From crystal lattice structure diffraction patterns, the crystalline solid-state composition of the samples was determined. The silver oxynitrate and other solid-state compounds were identified using XRD spectra deferring the Crystallography Open Database (COD) [21,23,29]. The relative compositions of the primary silver oxynitrate species and by-product or degradation products were approximated using the percent relative peak height of silver oxynitrate (36.3 °2Θ, COD Card 2310073) versus the additional solid-state compounds identified: AgO (32.4 °2Θ, COD Card 1509488), $Ag_2O$ (32.8 °2Θ, COD Card 4318188), Ag (38.0 °2Θ, COD Card 1100136), $AgNO_3$ (29.7 °2Θ, COD Card 1509468), and $Ag_2SO_4$ (28.1 °2Θ, COD Card 1509700). Over a four-month time course, the relative solid crystalline composition of each of the samples was observed. In this manner, the stability of the higher oxidation state silver compounds was evaluated under both ambient or room temperature and at elevated temperature, 40 °C, in a temperature regulated incubator.

*2.6. Aqueous Decomposition Studies*

The decomposition profile of silver oxynitrate and $Ag_7NO_{11}$:$SiO_2$ was evaluated in aqueous media over a course of seven days. In brief, approximately 1 g of silver oxynitrate or $Ag_7NO_{11}$:$SiO_2$ was added to a series of sealed glass vials denoted 2 h, 6 h, 24 h, 72 h, 120 h, and 168 h, each containing 10 mL water. Following addition of the higher oxidation state silver compounds, the vials were sealed and vortexed for 30 s to disperse the solids in solution then stored at room temperature away from direct light. At each specified time interval, the entire contents of the vial were quantitatively transferred to a Buchner funnel with Whatman 42 ashless filter paper. The grey-black solids were washed three times with water and was subsequently rinsed three times acetone. The solids were dried under air until a steady mass was obtained. At each specified time point, powder X-ray diffraction and scanning electron microscopy, as described above, was performed on the isolated solids.

*2.7. Planktonic Log Reduction Assay*

The planktonic antimicrobial activities of the higher oxidation state silver compounds were evaluated for their efficacy against *Staphylococcus aureus* (ATCC 6538) and *Pseudomonas aeruginosa* (ATCC 9027). Briefly, the equivalent of 10 mg silver (Ag) of the higher oxidation state silver compounds were weighed and added into a triplicate set of 15 mL sealed tubes containing 10 mL Mueller Hinton Broth (MHB) challenged with $1 \times 10^6$ CFU/mL inoculum of either *S. aureus* and *P. aeruginosa*. Testing included negative control tubes without any added silver compounds. Sterility control were also included; not containing any silver compounds and not inoculated. The test and control tubes were incubated at 37 °C for four (4) hours for *S. aureus* or one (1) hour for *P. aeruginosa*. After the final inoculation and at each required incubation time, contents of the reaction tubes were neutralized with 0.4 wt/wt% sodium thioglycolate solution (STS), serially diluted with 0.89 wt/wt% NaCl (Saline) and plated onto Mueller Hinton Agar (MHA). Plates were incubated for 18–24 hrs and then enumerated. Log reduction of the higher oxidation state silver compounds as compared to the control reaction tubes were calculated.

### 2.8. Biofilm Log Reduction Assay

Biofilm of *Staphylococcus aureus* (ATCC 6538) and *Pseudomonas aeruginosa* (ATCC 9027) were grown in a three-dimensional matrix to determine the efficacy of higher oxidation state silver compounds against bacterial biofilm. In, brief, three to five layers of sterile cotton gauze were placed in simulated wound fluid (SWF) in 6-well tissue culture plates for each strain tested. To each well, a $1 \times 10^6$ CFU/mL inoculum was added every 24 hrs for up to 72 hrs. During which time the plates were incubated at 37 °C with shaking at 200 rpm. After the incubation period, the gauze was removed from the liquid culture medium. The gauze biofilm was rinsed three times with sterile water to eliminate planktonic bacterium and then placed onto the surface of an MHA plate (Oxoid, Nepean, ON, Canada). The gauze biofilm was overlaid with additional MHA cooled to ca. 50 °C such that one-half of the biofilm was embedded in the agar and one-half was exposed. An equivalent of 10 mg silver (Ag) of the higher oxidation state silver compounds were weighted over-laid onto the biofilm and exposed to treatment at 37 °C for four (4) hours for *S. aureus* or two (2) hour for *P. aeruginosa*. After the exposure time, the dressings and biofilm (gauze pieces) were carefully removed from the plates and were placed into 10 mL of 0.4 wt/wt% STS. They were vortexed ($3 \times 1$ min) to disrupt the biofilm, then serially diluted in saline (0.89 wt/wt% saline) and spot-plated onto MHA for viable cell counts. Negative controls were made as follows: (i) gauze biofilm were grown and were overlaid with agar as described above; and (ii) no higher oxidation state silver compounds were overlaid on the biofilm in the same manner as the treatment conditions. Log reduction of the higher oxidation state silver compounds versus the control biofilm were calculated.

### 2.9. Statistical Analysis

Data are expressed as mean ± standard deviation of at least three independent experiments. Analysis of the data distribution was performed using Student's *t*-test to analyse the significance of differences between the treated group and the control group (without silver exposure). $p$ values of less than 0.05 were considered statistically significant.

## 3. Results

### 3.1. Characterization of $Ag_7NO_{11}$:$SiO_2$

Synthesis of higher oxidation states of metals, such as silver oxynitrate, may proceed through the addition of a soluble metal to a strong oxidizing agent [30]. Aqueous oxidization of silver nitrate by potassium persulfate, as per Equation (3), results in the formation or deposition of $Ag_7NO_{11}$.

$$\begin{array}{c} 7AgNO_{3\,(aq)} + K_2S_2O_{8\,(aq)} + 8H_2O_{(l)} \rightarrow \\ Ag_7O_8NO_{3\,(s)} + 6HNO_{3\,(aq)} + H_2SO_{4\,(aq)} + K_2SO_{4\,(aq)} + 4H_{2\,(g)} \end{array} \tag{3}$$

Based on the methods described herein, a series of silver oxynitrate-silica co-deposition products were prepared through the addition of an alkaline potassium silicate ($K_2SiO_3$) to a potassium persulfate solution prior to silver nitrate addition. Within the series of co-deposition products, the molar equivalents of silicon dioxide to silver were incrementally augmented following the series: 0:1, 0.1:1, 0.25:1, and 0.5:1 molar equivalents $SiO_2$:Ag; these co-deposition products are herein referred to as $Ag_7NO_{11}$:$SiO_2$. Subsequent to their isolation, the relative solid-state composition of the powders was determined by X-ray diffractometry (XRD). Principal diffraction patterns collected from each the powders, shown in Figure 1, were in agreement with the crystallographic parameters for $Ag_7NO_{11}$ [29]. Minor diffraction patterns were identified in each of the powder samples in agreement with the crystallographic parameters for silver sulfate. No additional diffraction patterns were identified in any of the powder samples. Silica gels formed from alkali silicates are commonly amorphous and poorly crystalline and therefore are expected to have limited or indiscernible diffraction patterns [31]. The relative diffraction peak area for silver sulfate $Ag_2SO_4$ (28.1 °2Θ) versus silver oxynitrate (36.3 °2Θ) was identified to increase with the relative molar ratio of $SiO_2$:Ag, as shown in Figure S1. Scanning electron microscopy (SEM) and transmission electron microscopy (TEM) were employed to observe the structural composition and distribution of silver oxynitrate and silica within the isolated products. SEM images of silver oxynitrate control and $Ag_7NO_{11}$:$SiO_2$ prepared at different silica concentrations ranging from 0:1 to 0.5:1 molar equivalents $SiO_2$:Ag are shown in Figure 2. It is observed from these images that distinct cuboctahedron structure of silver oxynitrate is conserved from 0:1 to 0.25:1 $SiO_2$:Ag, where the silica is observed as amorphous structures. At 0.5:1, $SiO_2$:Ag solid silica gel is the primary structure observed and the geometric silver oxynitrate structures are not visibly detected. Upon increasing the relative concentration of silica, it is also observed that the relative crystalline size of the silver oxynitrate decreases from 0:1 to 0.5:1 $SiO_2$:Ag. This reduction in crystalline size upon increasing silica concentration was also corroborated by XRD, employing the Debye-Scherrer equation, as shown in Figure S2 [32]. Complimentary to the SEM images, TEM provides a spatial orientation of the silver within the silica framework as the density disparity between silver oxynitrate and silica affords clear demarcation between the materials as shown in Figure 3. At a low relative concentration of silica, 0.1:1, molar equivalents $SiO_2$:Ag, the distinct cuboctahedra structure of silver oxynitrate is observed as a black silhouette, Figure 3A–C. Surrounding silver oxynitrate, the less electron dense structure of silica is observed on the structural faces of silver oxynitrate, Figure 3B, and encasing the silver oxynitrate particles in a framework of silica gel, Figure 3A. Increasing the relative concentration of silica to 0.5:1 molar equivalents $SiO_2$:Ag, as shown in Figure 3D–F, the distinct cuboctahedra structure of silver oxynitrate is partly conserved. Additionally, sub-micron sized areas of high electron density within the silica gel framework are observed, Figure 3D,E. Presence of silver in these areas of high electron density in $Ag_7NO_{11}$:$SiO_2$ containing 0.5:1 molar equivalents $SiO_2$:Ag was confirmed using energy dispersive X-ray spectroscopy as shown in Figure S3. Confirming that the silica framework contains 55 to 65 wt/wt% silver, in good agreement with quantitative silver titration results shown in Figure S4.

### 3.2. *Ambient and Accelerated Thermal Stability of $Ag_7NO_{11}$:$SiO_2$*

Diffraction patterns of silver oxynitrate and $Ag_7NO_{11}$:$SiO_2$ containing 0.1:1 $SiO_2$: were evaluated over a period of four months (16 weeks) under both ambient and elevated temperature storage conditions (40 °C). Within this timeframe, degradation of solid-state silver oxynitrate was evaluated by monitoring the depletion of the standard $Ag_7NO_{11}$ diffraction peaks: 31.3, 36.3, and 39.7 °2Θ and the onset of the first thermal decomposition product argentic oxide, Equation (4), following standard AgO diffraction patterns peaks: 32.2, 32.4, 37.3, and 38.6 °2Θ [21,23]. For the purposes of this study, the silver oxynitrate peak at 36.3 °2Θ peak was utilized to distinguish the presence/absence of silver oxynitrate as there are no overlapping impurity or degradation product diffraction peaks at this diffraction angle. At the study commencement, minor silver sulfate ($Ag_2SO_4$) impurities were additionally identified using standard diffraction patterns peaks: 28.1, 31.2, 33.9 °2Θ.

$$Ag_7NO_{11} \rightarrow 6AgO + AgNO_3 + O_2 \quad (4)$$

Following one month of silver oxynitrate storage under ambient conditions, Figure 4A, the primary solid-state compound observed was $Ag_7NO_{11}$ with minor peaks attributed to AgO and $Ag_2SO_4$. Following four months of storage at room temperature, AgO and $Ag_2SO_4$ alone were observed, consistent with the first thermal decomposition step shown in Equation (4). In parallel, the stability of $Ag_7NO_{11}$:$SiO_2$ containing 0.1:1 $SiO_2$:Ag was evaluated under ambient conditions as shown in Figure 4B. Silver oxynitrate was identified as the primary solid-state compound following one month of storage under ambient conditions with minor peaks attributed to AgO and $Ag_2SO_4$. Similarly, at four months of storage under ambient conditions, the primary solid-state compound observed was $Ag_7NO_{11}$ with only minor peaks attributed to AgO and $Ag_2SO_4$.

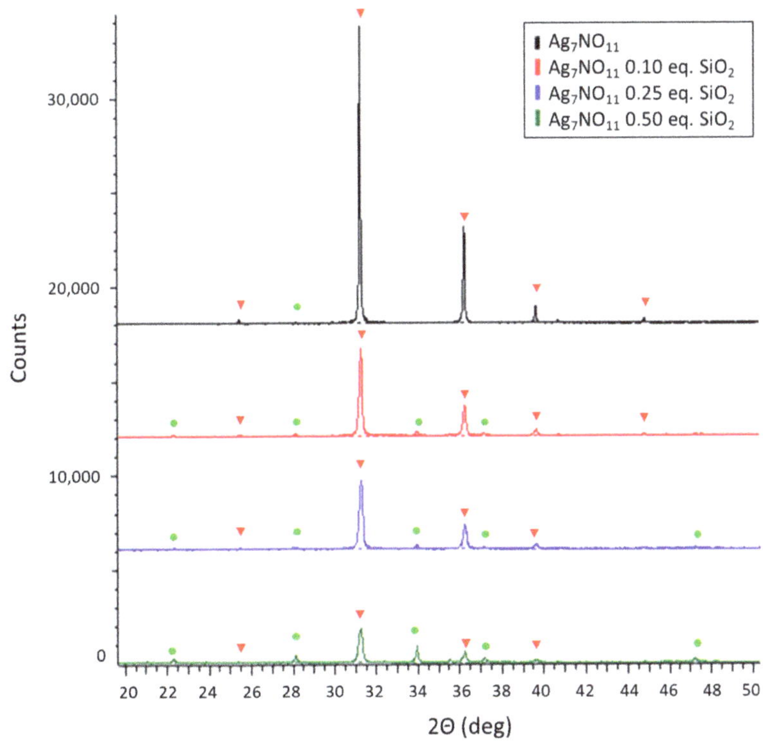


**Figure 1.** Powder X-ray diffraction of $Ag_7NO_{11}$:$SiO_2$ (0.0:1 to 0.5:1 molar equivalents $SiO_2$:Ag). Solid-state silver compounds were identified as silver oxynitrate ($Ag_7NO_{11}$, inverted red triangles) and silver sulfate ($Ag_2SO_4$, green circles).

Storage of silver oxynitrate under elevated temperatures, Figure 5A, was observed to result in multi-stage thermal decomposition processes as shown in Equations (5) to (7) [21,23]. Within one week of storage under 40 °C, the predominant solid-state compound was identified as silver nitrate, in good agreement with standard $AgNO_3$ diffraction peaks: 29.7, 32.8, and 35.3 °2Θ. Secondary solid-state compounds were identified as $Ag_7NO_{11}$, AgO, and $Ag_2SO_4$. Following 16 weeks of storage at elevated temperatures, the diffraction peak at 36.3 °2Θ for $Ag_7NO_{11}$ was not observed nor were any standard diffraction patterns for AgO. The primary solid-state diffraction patterns identified in the XRD pattern at 16 weeks were in agreement with $AgNO_3$, $Ag_2O$, and Ag (metallic silver), indicative of advanced thermal decomposition pathways as per Equations (5) to (7). In contrast, decomposition of $Ag_7NO_{11}$:$SiO_2$, containing 0.1:1 $SiO_2$:Ag, under elevated temperatures was less severe. Retaining $Ag_7NO_{11}$ as the predominant solid-state compound with only minor peaks attributed to AgO and $Ag_2SO_4$ after one week and conversion to the first thermal decomposition product, AgO, following 16 weeks of storage under 40 °C.

**Figure 2.** Scanning electron microscopy images of $Ag_7NO_{11}$:$SiO_2$ (0.0:1 to 0.5:1 molar equivalents $SiO_2$:Ag). (**A**,**B**) Silver oxynitrate ($Ag_7NO_{11}$); (**C**,**D**) $Ag_7NO_{11}$:$SiO_2$ containing 0.10:1, $SiO_2$:Ag; (**E**,**F**) $Ag_7NO_{11}$:$SiO_2$ containing 0.25:1, $SiO_2$:Ag; (**G**,**H**) $Ag_7NO_{11}$:$SiO_2$containing 0.50:1, $SiO_2$:Ag.

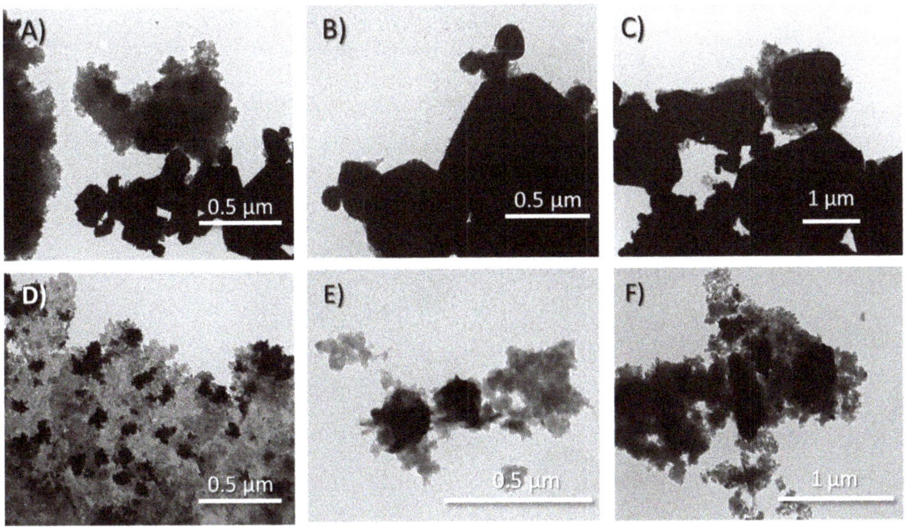


**Figure 3.** Transmission electron microscopy images of $Ag_7NO_{11}$:$SiO_2$ (0.0:1 to 0.5:1 molar equivalents $SiO_2$:Ag) prepared by the co-deposition synthetic process. (**A–C**) $Ag_7NO_{11}$:$SiO_2$ containing 0.10:1, $SiO_2$:Ag; (**D–F**) $Ag_7NO_{11}$:$SiO_2$ containing 0.50:1, $SiO_2$:Ag.

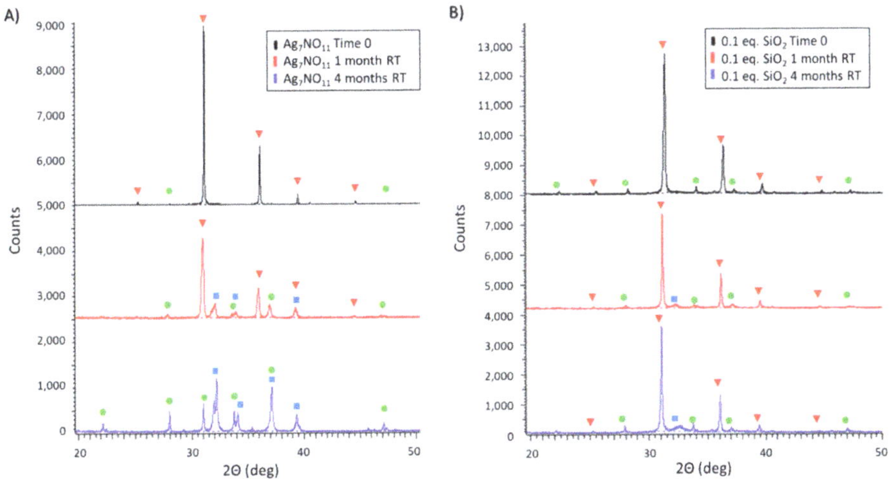


**Figure 4.** Powder X-ray diffraction patterns of (**A**) silver oxynitrate ($Ag_7NO_{11}$) and; (**B**) $Ag_7NO_{11}$:$SiO_2$ containing 0.10:1, $SiO_2$:Ag, equivalents silica over a four month period under storage at ambient room temperature (RT) conditions. Solid-state silver compounds were identified as silver oxynitrate ($Ag_7NO_{11}$, inverted red triangles), silver sulfate ($Ag_2SO_4$, green circles), and argentic oxide (AgO, blue squares).

$$6AgO + AgNO_3 \rightarrow 3Ag_2O + AgNO_3 + 1.5O_2 \quad (5)$$

$$3Ag_2O + AgNO_3 \rightarrow 3.5Ag_2O + NO_2 + 0.25O_2 \quad (6)$$

$$3.5Ag_2O \rightarrow 7Ag + 1.75O_2 \quad (7)$$

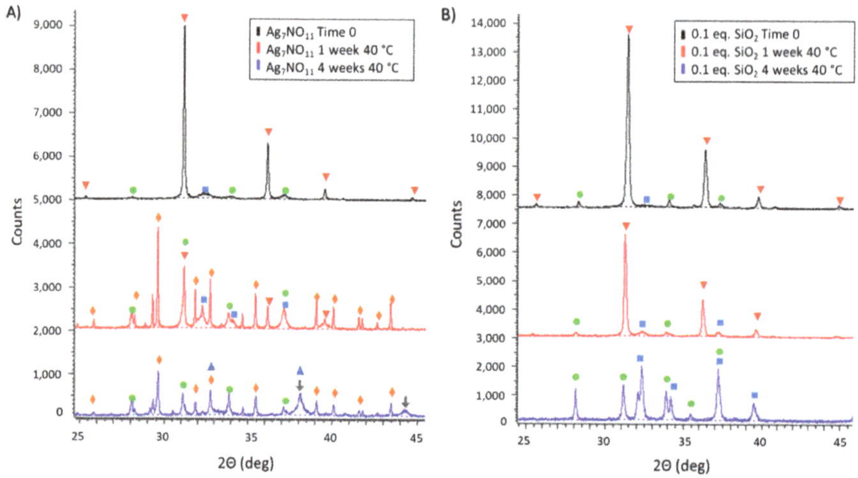


**Figure 5.** Powder X-ray diffraction patterns of (**A**) silver oxynitrate ($Ag_7NO_{11}$) and; (**B**) $Ag_7NO_{11}$:$SiO_2$ containing 0.10:1, $SiO_2$:Ag, equivalents silica over a four month period under accelerated storage in an incubator at 40 °C. Solid-state silver compounds were identified as silver oxynitrate ($Ag_7NO_{11}$, inverted red triangles), silver sulfate ($Ag_2SO_4$, green circles), argentic oxide (AgO, blue squares), silver nitrate (AgNO3, orange diamonds), silver oxide ($Ag_2O$, blue triangle), and metallic silver (Ag, grey arrows).

### 3.3. *Aqueous Decomposition of $Ag_7NO_{11}$:$SiO_2$*

The aqueous decomposition profiles of silver oxynitrate and $Ag_7NO_{11}$:$SiO_2$ containing 0.1:1 $SiO_2$:Ag were also investigated. In brief, the solid powder products were dispersed into aqueous media at room temperature and, at set time intervals over seven days, were evaluated by XRD and SEM. Following dispersion into aqueous media it was observed that, with increasing relative ratios of silica, the $Ag_7NO_{11}$:$SiO_2$ powders had an increased time of suspension as shown in Figure S5. The aqueous decomposition profiles were observed to be similar for silver oxynitrate and $Ag_7NO_{11}$:$SiO_2$ as seen in Figure 6. Within two hours of exposure to water, silver oxynitrate was confirmed as the primary solid-state silver species, with minor components attributed to AgO and $Ag_2SO_4$ confirmed by means of their respective standard diffraction patterns. Following 24 h in aqueous media, silver oxynitrate remained the primary solid-state compound, with minor components attributed to AgO and $Ag_2SO_4$. Relative counts for AgO versus $Ag_7NO_{11}$ were proportionally higher in $Ag_7NO_{11}$:$SiO_2$. At the terminus of the seven-day study, AgO was identified as the primary solid-state species for both silver oxynitrate and $Ag_7NO_{11}$:$SiO_2$, with secondary diffraction patterns attributed to $Ag_2SO_4$. The recovered solids at each time point were imaged by SEM. Typical geometric crystalline structures of silver oxynitrate were observed in Figure 7A,D for both silver oxynitrate and $Ag_7NO_{11}$:$SiO_2$. Subsequent to aqueous exposure, a crystalline morphology transformation was observed in Figure 7C,F previously identified to be associated with the formation of argentic oxide, AgO [22]. Retention of the silica framework in the co-deposition product was observed over the seven-day evaluation period as seen in Figure 7E,F.

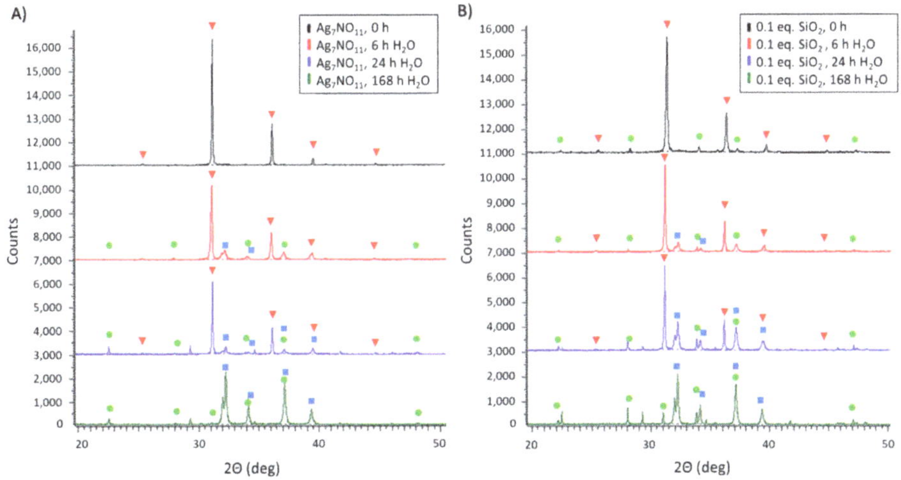


**Figure 6.** Powder X-ray diffraction patterns of (**A**) silver oxynitrate ($Ag_7NO_{11}$) and; (**B**) $Ag_7NO_{11}$:$SiO_2$ containing 0.10:1, $SiO_2$:Ag, equivalents silica over a seven-day period in an aqueous solution at room temperature. Solid-state silver compounds were identified as silver oxynitrate ($Ag_7NO_{11}$, inverted red triangles), silver sulfate ($Ag_2SO_4$, green circles), and argentic oxide (AgO, blue squares).

**Figure 7.** Scanning electron microscopy images of silver oxynitrate and $Ag_7NO_{11}$:$SiO_2$ containing 0.1:1, $SiO_2$:Ag, exposed to aqueous media for a period of seven days. Silver oxynitrate ($Ag_7NO_{11}$) (**A**) prior to aqueous exposure; (**B**) after 24 h of exposure to aqueous media; and (**C**) after 168 h of exposure to aqueous media. $Ag_7NO_{11}$:$SiO_2$ containing 0.1:1, $SiO_2$:Ag (**D**) prior to aqueous exposure; (**E**) after 24 h of exposure to aqueous media; and (**F**) after 168 h of exposure to aqueous media.

### *3.4. Antimicrobial Efficacy Evaluation of $Ag_7NO_{11}$:$SiO_2$*

Silver oxynitrate is known as a potent antimicrobial agent against planktonic, biofilm, and drug resistant bacterium [16–18]. Modification of the form or function of silver oxynitrate may result in an impairment of this antimicrobial activity. Therefore, it is imperative to determine the impact of synthetic modifications on the efficacy of silver oxynitrate. Towards this, a series of antimicrobial methodologies were employed to evaluate the efficacy of silver oxynitrate and $Ag_7NO_{11}$:$SiO_2$ against both planktonic and biofilm states of *Staphylococcus aureus* (ATCC 6538) and *Pseudomonas aeruginosa* (ATCC 9027).

Single time point log reduction of Gram-positive *S. aureus* (4 h) and Gram-negative *P. aeruginosa* (1 h) were evaluated in Mueller Hinton Broth (MHB) inoculated to a concentration of $1 \times 10^6$ CFU/mL. The efficacy of silver oxynitrate and $Ag_7NO_{11}$:$SiO_2$ containing 0.1:1 $SiO_2$:Ag were evaluated at normalized silver concentration. Following treatment, the remaining bacteria were numerated and log reduction quantified versus untreated control. The efficacy of $Ag_7NO_{11}$:$SiO_2$ was compared with silver oxynitrate and was found to be equivalent for both organisms tested ($p$ value > 0.05), as shown in Figure 8A.

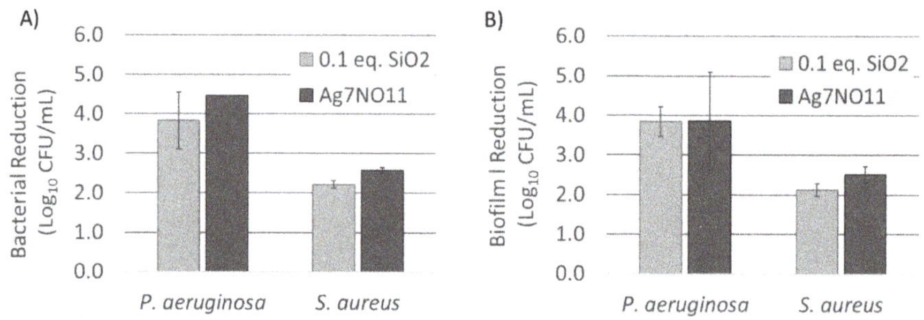


**Figure 8.** Antimicrobial efficacy of silver oxynitrate ($Ag_7NO_{11}$) and $Ag_7NO_{11}$:$SiO_2$ containing 0.1:1, $SiO_2$:Ag. (**A**) Single-time log reduction for 1-h treatment period against *P. aeruginosa* and 4-h treatment period against *S. aureus* challenged at $1 \times 10^6$ CFU/mL; (**B**) Biofilm log reduction values for 2-h treatment period against *P. aeruginosa* and 4-h treatment period against *S. aureus* established 3D biofilm. Results representing the average of triplicate data ($n$ = 3), error bars indicated represent standard deviations of the triplicate measurements.

The role of biofilm in infection and delayed wound healing is becoming increasingly evident [33–35]. Using a clinically relevant biofilm model, adapted from literature, we evaluated the capacity of silver oxynitrate and $Ag_7NO_{11}$:$SiO_2$ containing 0.1:1 $SiO_2$:Ag to target and reduce biofilm [17,36]. In brief, established biofilms of *P. aeruginosa* and *S. aureus* were treated with silver oxynitrate or $Ag_7NO_{11}$:$SiO_2$. Following a 2-h and 4-h treatment period for *P. aeruginosa* and *S. aureus* respectively, viable cells were recovered and numerated versus untreated control to determine log reduction values. Silver oxynitrate and $Ag_7NO_{11}$:$SiO_2$ were found to equivalently reduce viable biofilm counts ($p$ value > 0.05), as shown in Figure 8B. In comparison with silver oxynitrate, no hindrance of antibacterial or antibiofilm activity was observed with the incorporation of silver oxynitrate to the silicon dioxide framework through the described co-deposition procedures.

## 4. Discussion

Amorphous silicon dioxides are known to confer desirable attributes upon incorporation with active ingredients including enhanced biocompatibility, improved chemical stability, and improved thermal stability [7,9,37–39]. Modified Stöber methodologies employed by Cong et al. afford silica coatings on fluorescent dye-doped polymeric nanoparticles towards obtaining an insulating layer with improved physical stability [37]. Silver-silica core-shell structures, successfully prepared by Xu et al., use a modified Stöber process to confer chemical stability to metallic silver nanoparticles without hindering antimicrobial efficacy [9]. These methods are conducive to conventional organic drugs or weak oxidizing agents such as metallic silver, affording control over polymerization and deposition of silicon dioxide. However, in the presence of strong oxidizing agents such as silver oxynitrate, chemical incompatibilities restrict the successful application of these methodologies. Acid-catalyzed polymerization or deposition of alkali silicates are also known to generate amorphous silica [5,24,25,27]. Similar to the aforementioned silica-shell strategies, conventional acid precipitated silicate methodologies are reliant upon the use of surfactants and organic precursors as structure-directing agents and therefore pose challenges for use with highly oxidizing materials. In the present work, alternate methods for the in-situ polymerization of silica with higher oxidation state silvers are presented wherein the formation of the silica structure is self-directed through acidic by-products of silver oxynitrate synthetic reactions.

Silicate ions will polymerize to form silica in solutions having pH of less than about 10 ($H_4SiO_4$, $pK_{a1} = 9.8$) [40]. In the preparation of silver oxynitrate, Equation (3), it is known that sulfuric and nitric acids are generated as by-products of the silver oxidation reaction resulting in the reduction of pH of the solution from neutral to pH 1–2 (data not shown). These acids, as shown previously, are suitable to effect protonation of silicic acid at room temperature and formation of oligomeric silicic acids resulting in the precipitation of amorphous silica [24,25,40]. In the methodologies presented herein alkali silicates at various concentrations relative to silver are dispersed into an oxidizing solution of potassium persulfate to which a soluble silver solution is added resulting in the oxidation of silver and chemical deposition of silver oxynitrate, as verified by XRD in Figure 1. Finite nucleation locus of silver oxynitrate, a crystalline compound with the cubic space group Fm3m, generate an acid gradient in the immediate vicinity of silver oxynitrate nucleation sites, directing the spatial polymerization or precipitation of silica [41]. As observed in Figures 2 and 3, silver oxynitrate solids are observed to be impregnated in a silica framework. Absence of AgO or other degradation products of silver oxynitrate, as seen in in Figure 1, indicates that the co-deposition process does not impair the formation of silver oxynitrate. Formation of silver sulfate does not occur as a by-product of the reaction, as seen in Equation (3). Rather silver sulfate is an impurity forming due to incomplete conversion of soluble silver to silver oxynitrate. This impurity is a minor component in the silver oxynitrate control and observed at increasing relative quantities as silica ratios rise from 0.1:1 to 0.5:1, $SiO_2$:Ag, as shown by relative peak area in Figure S1.

Increasing the relative concentration of silica during the co-deposition reaction was also shown to be an effective method for crystalline size control of silver oxynitrate. Utilizing the FWHM for the four strongest reflections in the diffraction pattern for silver oxynitrate, observed at 31.2, 36.3, 52.3, 62,2 °2Θ, respectively (222), (400), (220) and (622) reflections, the relative size of the silver oxynitrate crystals were calculated as per the Debye-Scherrer equation [41,42]. Based upon the average crystal size determination from these primary reflections, the crystal size of silver oxynitrate decreases from 1032 ± 152 Å to 500 ± 75 Å as the relative ratio of $SiO_2$:Ag increases from 0.0:1 to 0.5:1. This same trend in silver oxynitrate crystalline size reduction upon increasing silica ratio may be observed the SEM and TEM images collected for the co-deposition products as shown in Figures 2 and 3.

As silica is known to confer enhanced stability of active ingredients, the stability of silver oxynitrate and $Ag_7NO_{11}$:$SiO_2$ were compared over the course of four months under ambient and 40 °C storage. The thermal degradation pathways for silver oxynitrate are well known [21,23,30]. The first two decomposition processes for silver oxynitrate, Equations (4) and (5) are exothermic with associated enthalpies of $\Delta H = -244$ KJ/mol and −75.8 KJ/mol and initiation temperatures of 80–85 °C and 115–120 °C respectively [21]. In contrast, the subsequent degradation processes, Equations (6) and (7), are endothermic with associated enthalpies of $\Delta H = 659$ KJ/mol and 338 KJ/mol and initiation temperatures of 350 °C and 425–435 °C respectively [21]. It is also known that the rate of decomposition of silver oxynitrate is accelerated in the presence of heat and water [30,41]. Under ambient conditions, silver oxynitrate is observed to proceed through the first decomposition process as shown in Figure 4A. The rate of this degradation process is retarded in $Ag_7NO_{11}$:$SiO_2$ as shown in Figure 4B. At elevated temperatures, a more rapid degradation of silver oxynitrate will result in increased thermal energy due to the exothermic nature of the first two degradation processes. In combination with the elevated storage temperature, initiation of subsequent endothermic degradation processes as shown in Equations (6) and (7) are observed to occur for silver oxynitrate as seen in Figure 5A. These advanced degradation processes are not observed in $Ag_7NO_{11}$:$SiO_2$. At elevated temperatures only the first degradation process, Equation (4), is observed for $Ag_7NO_{11}$:$SiO_2$ as seen in Figure 5B.

Silica is a thermal insulator with a very low thermal conductivity; beneficial in aerospace and construction applications [43–45]. As an aerogel, Reim et al. demonstrated silica granulates may afford a heat transfer coefficient of less than 0.4 W/($m^2$ K) [44]. Silica is also well known as a desiccant, capable of sequestering water from organic solvents to less than 100 ppm and significant reductions in water content in solid-state powder samples [46,47]. The results presented here suggest that the presence of an interspersed silica framework with thermally insulating and desiccating properties may obstruct heat transfer and restrict the step-wise degradation of silver oxynitrate. These properties enhance the long term and thermal stability of silver oxynitrate within the silica framework.

Silver oxynitrate has been previously shown to rapidly and effectively eliminate bacterial organisms in both free and biofilm states [16–18]. Silicate drug delivery systems, such as silica core-shell systems, may provide an enhanced stability and biocompatibility however, they are also known to modify drug release profiles [3,4]. Gaining understanding for how the silica co-deposition process may impact solution-phase behavior, a series of investigations were performed to evaluate silver oxynitrate and $Ag_7NO_{11}$:$SiO_2$ antimicrobial activities and degradation profiles in aqueous media. As described above, water is known to accelerate the rate of degradation of silver oxynitrate to silver oxide [30,41]. This effect is observed in Figure 6A where over the course of seven days silver oxynitrate proceeds through the degradation process described in Equation (4) to form argentic oxide: AgO. A similar degradation process is observed for $Ag_7NO_{11}$:$SiO_2$ as shown in Figure 6B. This solid-state transformation is visualized by the change in crystalline morphology by SEM where the cuboctahedra structure of silver oxynitrate observed prior to aqueous exposure, Figure 7A,D is replaced by geometric platelets of monoclinic silver (I, III) oxide over the seven day period [22,41]. Similarly, equivalent antimicrobial and antibiofilm log reduction values ($p$ value > 0.05) are observed for silver oxynitrate and $Ag_7NO_{11}$:$SiO_2$ against *P. aeruginosa* and *S. aureus* as shown in Figure 8. These results are in agreement with the silica co-deposition framework providing enhanced thermal stability without hindering aqueous degradation profiles or antimicrobial efficacy.

## 5. Conclusions

Methods described herein for the co-deposition of silver oxynitrate-silica in a one-pot synthesis yield a high purity silver oxynitrate within a three-dimensional silica framework. Spatially guided polymerization or deposition of silica, without the aid of alkyl silicates or surfactants, was achieved through the production of acidic by-product generated gradients from silver oxynitrate nucleation sites. Co-deposition of silica, from 0.1:1 to 0.5:1 molar equivalents $SiO_2$:Ag, were shown to influence control over the crystalline size of silver oxynitrate while providing enhanced long term and thermal stability of silver oxynitrate within the silica framework. It was shown that the enhanced thermal stability of silver oxynitrate within the silica co-deposition framework did not hinder aqueous degradation profiles or antimicrobial and antibiofilm activity of silver oxynitrate. The methods described herein will confer increased stability and versatility of silver oxynitrate when incorporated into inherently reducing materials such as natural fibers or hydrogels and during thermal processing as required for materials such as silicones, melt adhesives, or thermoplastics. Providing an opportunity to expand the application of silver oxynitrate into novel functional materials to address the growing need for anti-infective technologies.

**Supplementary Materials:** The following are available online at http://www.mdpi.com/2076-3417/9/23/5236/s1.

**Author Contributions:** C.J.S.: Conceptualization, Methodology, Formal Analysis, Writing—Original Draft, Visualization, Project Administration, Supervision, Funding Acquisition. R.L.: Methodology, Validation, Investigation. C.G.: Methodology, Formal Analysis, Investigation, Data Curation, Writing—Review & Editing. M.H.: Methodology, Validation, Investigation, Formal Analysis. R.P.: Writing—Review & Editing, Supervision, Project Administration, Funding Acquisition.

**Funding:** This research was supported by the BioTalent Canada, Student Work Placement Program (SWPP) between Carlie Gooodall, student of the University of Guelph, and Exciton Technologies Inc.

**Acknowledgments:** The authors gratefully acknowledge the support of the Medical Imaging Laboratories and the Centre for Nanostructured Imaging at the University of Toronto in Toronto, Ontario.

**Conflicts of Interest:** Carla J. Spina, Michelle Hay, Roohee Ladhani, Carlie Goodall, and Rod Precht were employees of Exciton Technologies Inc. at the time the work was completed.

## References


1. Qian, K.K.; Bogner, R.H. Application of mesoporous silicon dioxide and silicate in oral amorphous drug delivery systems. *J. Pharm. Sci.* **2012**, *101*, 444–463. [CrossRef]
2. BernArdos, A.; KouřimsKá, L. Applications of mesoporous silica materials in food-a review. *Czech J. Food Sci.* **2013**, *31*, 99–107. [CrossRef]
3. Tang, H.; Xiong, H.; Tang, S.; Zou, P. A starch-based biodegradable film modified by nano silicon dioxide. *J. Appl. Polym. Sci.* **2009**, *113*, 34–40. [CrossRef]
4. Santra, S.; Tapec, R.; Theodoropoulou, N.; Dobson, J.; Hebard, A.; Tan, W. Synthesis and characterization of silica-coated iron oxide nanoparticles in microemulsion: The effect of nonionic surfactants. *Langmuir* **2001**, *17*, 2900–2906. [CrossRef]
5. Chung, C.C.; Jean, J.H. Aqueous Synthesis of Y2O2S: Eu/Silica Core-Shell Particles. *J. Am. Ceram. Soc.* **2005**, *88*, 1341–1344. [CrossRef]
6. Stöber, W.; Fink, A.; Bohn, E. Controlled growth of monodisperse silica spheres in the micron size range. *J. Colloid Interface Sci.* **1968**, *26*, 62–69. [CrossRef]
7. Kobayashi, Y.; Katakami, H.; Mine, E.; Nagao, D.; Konno, M.; Liz-Marzán, L.M. Silica coating of silver nanoparticles using a modified Stöber method. *J. Colloid Interface Sci.* **2005**, *283*, 392–396. [CrossRef]
8. Zhu, Y.; Shi, J.; Shen, W.; Dong, X.; Feng, J.; Ruan, M.; Li, Y. Stimuli-responsive controlled drug release from a hollow mesoporous silica sphere/polyelectrolyte multilayer core–shell structure. *Angew. Chem.* **2005**, *117*, 5213–5217. [CrossRef]
9. Xu, K.; Wang, J.X.; Kang, X.L.; Chen, J.F. Fabrication of antibacterial monodispersed Ag–SiO2 core–shell nanoparticles with high concentration. *Mater. Lett.* **2009**, *63*, 31–33. [CrossRef]
10. Klasen, H.J. A historical review of the use of silver in the treatment of burns. II. Renewed interest for silver. *Burns* **2000**, *26*, 131–138. [CrossRef]
11. Barillo, D.J.; Marx, D.E. Silver in medicine: A brief history BC 335 to present. *Burns* **2014**, *40*, S3–S8. [CrossRef]
12. Alexander, J.W. History of the medical use of silver. *Surg. Infect.* **2009**, *10*, 289–292. [CrossRef]
13. Lansdown, A.B. Silver I: Its antibacterial properties and mechanism of action. *J. Wound Care* **2002**, *11*, 125–130. [CrossRef]
14. Jung, W.K.; Koo, H.C.; Kim, K.W.; Shin, S.; Kim, S.H.; Park, Y.H. Antibacterial activity and mechanism of action of the silver ion in Staphylococcus aureus and Escherichia coli. *Appl. Environ. Microbiol.* **2008**, *74*, 2171–2178. [CrossRef]
15. Holt, K.B.; Bard, A.J. Interaction of silver (I) ions with the respiratory chain of Escherichia coli: An electrochemical and scanning electrochemical microscopy study of the antimicrobial mechanism of micromolar Ag+. *Biochemistry* **2005**, *44*, 13214–13223. [CrossRef]
16. Lemire, J.A.; Kalan, L.; Bradu, A.; Turner, R.J. Silver oxynitrate, an unexplored silver compound with antimicrobial and antibiofilm activity. *Antimicrob. Agents Chemother.* **2015**, *59*, 4031–4039. [CrossRef]
17. Lemire, J.A.; Kalan, L.; Gugala, N.; Bradu, A.; Turner, R.J. Silver oxynitrate–an efficacious compound for the prevention and eradication of dual-species biofilms. *Biofouling* **2017**, *33*, 460–469. [CrossRef]
18. Kalan, L.R.; Pepin, D.M.; Ul-Haq, I.; Miller, S.B.; Hay, M.E.; Precht, R.J. Targeting biofilms of multidrug-resistant bacteria with silver oxynitrate. *Int. J. Antimicrob. Agents* **2017**, *49*, 719–726. [CrossRef]
19. Thomason, H.A.; Lovett, J.M.; Spina, C.J.; Stephenson, C.; McBain, A.J.; Hardman, M.J. Silver oxysalts promote cutaneous wound healing independent of infection. *Wound Repair Regen.* **2018**, *26*, 144–152. [CrossRef]
20. Vanysek, P. Electrochemical series. In *CRC Handbook of Chemistry and Physics*; CRC Press: Boca Raton, FL, USA, 2000; p. 8.
21. Jack, J.C.; Kennedy, T. The thermal analysis of "argentic oxynitrate" and silver oxides. *J. Anal.* **1971**, *3*, 25–33. [CrossRef]
22. Waterhouse, G.I.; Bowmaker, G.A.; Metson, J.B. The thermal decomposition of silver (I, III) oxide: A combined XRD, FT-IR and Raman spectroscopic study. *Phys. Chem. Chem. Phys.* **2001**, *3*, 3838–3845. [CrossRef]
23. Waterhouse, G.I.; Metson, J.B.; Bowmaker, G.A. Synthesis, vibrational spectra and thermal stability of Ag3O4 and related Ag7O8X salts (X = NO3-, ClO4-, HSO4-). *Polyhedron* **2007**, *26*, 3310–3322. [CrossRef]
24. Bałdyga, J.; Jasińska, M.; Jodko, K.; Petelski, P. Precipitation of amorphous colloidal silica from aqueous solutions—Aggregation problem. *Chem. Eng. Sci.* **2012**, *77*, 207–216. [CrossRef]

25. Musić, S.; Filipović-Vinceković, N.; Sekovanić, L. Precipitation of amorphous SiO2 particles and their properties. *Braz. J. Chem. Eng.* **2011**, *28*, 89–94. [CrossRef]
26. Belton, D.J.; Deschaume, O.; Perry, C.C. An overview of the fundamentals of the chemistry of silica with relevance to biosilicification and technological advances. *FEBS J.* **2012**, *279*, 1710–1720. [CrossRef]
27. Ray, R.C.; Ganguly, P.B. The optimum conditions for the formation of silica gels from alkali silicate solutions. *J. Phys. Chem.* **2002**, *34*, 352–358. [CrossRef]
28. Kobayashi, Y.; Sakuraba, T. Silica-coating of metallic copper nanoparticles in aqueous solution. *Colloids Surf. A Physicochem. Eng. Asp.* **2008**, *317*, 756–759. [CrossRef]
29. Naray-Szabo, I.; Argay, G.; Szabo, P. The structure of the Ag (I, III) oxide phases. *Acta Crystallogr.* **1965**, *19*, 180–184. [CrossRef]
30. Djokić, S.S. Deposition of silver oxysalts and their antimicrobial properties. *J. Electrochem. Soc.* **2004**, *151*, C359–C364. [CrossRef]
31. Hou, X.; Kirkpatrick, R.J.; Struble, L.J.; Monteiro, P.J. Structural investigations of alkali silicate gels. *J. Am. Ceram. Soc.* **2005**, *88*, 943–949. [CrossRef]
32. Klug, H.P.; Alexander, L.E. X-ray diffraction procedures: For polycrystalline and amorphous materials. In *X-Ray Diffraction Procedures: For Polycrystalline and Amorphous Materials*, 2nd ed.; Klug, H.P., Alexander, L.E., Eds.; Wiley-VCH: Hoboken, NJ, USA, 1974; p. 992. ISBN 0-471-49369-4.
33. Schierle, C.F.; De la Garza, M.; Mustoe, T.A.; Galiano, R.D. Staphylococcal biofilms impair wound healing by delaying reepithelialization in a murine cutaneous wound model. *Wound Repair Regen.* **2009**, *17*, 354–359. [CrossRef] [PubMed]
34. Metcalf, D.G.; Bowler, P.G. Biofilm delays wound healing: A review of the evidence. *Burn. Trauma* **2013**, *1*, 5. [CrossRef] [PubMed]
35. Wolcott, R.D.; Rhoads, D.D.; Dowd, S.E. Biofilms and chronic wound inflammation. *J. Wound Care* **2008**, *17*, 333–341. [CrossRef] [PubMed]
36. Seth, A.K.; Zhong, A.; Nguyen, K.T.; Hong, S.J.; Leung, K.P.; Galiano, R.D.; Mustoe, T.A. Impact of a novel, antimicrobial dressing on in vivo, Pseudomonas aeruginosa wound biofilm: Quantitative comparative analysis using a rabbit ear model. *Wound Repair Regen* **2014**, *22*, 712–719. [CrossRef]
37. Cong, L.; Takeda, M.; Hamanaka, Y.; Gonda, K.; Watanabe, M.; Kumasaka, M.; Kobayashi, Y.; Kobayashi, M.; Ohuchi, N. Uniform silica coated fluorescent nanoparticles: Synthetic method, improved light stability and application to visualize lymph network tracer. *PLoS ONE* **2010**, *5*, e13167. [CrossRef]
38. Tang, F.; Li, L.; Chen, D. Mesoporous silica nanoparticles: Synthesis, biocompatibility and drug delivery. *Adv. Mater.* **2012**, *24*, 1504–1534. [CrossRef]
39. Barbe, C.; Bartlett, J.; Kong, L.; Finnie, K.; Lin, H.Q.; Larkin, M.; Calleja, S.; Bush, A.; Calleja, G. Silica particles: A novel drug-delivery system. *Adv. Mater.* **2004**, *16*, 1959–1966. [CrossRef]
40. Makrides, A.C.; Turner, M.; Slaughter, J. Condensation of silica from supersaturated silicic acid solutions. *J. Colloid Interface Sci.* **1980**, *73*, 345–367. [CrossRef]
41. McMillan, J.A. Higher Oxidation States of Silver. *Chem. Rev.* **1962**, *62*, 65–80. [CrossRef]
42. Holzwarth, U.; Gibson, N. The Scherrer equation versus the'Debye-Scherrer equation'. *Nat. Nanotechnol.* **2011**, *6*, 534. [CrossRef]
43. Schmidt, M.; Schwertfeger, F. Applications for silica aerogel products. *J. Non-Cryst. Solids* **1998**, *225*, 364–368. [CrossRef]
44. Reim, M.; Körner, W.; Manara, J.; Korder, S.; Arduini-Schuster, M.; Ebert, H.P.; Fricke, J. Silica aerogel granulate material for thermal insulation and daylighting. *Sol. Energy* **2005**, *79*, 131–139. [CrossRef]
45. Sandberg, L.I.; Gao, T.; Jelle, B.P.; Gustavsen, A. Synthesis of hollow silica nanospheres by sacrificial polystyrene templates for thermal insulation applications. *Adv. Mater. Sci. Eng.* **2013**, *2013*, 483651. [CrossRef]
46. Williams, D.B.; Lawton, M. Drying of organic solvents: Quantitative evaluation of the efficiency of several desiccants. *J. Org. Chem.* **2010**, *75*, 8351–8354. [CrossRef]
47. Sherlock, G.; Block, W.; Benson, E.E. Thermal analysis of the plant encapsulation-dehydration cryopreservation protocol using silica gel as the desiccant. *CryoLetters* **2005**, *26*, 45–54.

applied sciences

MDPI

Article

# Enamine Barbiturates and Thiobarbiturates as a New Class of Bacterial Urease Inhibitors

**M. Ali [1], Assem Barakat [1,2,*], Ayman El-Faham [1,2,*], Abdullah Mohammed Al-Majid [1], Sammer Yousuf [3], Sajda Ashraf [4], Zaheer Ul-Haq [4], M. Iqbal Choudhary [3,4], Beatriz G. de la Torre [5] and Fernando Albericio [1,6,7,8]**

1 Department of Chemistry, College of Science, King Saud University, P.O. Box 2455, Riyadh 11451, Saudi Arabia; maly.c@ksu.edu.sa (M.A.); amajid@ksu.edu.sa (A.M.A.-M.); albericio@ub.edu (F.A.)
2 Department of Chemistry, Faculty of Science, Alexandria University, P.O. Box 426, Ibrahimia, Alexandria 21321, Egypt
3 Hussain Ebrahim Jamal Research Institute of Chemistry, International Center for Chemical and Biological Sciences, University of Karachi, Karachi 75270, Pakistan; dr.sammer.yousuf@gmail.com (S.Y.); iqbal.choudhary@iccs.edu (M.I.C.)
4 Panjwani Center for Molecular Medicine and Drug Research, International Center for Chemical and Biological Sciences, University of Karachi, Karachi 75270, Pakistan; sajda.ashraf@yahoo.com (S.A.); zaheer_qasmi@hotmail.com (Z.U.-H.)
5 KRISP—Kwazulu-Natal Research Innovation and Sequencing Platform, College of Health Sciences, University of KwaZulu-Natal, Westville, Durban 4001, South Africa; garciadelatorreb@ukzn.ac.za
6 School of Chemistry and Physics, University of KwaZulu-Natal, University Road, Westville, Durban 4001, South Africa
7 Biomedical Research Networking Center in Bioengineering, Biomaterials and Nanomedicine (CIBER-BBN), Networking Centre on Bioengineering, Biomaterials and Nanomedicine, PCB, Baldiri Reixac 10, 08028 Barcelona, Spain
8 Department of Organic Chemistry, University of Barcelona, Martí i Franqués 1-11, 08028 Barcelona, Spain
* Correspondence: ambarakat@ksu.edu.sa (A.B.); aelfaham@ksu.edu.sa (A.E.-F.); Tel.: +966-11467-5901 (A.B.); Fax: +966-11467-5992 (A.B.)

Received: 2 April 2020; Accepted: 15 May 2020; Published: 20 May 2020

**Abstract:** Urease is a therapeutic target associated with several important diseases and health problems. Based on our previous work on the inhibition of glucosidase and other enzymes and exploiting the privileged structure assigned to the (thio)barbiturate (pyrimidine) scaffold, here we tested the capacity of two (thio)barbiturate-based compound collections to inhibit urease. Several compounds showed more activity than acetohydroxamic acid as a standard tested compound. In addition, by means of a conformational study and using the Density Functional Theory (DFT) method, we identified energetically low-lying conformers. Finally, we undertook a docking study to explore the binding mechanism of these new pyrimidine derivatives as urease inhibitors.

**Keywords:** pyrimidine-trione; barbituric; thiobarbituric; urease inhibitors; DFT

## 1. Introduction

Many microorganisms use nitrogen from urease for growth (urea amidohydrolase, EC: 3.5.1.5). Urease is an enzyme that catalyzes the hydrolysis of urea into ammonia and carbon dioxide. There is a link between health complications and ammonia production by urease [1]. In animals and humans, low pH of the stomach allows microbial strains to survive, multiply, and grow, resulting in pyelonephritis, hepatic coma, gastric carcinoma, gastric lymphoma, kidney stones, and peptic ulcer complications [2,3]. The treatment of bacterial infection with therapeutics has often proven ineffective due to drug resistance. Thus, there is a clear need for alternatives or novel treatments. Given the involvement of ureases in various diseases, pharmaceutical research has channeled considerable efforts into discovering potent and safe urease inhibitors [4–7].

Molecules with a binding site that can chelate metals are an interesting challenge and could be a promising line of action to prevent the adverse effects of ureolytic bacterial infections in humans [8]. In this regard, many types of potent urease inhibitors have been designed [9], such as dihydropyrimidines [10], urea derivatives [11], semicarbazones [12], Schiff bases [13], hydroxamic acid derivatives [14], piperazines [15], biscoumarines [16], benzimidazoles [17], and sulfonamides [18].

Enaminone molecular scaffolds have been found in many synthetic drugs and natural products [19,20]. Specially, pyrimidine-based enamines have demonstrated an array of biological activities owing to the presence of the alkenylamine moiety ($R_2NCH = CH-$) in their structures, and are capable of strong bonding with metal ion chelates in biological systems [21,22].

The barbituric and thiobarbituric acids [(thio)pyrimdine trione analog derivatives] have been reported urease inhibitors [12]. These can be considered privileged structures because they have antifungal [23], antimicrobial [24,25], anti-adiponectin [26], anti-sclerosis [27], anti-convulsing [28,29], antiglycation [30], and α-glucosidase inhibitory [30] properties. Several compounds **A, B,** and **C** have been reported, and showed anti-urease activity comparable to acetohydroxamic acid (Figure 1) [31,32].

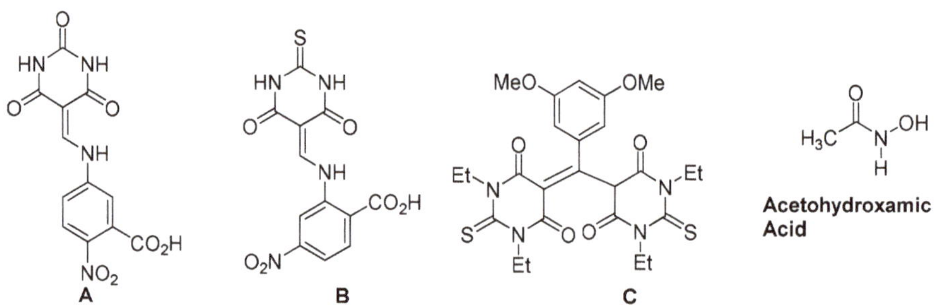


**Figure 1.** Some examples of bacterial urease inhibitors.

In this context, and as a continuation of our extended medicinal chemistry program based on the barbituric and thiobarbituric acid moieties [30,33–44], here we examined the in vitro anti-urease activity of a collection of 1,3-dimethylbarbiturate-enamine compounds and their analogs, thiobarbiturate derivatives. In addition, molecular docking studies were performed to evaluate the molecular interactions of the newly synthesized compounds with selected drug targets (PDB ID 4GY7).

## 2. Materials and Methods

The synthesis and the full characterization of compounds **3a**, **3d**, **3h**, **3j–p**, **4**, and **5** (Table 1) have been previously described by our group [30,33]. The rest of the compounds studied were prepared following the previously described method. The yields and full characterization are provided in the Supplementary materials.

### 2.1. *In Vitro Urease Inhibition Assay Protocol*

The urease inhibition assay was performed spectrophotometrically following the manufacturer's instructions [45–47]. The source of urease is Jack bean Urease and the full protocol provided in the supplementary materials. In the present study, urease was preincubated with the inhibitors for a period of 15 min, which proved to be sufficient in our studies.

**Table 1.** Urease inhibition capacity of compounds **3a–p**, **4**, and **5**.

| Compound | Structure | Urease Inhibition $IC_{50} \pm SEM$ [μM] |
|---|---|---|
| **3a** | 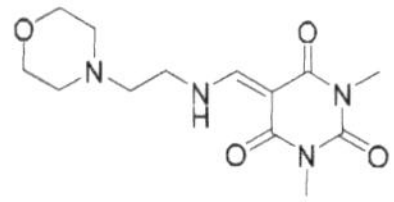 | 9 ± 2 |
| **3b** | 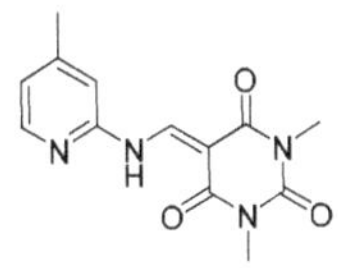 | NA |
| **3c** | 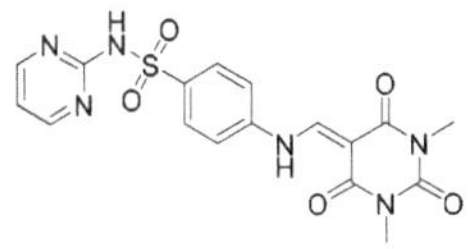 | 26 ± 1 |
| **3d** | 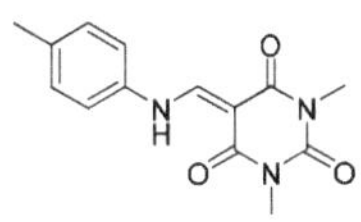 | 8 ± 0.3 |
| **3e** | 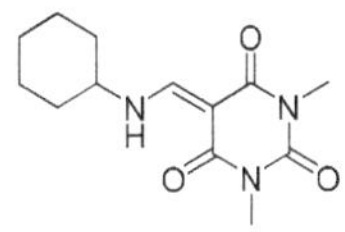 | 11 ± 0.3 |
| **3f** | 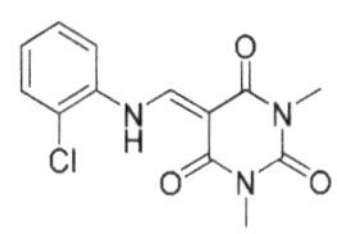 | 10 ± 0.6 |
| **3g** | 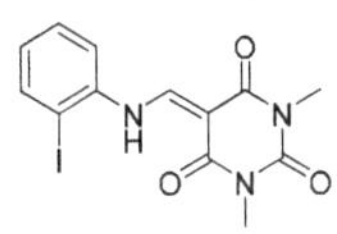 | NA |
| **3h** | 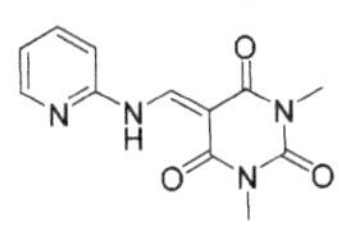 | 6.4 ± 0.3 |
| **3i** | 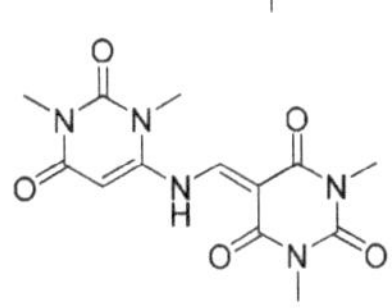 | 66 ± 2.4 |
| **3j** | 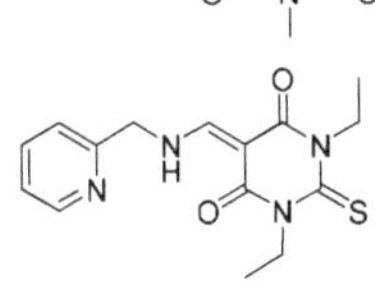 | 10 ± 0.9 |
| **3k** | 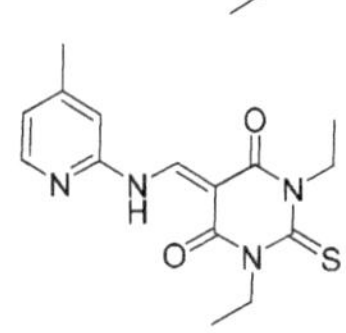 | 11 ± 1.2 |

**Table 1.** *Cont.*

| Compound | Structure | Urease Inhibition $IC_{50}$ ± SEM [µM] |
|---|---|---|
| **3l** | 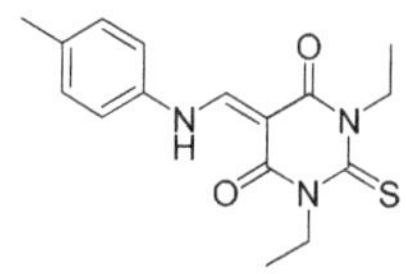 | 8 ± 0.4 |
| **3m** | 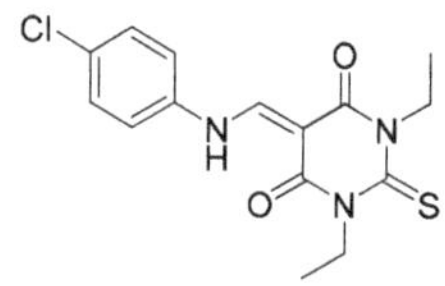 | 15 ± 1.2 |
| **3n** | 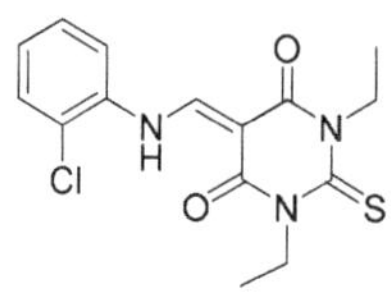 | 11 ± 1.1 |
| **3o** | 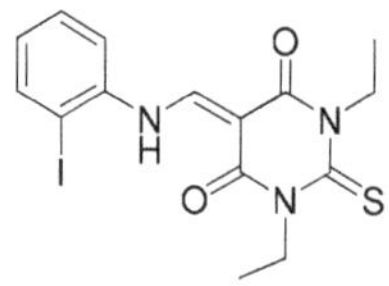 | 9 ± 0.3 |
| **3p** | 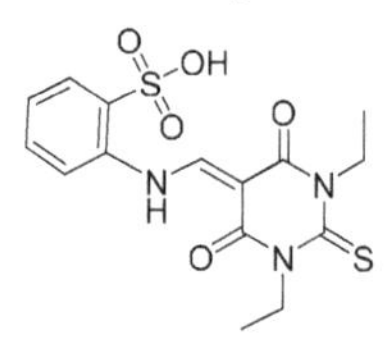 | 22 ± 0.8 |
| **4** | 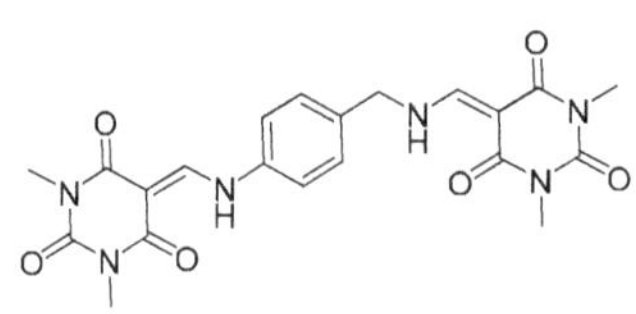 | 10 ± 1.2 |
| **5** | 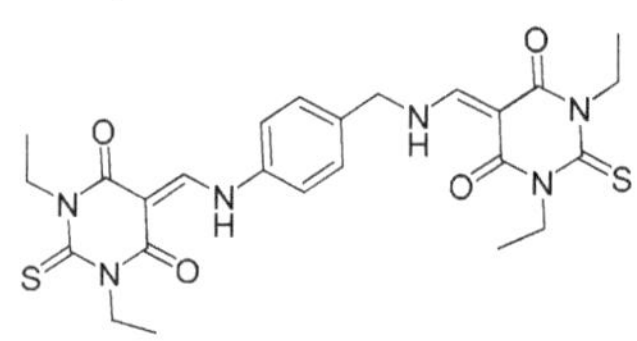 | 42 ± 2.3 |
| **STD** | **Acetohydroxamic Acid (AHA)** | 20 ± 0.4 |

## *2.2. DFT Calculation*

The molecular building and geometric optimization of all 18 compounds was performed by DFT using Gaussian 03W software package with basis set method B3LYP/6-31 + G(d,p). The Gauss-view program was used for molecular visualization. Geometrical, energetic and electronic parameters were calculated using structures optimized at 298 K and 1 atm. The Gibbs free energy equation ($\Delta G = -RT \ln K$) was used to search conformational space.

### 2.3. Molecular Docking

Molecular docking studies were conducted to rationalize the binding mode of the pyrimidine derivatives using the MOE package [Molecular Operating Environment, 2016.0810; Chemical Computing Group ULC, 1010 Sherbrooke St. West, Suite #910, Montreal, QC, Canada, H3A 2R7, 2017]. The three-dimensional structure of the urease protein was retrieved from the Protein Data Bank (PDB ID 4GY7) [48]. Water molecules were removed, and the protonation step was performed. Energy minimization using default parameters was performed to achieve a stable conformation. The optimized geometries of the pyrimidine derivatives were docked to the active site of the urease protein. For docking studies, an induced fit protocol with the triangular matcher placement method and London dG scoring function were used. The interactive docking procedure was performed for all the optimized conformers of each compound in the active site. A score was assigned to each docked compound based on its fitting in the binding cavity and its binding mode (docking study protocol are provided in the Supplementary materials).

## 3. Results and Discussion

### 3.1. Synthesis

Derivatives **3a–p**, **4**, and **5** (Table 1) were straightforwardly synthesized by nucleophilic substitution of the corresponding amines with enaminone scaffolds **1** [49] and **2** [50] in methanolic solution, following the reported method [30,33] (Scheme 1). Due to the potential tautomerism of these families of compounds, X-ray crystallography was carried out on two compounds (**3b** and **3k**) to confirm the structures. ($^1$H-NMR and $^{13}$C-NMR spectra, Figures S1–S21, and X-ray data of compounds **3b** and **3k** are provided in the Supplementary materials).

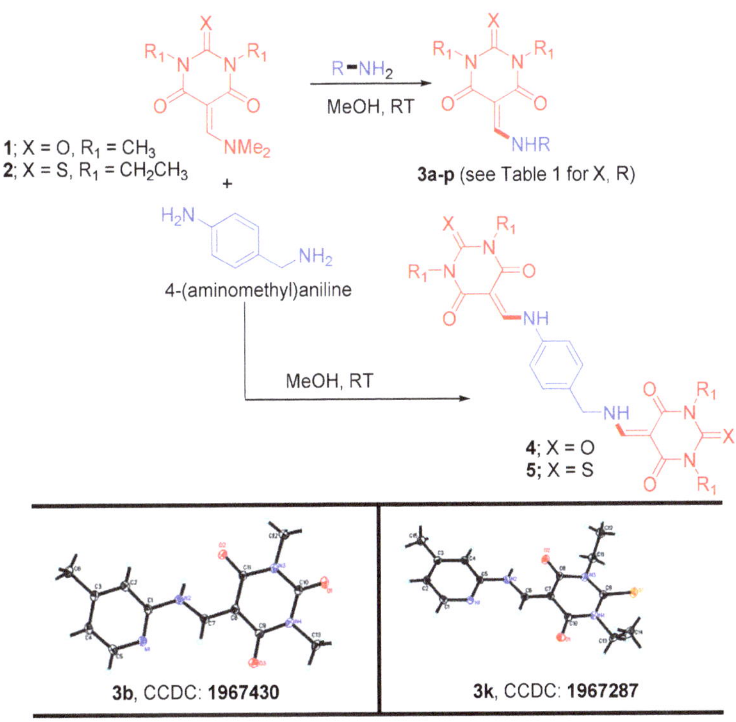


**Scheme 1.** Synthesis of **3a–p**, **4**, and **5**.

### 3.2. *In Vitro Evaluation of Urease Inhibition*

The capacity of synthesized derivatives of *N,N*-dimethylpyrimidine trione (**3a–i** and **4**) and *N,N*-diethyl-thio pyrimidine dione (**3j–p** and **5**) to inhibit urease in vitro was examined and compared with that of acetohydroxamic acid **(AHA)** ($IC_{50}$ = 20± 0.4 μM), as a standard compound (Table 1).

Of note, the tested compounds showed potent urease inhibition activity, twelve of them (**3a, 3d, 3e, 3f, 3h, 3j, 3k, 3l, 3m, 3n, 3o,** and **4**) exerting greater activity than the reference acetohydroxamic acid. The most active member between this series is compound **3h** which the structure attributed the pyridine moiety. Separation the pyridine ring from the enaminone functionality in compound **3j** with $CH_2$ group, the activity has been decreased compared to the most active member **3h**. On the other hands, compounds **3k**, and **3o** showed superior activity to their counterparts **3b** and **3g**. Indeed, **3l** and **3o** showed activity similar to that of **3d**, and **3f**. The chloro atom in *para*-position in compound **3m**, the activity has been decreased compared to the *ortho* position of compounds **3f**, and **3n**. Compounds **3c**, **3i** and **3p** with sulfonic, cyclic amide or sulfonamide functionalities shown less activity compared to standard drug. The cyclohexyl group in compound **3e** increases the activity with 1.8 folds than the control. it was observed that in case of compounds **4** and **5** the presence of the *N,N*′-diethyl group and the S atom in the thiobarbituric derivative **5** make them more voluminous compared to its analogs *N,N*′-dimethyl and the "O" atom in the barbituric derivatives **4**. Finally, compounds **3b**, and **3g** were not active.

However, there are some discrepancies regarding whether AHA is or is not a competitive inhibitor to the active site of the urease enzyme [51,52]. Therefore, some limitations could be applied, and we think that comparisons are important.

### 3.3. *Docking Study*

Molecular docking studies have the potential to identify novel drug-like molecules that display high binding affinity for the target protein, and they can facilitate the understanding of biological activity data with the purpose of designing new compounds with improved activity. Hence, we extended our study to explore the conformational space and binding orientation of the synthesized pyrimidine derivatives.

The binding conformations of these compounds were explored by MOE-Dock module implemented in the MOE program [53]. The optimized conformers obtained from the Gaussian program were docked to the active site of urease. The docking results indicated that all the conformers were well accommodated inside the active site and were stabilized by various hydrophilic, hydrophobic and van der Waals interactions. The residues involved in these interactions were Arg439, Ala440, His492, Asp494, His593, His594, Asp633 Ala636 and Met637. Compound **3h** ($IC_{50}$ = 6 ± 0.3 μM) showed strong inhibitory potential against urease as compared to **3b** (inactive) due to methyl substitution at the *para* position of the pyridine ring. Compound **3h** showed good interactions with the residues of the active site by coordinating with the bi-nickel center, via its carbonyl group at the pyrimidine ring and anchoring itself in a way that permitted strong interaction of the carbonyl group with the two nickel ions. Additionally, the two carbonyl groups on the pyrimidine ring acted as hydrogen bond acceptors and mediated hydrogen bond interaction with His492 (2.3 Å) and Arg439 (2.04 Å) of the binding pocket (Figure 2b).

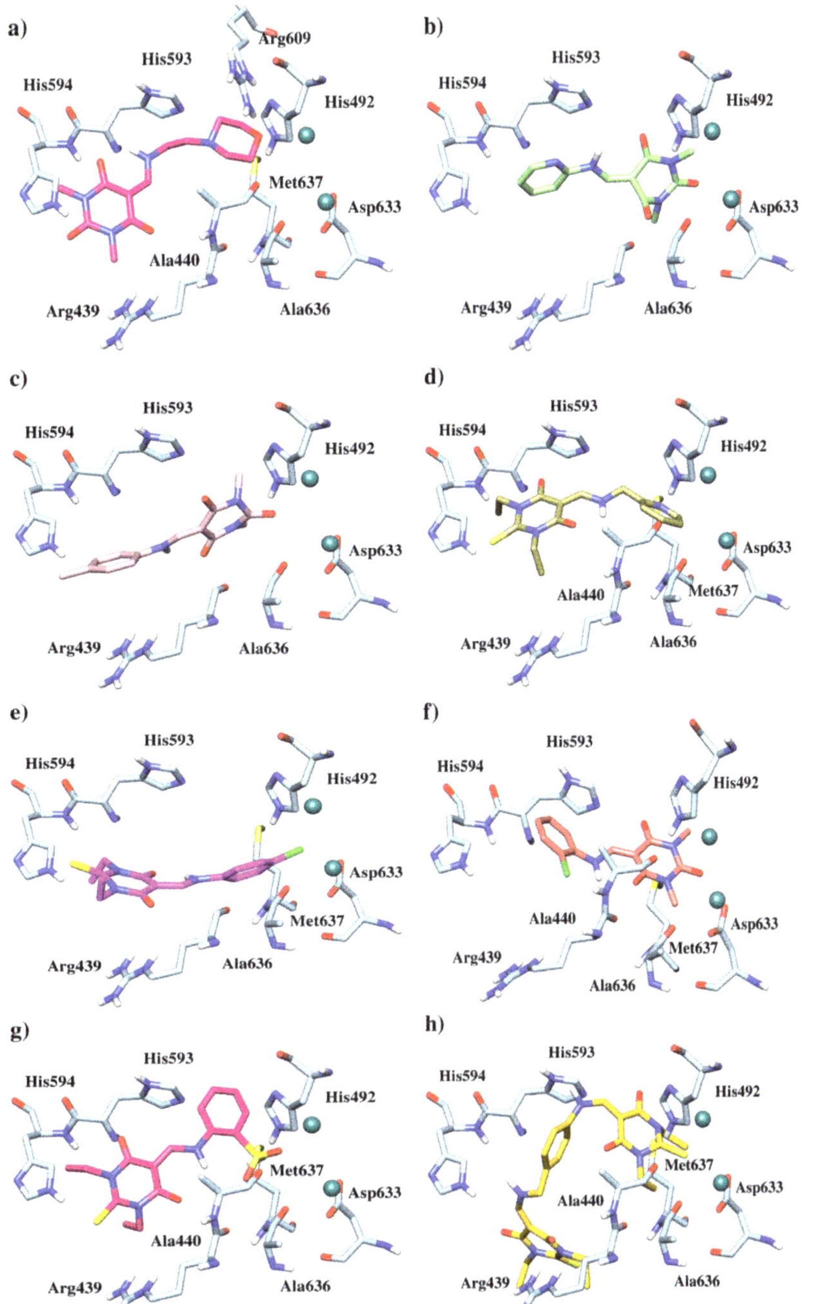


**Figure 2.** Depiction of the docking results of low energy conformer: (**a**) **3a**, (**b**) **3h**, (**c**) **3d**, (**d**) **3j**, (**e**) **3m**, (**f**) **3f**, (**g**) **3p**, and (**h**) **5**. The key residues are presented as sticks models and nickel atoms are shown as green circles.

Compound **3a** ($IC_{50}$ = 9 ± 2 μM), another active compound of the series, established various potential interactions with the active site of urease, as well as with nickel ions. The oxygen atom of the morpholine ring was involved in the chelation process with the nickel center in the catalytic site. This compound was further stabilized by two typical hydrogen bonds with His492 (2.8 Å) and Arg609 (3.0 Å). Apart from hydrogen bond interactions, a tetramine of the compound participated in the salt bridge interaction with the negatively charged Asp494 residue (Figure 2a). Compounds **3d** and **3l** ($IC_{50}$ = 8 ± 0.3 μM) had almost similar structures, biological activities and patterns of binding interactions with the active site residues. The only difference was the interaction with metal ions. The lone pair of electrons for sulfur (C = S) of **3l** mediated strong interaction with the two nickel ions and hydrogen bond interaction with Arg609 (2.4 Å) and Met637 (2.8 Å), while in case of **3d**, the carbonyl group on the pyrimidine ring was involved in the interaction with the nickel ions as depicted in Figure 2c.

Compounds **3j** ($IC_{50}$ =10 ± 0.9 μM) and **3k** ($IC_{50}$ =11 ± 1 μM), which showed good biological activity, presented similar types of interactions. The additional effectiveness of **3j** compared to **3k** was due to the absence of a methyl group at the *meta* position of the benzene ring, which allowed the compound to establish close contacts with the active site residues (Figure 2d). The ring nitrogen of the pyridine is involved in two productive hydrogen bond interaction with His492 and Met637 at 2.54 and 3.4 Å, respectively. Moreover, the C = O of pyrimidine rind mediated a potential hydrogen bond interaction with the NH of His593 at 2.59 Å, while the compound was further stabilized through hydrophobic interactions with Ala440, Cme592, and His594.

The compounds with electron-withdrawing substitution, especially halogens, showed noteworthy inhibitory potential. For **3m**, **3n**, and **3o**, the presence of halogen atoms with their respective position on the benzene ring affected the binding pattern and orientation of the compound within the pocket (Figure 2e). Compounds **3f** (*o*-chloro phenyl) and **3o** (*o*-iodo phenyl) with $IC_{50}$ = 10 ± 0.6 μM showed two hydrogen bond interactions with His492 and Ala440 (Figure 2f), while 3m and 3n were stabilized by a single hydrogen bond with His492. These compounds also displayed hydrophobic interaction with Ala440, His593 and Met637. Compounds **3c** ($IC_{50}$ = 26 ± 1μM) and **3p** ($IC_{50}$ = 22 ± 0.8 μM), both with moderate biological activity, established two hydrogen bond interactions with His492 and Ala440. However, only one interaction with the metallocenter was observed (Figure 2g). The compounds with low biological activity, as compared to the standard and other pyrimidine derivatives such as **3i** ($IC_{50}$= 66 ± 2.4 μM) and **5** ($IC_{50}$= 42 ± 2.3 μM), established single hydrogen bond interaction with the active site residue Arg439 while compound mostly established by hydrophobic interaction with the active site residues (Figure 2h). However, no interaction with nickel was observed for **3i**.

Indeed, the urease inhibition capacity of the synthesized pyrimidine derivatives is attributed to the mutual contribution of the distinct substitutions they bear. However, interaction with the metallocenter and hydrophilic interaction with Ala440, His492, His593, and Met637 are found to be crucial for the activity of these compounds.

### 3.4. Density Functional Theory (DFT)

The physicochemical properties and frontier molecular orbitals (FMOs) of the new enaminone compounds play a crucial role in enhancing bioactivity. Khon-Sham's DFT approach with the B3LYP method was used for geometry optimization [54]. The electron-donating and -withdrawing ability of a compound can be explained by its HOMO and LUMO. The higher the energy value of HOMO, the greater the electron-contributing ability of the compound. The energy difference between HOMO and LUMO is an established parameter to measure the electron conductivity or degree of intermolecular charge transfer, which also affects bioactivity [55]. In this study, to examine the urease inhibition capacity of the pyrimidine derivatives, we randomly selected compounds for comparison with DFT results.

The results of HOMO-LUMO energies were plotted against the biological activity of the pyrimidine derivatives (Table 2). Good correlation was observed between biological activity and the energies of the LUMO orbitals. Compounds **3h**, **3k**, **3f**, and **3o** showed significant LUMO energy of −1.89, −2.24, −1.96 and −2.26 eV, respectively.

**Table 2.** Theoretical results of pyrimidine derivatives.

| Compound | $IC_{50}$ | HOMO (eV) | LUMO (eV) | $E_{gap}$ (eV) |
|---|---|---|---|---|
| **3a** | 9 ± 2 | −5.79330622969 | −1.21770997548 | −4.57559625421 |
| **3d** | 8 ± 0.3 | −5.88881823 | −1.70370550982 | −4.76988562237 |
| **3f** | 10 ± 0.6 | −6.2202530345 | −1.96248588673 | −4.25776714777 |
| **3h** | 6 ± 0.3 | −6.24583174683 | −1.88955934518 | −4.35627240165 |
| **3k** | 11 ± 1 | −5.70922301574 | −2.24439595033 | −3.46482706541 |
| **3o** | 9 ± 0.3 | −5.75357759138 | −2.25963433214 | −3.49394325924 |

The visualization of the HOMO-LUMO orbitals of **3h** reflects the localization of FMO (frontier molecular orbital) (Figure 3). The negative and positive phases of orbitals are shown in green and red, respectively. For **3h**, HOMO is localized on the pyrimidine ring, distal pyridine moiety and carbonyl group, whereas LUMO is on the pi-bond adjacent to the pyrimidine ring. In contrast, for **3o**, electrons are delocalized mainly on the benzene ring with iodine group as a substituent. The influence of LUMO energy on the inhibitory activity of the compounds might be due to the presence of halogen substitution and the pyridine ring. The halogen substitution at *ortho* and *para* positions made a great contribution to the urease inhibition capacity of the derivatives. Notably, neither HOMO nor LUMO were located on the methyl group, which would explain why this group contributed the least to binding with the protein.

The energy gap for the most active compound **3h** of the series was −4.35. A similar energy difference was observed for **3k**, **3f**, and **3o**. The smaller the difference in HOMO-LUMO energies, the greater the chemical reactivity. For any potential interaction, electron transfer from high-lying HOMO to low-lying LUMO is always energetically favorable. Given this consideration, **3h**, **3k**, **3f**, and **3o** possess good activity, which correlated well with urease inhibitory activity.

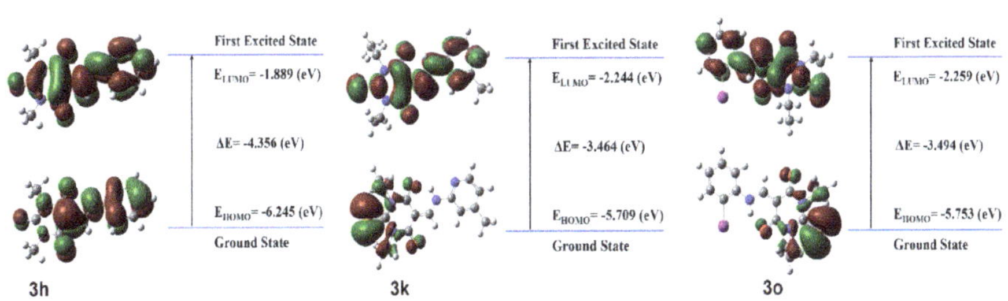


**Figure 3.** The optimized geometries and surfaces of the HOMO (Highest occupied molecular orbital)-LUMO (Lowest unoccupied molecular orbital) of **3h**, **3k** and **3o** obtained at the B3LYP/6-31G (d, p) level.

Molecular electrostatic potentials (MEP) were run for compounds **3h**, **3k**, and **3o** at B3LYP/6-31G (d, p), providing information about the sites reactive towards nucleophilic and electrophilic attack, together with hydrogen-bond interactions. The potential of electrostatic interaction at the surface is shown by blue, red and green, representing the sites for positive, negative and no electrostatic potential, respectively (Figure 4). Moreover, the positive and negative regions of the maps were responsible for the nucleophilic and electrophilic reactivity of the compounds.

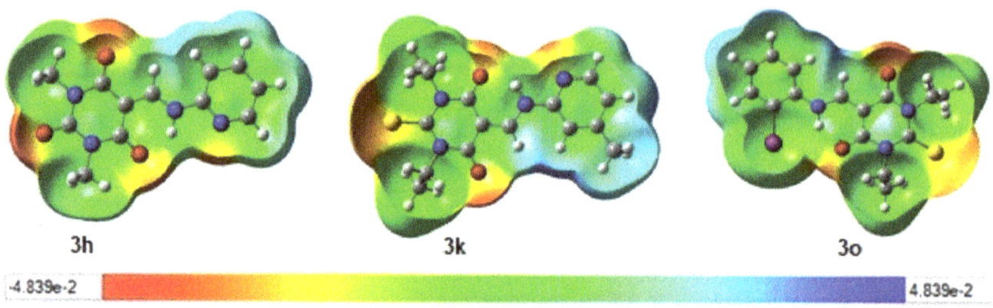


**Figure 4.** The molecular electrostatic potential (MEP) of **3h**, **3k**, and **3o**.

## 4. Conclusions

Based on our previous work addressing the inhibition of glucosidase and taking advantage of the privileged structure associated with the (thio)barbiturate (pyrimidiene) scaffold, here we have identified compounds with high in vitro anti-urease activity. All conformations of the pyrimidine derivatives were optimized with the B3LYP/6–31G method. The DFT results for the compounds correlated well with the experimental data, which indicated that the presence of the pyridine ring and electron-withdrawing groups, especially halogens (compounds **3m**, **3n**, and **3o**), play an important role in conferring urease inhibition activity. Moreover, the molecular docking studies further helped to explain the experimental results. Because of our findings, compounds **3a**, **3d**, **3h**, **3k**, and **3o** emerge as potential leads for the development of urease inhibitors. In addition, the obtained results confirm that the use of privileged structures for medicinal chemistry programs is an excellent strategy for identifying hits and leads. Further investigation will be carried out in the future for urease enzyme competitive inhibition.

**Supplementary Materials:** The following are available online at http://www.mdpi.com/2076-3417/10/10/3523/s1. The details of the synthesis and characterization of the studied compounds; X-ray data of compounds **3b** and **3k**; biological activity assay; docking study protocol and copies of NMR spectra (Figures S1–S21) of the synthesized target compounds are provided as Supplementary materials.

**Author Contributions:** Conceptualization, A.B. and A.E.-F.; Data curation, S.A. and Z.U.-H.; Funding acquisition, A.M.A.-M.; Investigation, M.A. and A.B.; Methodology, M.A.; Software, S.Y., S.A. and Z.U.-H.; Supervision, A.E.-F., M.I.C. and F.A.; Visualization, A.M.A.-M., M.I.C. and B.G.d.l.T.; Writing—original draft, A.B.; Writing—review & editing, A.B., M.I.C., B.G.d.l.T. and F.A. All authors have read and agreed to the published version of the manuscript.

**Funding:** Deanship of Scientific Research at King Saud University, Saudi Arabia, research group No. (RGP-044).

**Acknowledgments:** The authors would like to extend their sincere appreciation to the Deanship of Scientific Research at King Saud University for supporting and funding the research group No. (RGP-044).

**Conflicts of Interest:** The authors declare no conflict of interest.

## References

1. Mobley, H.L.; Hausinger, R.P. Microbial ureases: Significance, regulation, and molecular characterization. *Microbiol. Mol. Biol. Rev.* **1989**, *53*, 85–108. [CrossRef]
2. Zonia, L.E.; Stebbins, N.E.; Polacco, J.C. Essential role of urease in germination of nitrogen-limited Arabidopsis thaliana seeds. *Plant Physiol.* **1995**, *107*, 1097–1103. [CrossRef] [PubMed]
3. Mulvaney, R.L.; Bremner, J.M. *Soil Biochemistry*; Paul, E.A., Ladd, J.N., Eds.; Marcel Dekker, Inc.: New York, NY, USA, 1981; pp. 153–196.
4. Kataria, R.; Khatkar, A. In-silico Designing, ADMET Analysis, Synthesis and Biological Evaluation of Novel Derivatives of Diosmin Against Urease Protein and Helicobacter pylori Bacterium. *Curr. Top. Med. Chem.* **2019**, *19*, 2658–2675. [CrossRef] [PubMed]
5. Li, W.Y.; Ni, W.W.; Ye, Y.X.; Fang, H.L.; Pan, X.M.; He, J.L.; Zhou, T.-L.; Yi, J.; Liu, S.-S.; Zhou, M.; et al. N-monoarylacetothioureas as potent urease inhibitors: Synthesis, SAR, and biological evaluation. *J. Enzyme Inhib. Med. Chem.* **2020**, *35*, 404–413. [CrossRef]

6. Qamar, N.; Sultan, H.; Raheel, A.; Ashfaq, M.; Azmat, R.; Naz, R.; Raheela, L.M.; Khan, K.M.; Arshad, T. Heterochelates of metals as an effective anti-Urease agents couple with their docking studies. *Pak. J. Pharm. Sci.* **2019**, *32*, 1179–1183.
7. Shah, S.R.; Shah, Z.; Khiat, M.; Khan, A.; Hill, L.R.; Khan, S.; Hussain, J.; Csuk, R.; Anwar, M.U.; Al-Harrasi, A. Complexes of N-and O-Donor Ligands as Potential Urease Inhibitors. *ACS Omega* **2020**, *17*, 10200–10206. [CrossRef]
8. Hanif, M.; Saleem, M.; Hussain, M.T.; Rama, N.H.; Zaib, S.; Aslam MA, M.; Iqbal, J. Synthesis, urease inhibition, antioxidant and antibacterial studies of some 4-amino-5-aryl-3H-1,2,4-triazole-3-thiones and their 3,6-disubstituted 1,2,4-triazolo[3,4-b]1,3,4-thiadiazole derivatives. *J. Braz. Chem. Soc.* **2012**, *23*, 854–860. [CrossRef]
9. Kafarski, P.; Talma, M. Recent advances in design of new urease inhibitors: A review. *J. Adv. Res.* **2018**, *13*, 101–112. [CrossRef]
10. Shamim, S.; Khan, K.M.; Salar, U.; Ali, F.; Lodhi, M.A.; Taha, M.; Perveen, S. 5-Acetyl-6-methyl-4-aryl-3,4-dihydropyrimidin-2(1H)-ones: As potent urease inhibitors; synthesis, in vitro screening, and molecular modeling study. *Bioorg. Chem.* **2018**, *76*, 37–52. [CrossRef]
11. Perveen, S.; Khan, K.M.; Lodhi, M.A.; Choudhary, M.I.; Voelter, W. Urease and α-chymotrypsin inhibitory effects of selected urea derivatives. *Lett. Drug Des. Discov.* **2008**, *5*, 401–405. [CrossRef]
12. Pervez, H.; Chohan, Z.H.; Ramzan, M.; Nasim FU, H.; Khan, K.M. Synthesis and biological evaluation of some new N(4)-substituted isatin-3-thiosemicarbazones. *J. Enzyme Inhib. Med. Chem.* **2009**, *24*, 437–446. [CrossRef] [PubMed]
13. Aslam MA, S.; Mahmood, S.U.; Shahid, M.; Saeed, A.; Iqbal, J. Synthesis, biological assay in vitro and molecular docking studies of new Schiff base derivatives as potential urease inhibitors. *Eur. J. Med. Chem.* **2011**, *46*, 5473–5479. [CrossRef] [PubMed]
14. Arora, R.; Issar, U.; Kakkar, R. In silico study of the active site of Helicobacter pylori urease and its inhibition by hydroxamic acids. *J. Mol. Graph. Model.* **2018**, *83*, 64–73. [CrossRef] [PubMed]
15. Taha, M.; Wadood, A. Synthesis and molecular docking study of piperazine derivatives as potent urease inhibitors. *Bioorg. Chem.* **2018**, *78*, 411–417.
16. Khan, K.M.; Iqbal, S.; Lodhi, M.A.; Maharvi, G.M.; Choudhary, M.I.; Perveen, S. Biscoumarin: New class of urease inhibitors; economical synthesis and activity. *Bioorg. Med. Chem.* **2004**, *12*, 1963–1968. [CrossRef]
17. Menteşe, E.; Bektaş, H.; Sokmen, B.B.; Emirik, M.; Çakır, D.; Kahveci, B. Synthesis and molecular docking study of some 5,6-dichloro-2-cyclopropyl-1H-benzimidazole derivatives bearing triazole, oxadiazole, and imine functionalities as potent inhibitors of urease. *Bioorg. Med. Chem. Lett.* **2017**, *27*, 3014–3018.
18. Saeed, A.; Mahmood, S.U.; Rafiq, M.; Ashraf, Z.; Jabeen, F.; Seo, S.Y. Iminothiazoline-sulfonamide hybrids as Jack bean urease inhibitors; synthesis, kinetic mechanism and computational molecular modeling. *Chem. Biol. Drug. Des.* **2016**, *87*, 434–443. [CrossRef]
19. Mabkhot, Y.N.; Barakat, A.; Yousuf, S. Substituted thieno [2, 3-b] thiophenes and related congeners: Synthesis, β-glucuronidase inhibition activity, crystal structure, and POM analyses. *Bioorg. Med. Chem.* **2014**, *22*, 6715–6725. [CrossRef]
20. Mabkhot, Y.N.; Al-Majid, A.M.; Barakat, A.; Alshahrani, S.; Siddiqui, Y. 1, 1′-(3-Methyl-4-phenylthieno [2–b] thiophene-2, 5-diyl) diethanone as a building block in heterocyclic synthesis. novel synthesis of some pyrazole and pyrimidine derivatives. *Molecules* **2011**, *16*, 6502–6511. [CrossRef]
21. Fırıncı, R.; Fırıncı, E.; Başbülbül, G.; Dabanca, M.B.; Celepci, D.B.; Günay, M.E. Enamines of 1, 3-dimethylbarbiturates and their symmetrical palladium (II) complexes: Synthesis, characterization and biological activity. *Trans. Met. Chem.* **2019**, *44*, 391–397.
22. Cox, D.S.; Scott, K.R.; Gao, H.; Eddington, N.D. Effect of P-glycoprotein on the pharmacokinetics and tissue distribution of enaminone anticonvulsants: Analysis by population and physiological approaches. *J. Pharmacol. Exp. Ther.* **2002**, *302*, 1096–1104. [CrossRef] [PubMed]
23. Kidwai, M.; Thakur, R.; Mohan, R. Ecofriendly synthesis of novel antifungal (thio)- barbituric acid derivatives. *Acta Chim. Solv.* **2005**, *52*, 88–92.
24. Figueiredo, J.; Serrano, J.L.; Cavalheiro, E.; Keurulainen, L.; Yli-Kauhaluoma, J.; Moreira, V.M.; Almeida, P. Trisubstituted barbiturates and thiobarbiturates: Synthesis and biological evaluation as xanthine oxidase inhibitors, antioxidants, antibacterial and anti-proliferative agents. *Eur. J. Med. Chem.* **2018**, *143*, 829–842. [CrossRef] [PubMed]

25. Dabholkar, V.V.; Ravi, T.D. Synthesis of Biginelli products of thiobarbituric acids and their antimicrobial activity. *J. Serb. Chem. Soc.* **2010**, *75*, 1033–1040. [CrossRef]
26. Ma, L.; Li, S.; Zheng, H. Synthesis and biological activity of novel barbituric and thiobarbituric acid derivatives against non-alcoholic fatty liver disease. *Eur. J. Med. Chem.* **2011**, *46*, 2003. [CrossRef] [PubMed]
27. Khan, K.M.; Ali, M.; Farooqui, T.A.; Khan, M.; Taha, M.; Perveen, S. An improved method for the synthesis of 5- arylidene barbiturates using $BiCl_3$. *J. Chem. Soc. Pak.* **2009**, *31*, 823–828.
28. Archana, S.V.K.; Kumar, A. Synthesis of some newer derivatives of substitute quinazolinonyl-2-oxo/thiobarbituric acid as potent anticonvulsant agents. *Bioorg. Med. Chem.* **2004**, *12*, 1257–1264. [CrossRef]
29. Hassan, M.; Khan Faidallah, K.A. Synthesis and biological evaluation of new barbituric and thiobarbituric acid fluoro analogs of benzenesulfonamides as antidiabetic and antibacterial agents. *J. Fluorine Chem.* **2012**, *142*, 96–104.
30. Ali, M.; Barakat, A.; El-Faham, A.; Al-Rasheed, H.H.; Dahlous, K.; Al-Majid, A.M.; Choudhary, M.I. Synthesis and characterisation of thiobarbituric acid enamine derivatives, and evaluation of their α-glucosidase inhibitory and anti-glycation activity. *Enzyme Inhib. Med. Chem.* **2020**, *35*, 692–701. [CrossRef]
31. Puerta, D.T.; Cohen, S.M.A. A bioinorganic perspective on matrix metalloproteinase inhibition. *Curr. Topic. Med. Chem.* **2004**, *4*, 1551–1573. [CrossRef]
32. Hameed, A.; Al-Rashida, M.; Uroos, M. A patent update on therapeutic applications of urease inhibitors (2012–2018). *Expert Opin. Rapeutic Pat.* **2019**, *29*, 181–189. [CrossRef] [PubMed]
33. Barakat, A.; Soliman, S.M.; Ali, M.; Elmarghany, A.; Al-Majid, A.M.; Yousuf, S.; El-Faham, A. Synthesis, crystal structure, evaluation of urease inhibition potential and the docking studies of cobalt(III) complex based on barbituric acid Schiff base ligand. *Inorg. Chim. Acta.* **2020**, *503*, 119405. [CrossRef]
34. Barakat, A.; Ali, M.; Al-Majid, A.M.; Yousuf, S.; Choudhary, M.I.; Khalil, R.; Ul-Haq, Z. Synthesis of thiobarbituric acid derivatives: In vitro α-glucosidase inhibition and molecular docking studies. *Bioorg. Chem.* **2017**, *75*, 99–105. [CrossRef] [PubMed]
35. Barakat, A.; Al-Majid, A.M.; Al-Najjar, H.J.; Mabkhot, Y.N.; Javaid, S.; Yousuf, S.; Choudhary, M.I. Zwitterionic pyrimidinium adducts as antioxidants with therapeutic potential as nitric oxide scavenger. *Eur. J. Med. Chem.* **2014**, *84*, 146. [CrossRef] [PubMed]
36. Barakat, A.; Islam, M.S.; Al-Majid, A.M.; Ghabbour, H.A.; Fun, H.K.; Javed, K.; Wadood, A. Synthesis, in vitro biological activities and in silico study of dihydropyrimidines derivatives. *Bioorg.Med. Chem.* **2015**, *23*, 6740. [CrossRef] [PubMed]
37. Barakat, A.; Al-Majid, A.M.; Soliman, S.M.; Lotfy, G.; Ghabbour, H.A.; Fun, H.K.; Sloop, J.C. New diethyl ammonium salt of thiobarbituric acid derivative: Synthesis, molecular structure investigations and docking studies. *Molecules* **2015**, *20*, 20642–20658. [CrossRef] [PubMed]
38. Barakat, A.; Al-Majid, A.M.; Lotfy, G.; Arshad, F.; Yousuf, S.; Choudhary, M.I.; Ul-Haq, Z. Synthesis and Dynamics Studies of Barbituric Acid Derivatives as Urease Inhibitors. *Chem. Cent. J.* **2015**, *9*, 63. [CrossRef]
39. Badria, F.A.; Atef, S.; Al-Majid, A.M.; Ali, M.; Elshaier, Y.A.; Ghabbour, H.A.; Barakat, A. Synthesis and inhibitory effect of some indole-pyrimidine-based hybrid heterocycles on α-glucosidase and α-amylase as potential hypoglycemic agents. *ChemistryOpen* **2019**, *8*, 1288–1297. [CrossRef]
40. Barakat, A.; Islam, M.S.; Al-Majid, A.M.; Ghabbour, H.A.; Yousuf, S.; Ashraf, M.; Ul-Haq, Z. Synthesis of pyrimidine-2,4,6-trione derivatives: Anti-oxidant, Anti-cancer, α-glucosidase, ß-glucuronidase inhibiton and their molecular docking studies. *Bioorg. Chem.* **2016**, *86*, 72–79. [CrossRef]
41. Barakat, A.; Soliman, S.M.; Elshaier, Y.A.M.M. Molecular structure investigation, hypoglycemic/anticancer and docking studies of 5-((4-fluorophenyl)(2-hydroxy-6-oxocyclohex-1-en-1-yl)methyl)-6-hydroxy-1,3-dimethylpyrimidine-2,4(1H,3H)-dione. *J. Mol. Struc.* **2017**, *1134*, 99–111. [CrossRef]
42. Ul-Haq, Z.; Ashraf, S.; Al-Majid, A.M.; Barakat, A. 3D-QSAR Studies on Barbituric Acid Derivatives as Urease Inhibitors and the Effect of Charges on the Quality of a Model. *Int. J. Mol. Sci.* **2016**, *17*, 657. [CrossRef] [PubMed]
43. Khan, K.M.; Rahim, F.; Khan, A. Synthesis and structure–activity relationship of thiobarbituric acid derivatives as potent inhibitors of urease. *Bioorg. Med. Chem.* **2014**, *22*, 4119. [CrossRef] [PubMed]
44. Altowyan, M.S.; Barakat, A.; Soliman, S.M.; Al Majid, A.M.; Ali, M.; Elshaier, Y.A.; Ghabbour, H.A. A New Barbituric Acid Derivatives as Reactive Oxygen Scavenger: Experimental and Theoretical Investigations. *J. Mol. Struc.* **2019**, *1175*, 524–535. [CrossRef]

45. Weatherburn, M.W. Phenol-hypochlorite reaction for determination of ammonia. *Anal. Chem.* **1967**, *39*, 971. [CrossRef]
46. Mabkhot, Y.N.; Aldawsari, F.D.; Al-Showiman, S.S.; Barakat, A.; Soliman, S.M.; Choudhary, M.I.; Hadda, T.B. Novel enaminone derived from thieno [2–b] thiene: Synthesis, x-ray crystal structure, HOMO, LUMO, NBO analyses and biological activity. *Chem. Cent. J.* **2015**, *9*, 24. [CrossRef]
47. Khan, M.K.; Saify, Z.S.; Arif Lodhi, M.; Butt, N.; Perveen, S.; Murtaza Maharvi, G.; Atta-Ur-Rahman. Piperidines: A new class of urease inhibitors. *Nat. Prod. Res.* **2006**, *20*, 523. [CrossRef]
48. Begum, A.; Banumathi, S.; Choudhary, M.I.; Betzel, C. Crystallographic structure analysis of urease from Jack bean (Canavalia ensiformis) at 1.49 A Resolution. *Macromolecules* **2012**. [CrossRef]
49. Kulinich, A.V.; Derevyanko, N.A.; Ishchenko, A.A. Synthesis and spectral properties of cyanine dyes-derivatives of 10,10-dimethyl-7,8,9,10-tetrahydro-6H-pyrido[1–a]indolium. *J. Photochem. Photobiol.* **2008**, *198*, 119–125. [CrossRef]
50. Gorobets, N.Y.; Yousefi, B.H.; Belaj, F.; Kappe, C.O. Rapid microwave-assisted solution phase synthesis of substituted 2-pyridone libraries. *Tetrahedron* **2004**, *60*, 8633–8644. [CrossRef]
51. Kumar, S.; Kayastha, A.M. Acetohydroxamic Acid—A Competitive Inhibitor of Urease from Soybean "Glycine max". *J. Proteins Proteom.* **2013**, *1*, 3–8.
52. Krajewska, B.; Zaborska, W. Jack bean urease: The effect of active-site binding inhibitors on the reactivity of enzyme thiol groups. *Bioorg. Chem.* **2007**, *35*, 355–365. [CrossRef] [PubMed]
53. Chemical Computing Group. *Molecular Operating Environment (MOE) 2016.0810*; Chemical Computing Group Inc.: Montreal, QC, Canada, 2016. Available online: https://www.chemcomp.com/Products.htm (accessed on 19 May 2020).
54. Rauf, A.; Shahzad, S.; Bajda, M.; Yar, M.; Ahmed, F.; Hussain, N.; Jończyk, J. Design and synthesis of new barbituric-and thiobarbituric acid derivatives as potent urease inhibitors: Structure activity relationship and molecular modeling studies. *Bioorg. Med. Chem.* **2015**, *23*, 6049–6058. [CrossRef] [PubMed]
55. Stephens, P.J.; Devlin, F.J.; Chabalowski, C.F.N.; Frisch, M.J. Ab initio calculation of vibrational absorption and circular dichroism spectra using density functional force fields. *J. Phys. Chem.* **1994**, *98*, 11623–11627. [CrossRef]

applied sciences

MDPI

Article

# Synthesis of Novel Potent Biologically Active *N*-Benzylisatin-Aryl Hydrazones in Comparison with Lung Cancer Drug 'Gefitinib'

**Huda S. Al-Salem [1,*], Hatem A. Abuelizz [1], Iman S. Issa [1], Amany Z. Mahmoud [2,3], Ali AlHoshani [4], Md Arifuzzaman [5] and A. F. M. Motiur Rahman [1,*]**

1 Department of Pharmaceutical Chemistry, College of Pharmacy, King Saud University, Riyadh 11451, Saudi Arabia; habuelizz@ksu.edu.sa (H.A.A.); iman_issa69@yahoo.com (I.S.I.)
2 Departments of Pharmaceutics, College of Pharmacy, King Saud University, Riyadh 11451, Saudi Arabia; azmahmoud@ksu.edu.sa
3 Department of Pharmaceutical Medicinal Chemistry, Faculty of Pharmacy, Assiut University, Assiut 71516, Egypt
4 Departments of Pharmacology and Toxicology, College of Pharmacy, King Saud University, Riyadh 11451, Saudi Arabia; ahoshani@ksu.edu.sa
5 College of Pharmacy, Yeungnam University, Gyeongsan 38541, Korea; arifmilon2016@gmail.com
* Correspondence: hhalsalem@ksu.edu.sa (H.S.A.-S.); afmrahman@ksu.edu.sa (A.F.M.M.R.); Tel.: +966-11-29-52740 (H.S.A.); +966-11-46-70237 (A.F.M.M.R.)

Received: 4 April 2020; Accepted: 13 May 2020; Published: 26 May 2020

**Featured Application: Excellent potential scaffold for the development of anticancer therapeutics.**

**Abstract:** Developing anticancer therapeutics with no/few side effects is a challenge for medicinal chemists. The absence of antibacterial activity of an anticancer drug removes its detrimental effect toward intestinal flora and therefore leads to reduced side effects. Here, a series of novel *N*-benzylisatin-aryl-hydrazones was designed, synthesized and evaluated for their antimicrobial and antiproliferative activities with SAR and ADME studies, aiming to develop anticancer drugs with no antimicrobial, yet high antiproliferative activities. The results were then compared with the effects of first-line treatments for lung cancer drug Gefitinib. Novel *N*-benzylisatin-aryl-hydrazones were synthesized from isatin and benzyl bromide in three steps with good to moderate yields. Antimicrobial activity was tested with six Gram-positive/negative bacterial strains, antifungal activity with a fungal strain and antiproliferative activity against 'A549' and 'HeLa cell lines', respectively. As expected, synthesized hydrazones reveled no effects on any of the strains of bacteria and fungi up to 100-μg/disc concentration. However, four compounds showed two-to-four fold antiproliferative activity over Gefitinib. For instance, $IC_{50}$ of a compound (**6c**) shows concentration of 4.35 μM, whereas gefitinib shows 15.23 μM against 'A549.' ADME predicted studies reveled that our synthesized hydrazones exhibited higher ADME values than the Gefitinib. Therefore, our synthesized hydrazones can be an excellent scaffold for the development of anticancer therapeutics after considering further investigations.

**Keywords:** *N*-benzylisatin-aryl hydrazones; gefitinib; cancer; A549 cell lines

## 1. Introduction

Cancer is a life-threatening disease; one of the major challenges for relieving its burden is to develop highly effective drugs with specificity for cancer, but few to no side effects on normal mammalian cells [1]. Gefitinib, the epidermal growth factor receptor (EGFR) inhibitor, was added recently to the

Food and Drug Administration (FDA) list of the recommended first-line treatments for lung cancer [2]. Gefitinib is particularly recommended for non-small cell lung cancer (NSCLC) type [3]. Although that erlotinib and afatinib are also included in this list, gefitinib may be the most tolerable [4]. The most common adverse effects of this group of drugs were gastrointestinal diarrhea, nausea and/or vomiting which is due to the toxic effect of the drug on normal flora [5,6]. The absence of antibacterial activity of the anticancer drug removes its detrimental effect against intestinal flora, suggesting a highly promising new strategy for the development of anticancer drugs with reduced side effects [7]. In 2008, O'Shea and Moser reported [8] that molecular weight is an important factor for antimicrobial activity. They did experiment on 147 antibacterial and 4623 non-antibacterial compounds and found average molecular weight for an antibacterial compound were 812 Da (Gram '+' positive) and 414 Da (Gram '−' negative). In addition, another important factor for becoming an antimicrobially active compound is polarity. According to the Lipinski's rule of five, also known as Pfizer's rule of five or simply the rule of five (RO5) [9], 70.4% of the antibacterial active compounds showed logP values ranges from 0–5. Therefore, in order to target a compound with no/little antimicrobial activity, lower molecular weight and with more logP values should be taken into consideration, isatin and its derivatives with low molecular weight could be a choice of interests [10–12]. Isatin, a natural compound, is known for more than a century and still being used extensively in medicinal compound synthesis [13–19]. It has been reported that, various substituents on isatin nucleus displayed numerous biological activities [12,20]. In recent years, number of isatin derivatives were reported with extensive biological activities [21–25], included EGFR activity [26]. *N*-benzylisatin hydrazones of fluorescein had showed antiproliferative activity as well as topo II inhibitory activity [27]. Moreover, two series of hydrazone derivatives has been reported recently with antiproliferative activity [28,29]. A number of marketed drugs and potential anticancer agents having isatin moiety, depicted in Figure 1, inspired us to synthesize a series of isatin hydrazones having *N*-benzyl protection at 1-position of isatin and hydrazone formation at 3-position with various aryl substituents. We thus anticipating that simplification of isatin molecule would prevent its inhibitory effect on different microbes, providing more selectivity in the action on cancerous cell lines with less/no toxicity on the gastrointestinal tract (GIT) lining cells. Considering the above points and the importance of the development of anticancer therapeutics with no/few side effects, we therefore designed and synthesized a series of *N*-protected *N*-benzylisatin-aryl hydrazones with a lower molecular weight (353–418 Da), evaluated their antibacterial activity against two Gram-positive, four Gram-negative bacterial strains and antifungal activity against *Candida albicans* NRRL Y-477, and antiproliferative activities against non-small cell lung cancer cell lines 'A549', as well as human cervical cancer cells lines 'HeLa.' In addition, for comparing the potency of the synthesized compounds, "gefitinib" was used as a positive control for antiproliferative activity evaluation.

In order to consider a compound as drug molecule, it is necessary to test their drug likeness properties as well as the analysis of physiological descriptors such as absorption, distribution, metabolism and excretion (ADME). ADME are important physiological descriptors of chemical compounds for selecting highly potential drug targets. However, testing a wide range of compounds directly in clinical or pre-clinical phase is extensively time consuming and costly. Moreover, ADME was considered as the last step of drug development where many drugs (approximately 60%) were failed after all the procedures. To solve these problems, recent experiments utilizes in silico ADME tools as the first step to shorten the amount of target compounds, by calculating predicted ADME properties and by discarding the compounds with unsatisfactory ADME values from the drug designing pipe line [30]. This prediction enabled us to identify potent drug candidate by analyzing the properties of the designed compounds. Therefore, we studied the ADME predicted parameters of the synthesized compounds (**6a–j**) using in silico ADME tools and compared them with that of "gefitinib".

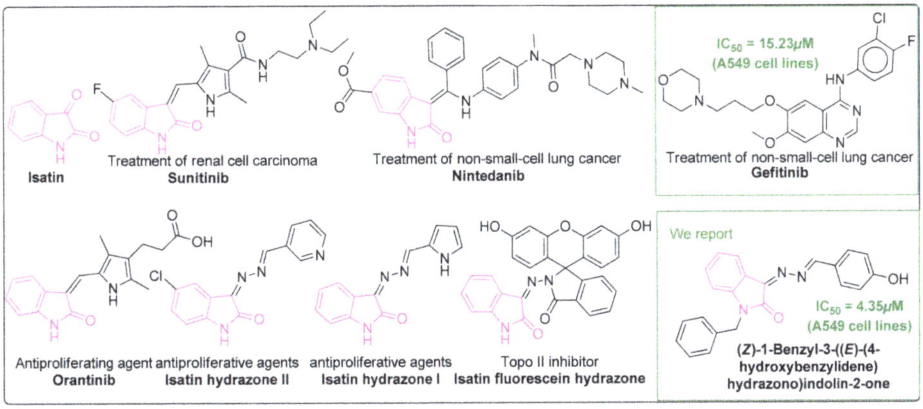


**Figure 1.** Isatin moiety containing active & potential drugs and gefitinib.

## 2. Materials and Methods

### 2.1. Chemicals and Solvents Were of Commercial Reagent Grade (Sigma-Aldrich, St. Louis, MO, USA) and Used without Further Purification

The progress of reactions and purity of reactants and products were checked using pre-coated silica gel 60 aluminum TLC sheets with fluorescent indicator UV254 of Macherey-Nagel, and detection was carried out with ultraviolet light (254 nm). Melting points were determined on a Fisher ScientificTM digital melting point apparatus (model number IA9100) and are uncorrected. Electrospray ionization (ESI) mass spectrometry (MS) experiments were performed using an Agilent high performance liquid chromatography (HPLC) 1200 connected to an Agilent 6320 ion trap mass spectrometer fitted with an electrospray ionization (ESI) ion source (Agilent Technologies, Palo Alto, CA, USA). Infrared spectra were recorded as KBr disks using the Fourier Transform Infrared Spectrophotometer of Shimadzu; model: IR affinity-1S (Shimadzu, Tokyo, Japan). NMR spectra were taken on Agilent-NMR-VNMRS 600 MHz spectrometer (Agilent Technologies, Palo Alto, CA, USA) and DMSO-d6 was used as solvent.

### 2.2. Preparation and Characterization of Target Compounds

#### 2.2.1. 1-Benzylindoline-2,3-dione (**3**)

Orange powder (95%) mp = 130–131 °C (Lit. [31] mp. = 125–126 °C).

#### 2.2.2. (Z)-1-Benzyl-3-hydrazonoindolin-2-one (**4**)

Yellow powder (90%) mp = 124.5–126 °C (Lit. [32,33] mp. = 125–126 °C). ESI mass m/z = 252 $[M + H]^+$.

#### 2.2.3. General Procedure for the Synthesis of **6a–j**

To a mixture of isatin monohydrazone (1 mmol) and substituted aryl aldehyde (1 mmol) in 10 mL ethanol, a few drops of glacial acetic acid was added. The reaction mixture was refluxed for 4 h. The precipitate solid was filtered, washed with cold ethanol and air dried to obtain the target compounds (**6a–j**), which was then further purified by recrystallization using methanol. $^{13}$CNMR spectra for **6a-j** is available in the Supplementary Materials file.

1-Benzyl-3-((4-(dimethylamino) benzylidene)hydrazono)indolin-2-one (**6a**)

Dark red solid (53%) mp = 183–184 °C. IR (KBr): $\nu_{max}$ (cm$^{-1}$): 2926 (C-H aliphatic), 1707 (C=O), 1606 (C=N), 1558 (C=N). $^1$H NMR (DMSO-d6, 600 MHz: δ 3.06 (s, 6H, $N(CH_3)_2$), 4.97 (s, 2H, N-$CH_2$-Ph),

6.84 (d, $J$ = 7.8, 2H, ArH), 6.98 (d, $J$ = 7.2, 1H, ArH), 7.15 (t, 1H, ArH), 7.26 (s, 1H, ArH), 7.34 (m, 5H, ArH), 7.85 (d, $J$ = 7.8, 2H, ArH), 8.30 (d, $J$ = 7.2, 1H, ArH), 8.67 (s, 1H, N=CH) ppm. $^{13}$C NMR (DMSO-d6, 150 MHz: δ 166.41, 164.43, 153.76, 149.56, 144.80, 136.70, 133.09, 131.97, 129.18, 129.08, 127.99, 127.68, 123.35, 120.76, 117.04, 112.30, 110.22, 43.11 and 40.14 ppm. ESI mass m/z = 383 $[M + H]^+$.

1-Benzyl-3-(((4-chlorocyclohexa$^{-1}$, 5-dien-1-yl) methylene)hydrazono)indolin-2-one (**6b**)

Dark red solid (30%) mp = 210–211 °C. IR (KBr): $\nu_{max}$ (cm$^{-1}$): 2918 (C-H aliphatic), 1718 (C=O), 1591 (C=N). $^1$H NMR (DMSO-d6, 600 MHz: δ 4.97 (s, 2H, N-$CH_2$-Ph), 7.00 (d, $J$ = 7.8, 1H, ArH), 7.08 (t, 1H, ArH), 7.24–7.44 (m, 6H, ArH), 7.64 (d, $J$ = 7.8, 2H, ArH), 7.92 (d, $J$ = 6.6, 1H, ArH), 8.05 (d, $J$ = 7.8, 2H, ArH), 8.66 (s, 1H, N=CH) ppm. $^{13}$C NMR (DMSO-d6, 150 MHz: δ 163.69, 159.86, 150.00, 145.57, 137.31, 136.45, 134.18, 132.66, 131.03, 129.87, 129.20, 128.00, 127.72, 123.60, 116.30, 110.63 and 43.23 ppm. ESI mass m/z = 374 $[M(^{35}Cl) + H]^+$, 376 $[M(^{37}Cl) + H]^+$.

1-Benzyl-3-((4-hydroxybenzylidene)hydrazono)indolin-2-one (**6c**)

Orange solid (50%) mp = 235–236 °C. IR (KBr): $\nu_{max}$ (cm$^{-1}$): 3329 (O-H), 3022 (C-H aliphatic), 1705 (C=O), 1608 (C=N), 1589 (C=N). $^1$H NMR (DMSO-d6, 600 MHz: δ 4.97 (s, 2H, N-$CH_2$-Ph), 6.92 (d, $J$ = 7.8, 2H, ArH), 6.98 (d, $J$ = 7.2, 1H, ArH), 7.08 (t, 1H, ArH), 7.22–7.4 (m, 6H, ArH), 7.73 (bs, 1H, OH), 7.86 (d, $J$ = 7.2, 2H, ArH), 8.15 (d, $J$ = 7.8, 1H, ArH), 8.65 (s, 1H, N=CH) ppm. $^{13}$C NMR (DMSO-d6, 150 MHz: δ 164.32, 164.11, 162.32, 150.79, 150.31, 145.17, 136.56, 133.62, 132.08, 129.29, 129.18, 127.96, 127.49, 124.88, 123.49 and 43.14 ppm. ESI mass m/z = 356 $[M + H]^+$.

1-Benzyl-3-((2-bromobenzylidene)hydrazono)indolin-2-one (**6d**)

Orange solid (65%) mp = 173–174 °C. IR (KBr): $\nu_{max}$ (cm$^{-1}$): 2935(C-H aliphatic), 1722 (C=O), 1606 (C=N). $^1$H NMR (DMSO-d6, 600 MHz: δ 4.95 (s, 2H, N-$CH_2$-Ph), 6.99 (d, $J$ = 4.8, 1H, ArH), 7.06 (t, 1H, ArH), 7.25 (m, 1H, ArH), 7.3–7.37 (m, 4H, ArH), 7.4 (t, 1H, ArH), 7.51 (t, 1H, ArH), 7.58 (t, 1H, ArH), 7.78 (d, $J$ = 8.4, 1H, ArH), 7.86 (d, $J$ = 7.2, 1H, ArH), 8.22 (d, $J$ = 7.2, 1H, ArH), 8.75 (s, 1H, N=CH) ppm. $^{13}$C NMR (DMSO-d6, 150 MHz: δ 163.57, 158.38, 149.98, 145.68, 136.37, 134.37, 134.28, 134.15, 132.16, 129.31, 129.19, 129.15, 129.10, 128.00, 127.71, 125.68, 123.64, 116.16, 116.16, 110.68 and 43.22 ppm. ESI mass m/z = 418 $[M(^{79}Br) + H]^+$, 420 $[M(^{81}Br) + H]^+$.

1-Benzyl-3-((E)-(3-bromobenzylidene)hydrazono)indolin-2-one (**6e**)

Orange solid (52%) mp = 173–174 °C. IR (KBr): $\nu_{max}$ (cm$^{-1}$): 2920 (C-H aliphatic), 1720 (C=O), 1624 (C=N), 1573 (C=N). $^1$H NMR (DMSO-d6, 600 MHz: δ 4.97 (s, 2H, N-$CH_2$-Ph), 7.01 (d, $J$ = 8.4, 1H, ArH), 7.09 (t, 1H, ArH), 7.25 (m, 1H, ArH), 7.24–7.38 (m, 4H, ArH), 7.41 (t, 1H, ArH), 7.54 (t, 1H, ArH), 7.79 (d, $J$ = 8.4, 1H, ArH), 7.86 (d, $J$ = 7.8, 1H, ArH), 8.02 (d, $J$ = 7.8, 1H, ArH), 8.12 (s, 1H, ArH), 8.60 (s, 1H, N=CH) ppm. $^{13}$C NMR (DMSO-d6, 150 MHz: δ 163.60, 158.67, 149.59, 145.61, 136.43, 136.10, 135.22, 134.12, 132.07, 131.88, 129.21, 129.10, 128.01, 127.73, 127.66, 123.65, 122.86, 116.20, 110.67 and 43.25 ppm. ESI mass m/z = 418 $[M(^{79}Br) + H]^+$, 420 $[M(^{81}Br) + H]^+$.

1-Benzyl-3-((4-bromobenzylidene)hydrazono)indolin-2-one (**6f**)

Orange solid (68%) mp = 216–217 °C. IR (KBr): $\nu_{max}$ (cm$^{-1}$): 2922 (C-H aliphatic), 1720 (C=O), 1602 (C=N). $^1$H NMR (DMSO-d6, 600 MHz: δ 4.95 (s, 2H, N-$CH_2$-Ph), 6.85 (d, $J$ = 7.2, 1H, ArH), 7.05 (t, 1H, ArH), 7.22–7.42 (m, 6H, ArH), 7.76 (d, $J$ = 7.8, 2H, ArH), 7.9 (d, $J$ = 7.8, 3H, ArH), 8.60 (s, 1H, N=CH) ppm. $^{13}$C NMR (DMSO-d6, 150 MHz: δ 163.67, 159.89, 149.91, 145.53, 136.42, 134.18, 132.94, 132.77, 131.15, 129.19, 128.01, 127.70, 126.34, 123.60, 116.26, 110.61 and 43.20 ppm. ESI mass m/z = 418 $[M(^{79}Br) + H]^+$, 420 $[M(^{81}Br) + H]^+$.

1-Benzyl-3-((2-methyl benzylidene)hydrazono)indolin-2-one (**6g**)

Orange solid (60%) mp = 176–177 °C. IR (KBr): $\nu_{max}$ ($cm^{-1}$): 2920 (C-H aliphatic), 1716 (C=O), 1624 (C=N). $^1H$ NMR (DMSO-d6, 600 MHz: δ 2.53 (t, 3H, Ph-$CH_3$), 4.96 (s, 2H, N-$CH_2$-Ph), 7.0 (d, *J* = 7.8, 1H, ArH), 7.07 (t, 1H, ArH), 7.23–7.41 (m, 8H, ArH), 7.45 (t, 1H, ArH), 7.95 (d, *J* = 7.8, 1H, ArH), 8.08 (d, *J* = 7.2, 1H, ArH), 8.85 (s, 1H, N=CH) ppm. $^{13}C$ NMR (DMSO-d6, 150 MHz: δ 163.80, 159.92, 149.81, 145.45, 139.86, 136.46, 134.04, 132.44, 131.80, 131.72, 129.19, 128.91, 128.07, 128.00, 127.71, 127.09, 123.54, 116.33, 110.59, 43.19 and 19.59 ppm. ESI mass m/z = 354 $[M + H]^+$.

1-Benzyl-3-((3-methylbenzylidene)hydrazono)indolin-2-one (**6h**)

Pale orange solid (60%) mp = 138–139 °C. IR (KBr): $\nu_{max}$ ($cm^{-1}$): 2920 (C-H aliphatic), 1718 (C=O), 1670 (C=N), 1606 (C=N). $^1H$ NMR (DMSO-d6, 600 MHz: δ 2.38 (s, 3H, Ph-$CH_3$), 4.95 (s, 2H, N-$CH_2$-Ph), 7.0 (d, *J* = 8.4, 1H, ArH), 7.07 (t, 1H, ArH), 7.22–7.42 (m, 7H, ArH), 7.45 (t, 1H, ArH), 7.76 (s, 1H, ArH), 7.78 (d, *J* = 7.8, 1H, ArH), 7.95 (d, *J* = 7.2, 1H, ArH), 8.85 (s, 1H, N=CH) ppm. $^{13}C$ NMR (DMSO-d6, 150 MHz: δ 163.74, 161.14, 149.83, 145.45, 139.06, 136.46, 134.03, 133.75, 133.43, 129.98, 129.62, 129.19, 129.13, 129.10, 127.99, 127.71, 126.45, 123.60, 117.81, 116.34, 110.57, 43.19 and 21.35 ppm. ESI mass m/z = 354 $[M + H]^+$.

1-Benzyl-3-((4-methylbenzylidene)hydrazono)indolin-2-one (**6i**)

Pale red solid (82%) mp = 191–192 °C. IR (KBr): $\nu_{max}$ ($cm^{-1}$): 2920 (C-H aliphatic), 1722 (C=O), 1683 (C=N), 1602 (C=N). $^1H$ NMR (DMSO-d6, 600 MHz: δ 2.38 (s, 3H, Ph-$CH_3$), 4.95 (s, 2H, N-$CH_2$-Ph), 6.99 (d, *J* = 7.8, 1H, ArH), 7.07 (t, 1H, ArH), 7.22–7.41 (m, 8H, ArH), 7.87 (d, *J* = 8.4, 2H, ArH), 8.00 (d, *J* = 7.2, 1H, ArH), 8.65 (s, 1H, N=CH) ppm. $^{13}C$ NMR (DMSO-d6, 150 MHz: δ 163.83, 161.88, 150.09, 145.40, 143.17, 136.48, 133.96, 131.14, 130.33, 129.53, 129.19, 127.98, 127.70, 123.55, 116.43, 110.54, 43.18 and 21.74 ppm. ESI mass m/z = 354 $[M + H]^+$.

1-Benzyl-3-((4-(methylthio)benzylidene)hydrazono)indolin-2-one (**6j**)

Orange solid (65%) mp = 177–178 °C. IR (KBr): $\nu_{max}$ ($cm^{-1}$): 2926 (C-H aliphatic), 1720 (C=O), 1606 (C=N). $^1H$ NMR (DMSO-d6, 600 MHz: δ 2.49 (s, 3H, S-$CH_3$), 4.97 (s, 2H, N-$CH_2$-Ph), 7.01 (d, *J* = 7.8, 1H, ArH), 7.09 (t, 1H, ArH), 7.27 (m, 1H, ArH) 7.3–7.42 (m, 5H, ArH), 7.43 (d, *J* = 8.4, 2H, ArH), 7.92 (d, *J* = 7.8, 2H, ArH), 8.04 (d, *J* = 7.2, 1H, ArH), 8.70 (s, 1H, N=CH) ppm. $^{13}C$ NMR (DMSO-d6, 150 MHz: δ 163.89, 162.07, 150.20, 145.41, 144.96, 136.51, 133.95, 130.01, 129.88, 129.20, 127.99, 127.72, 126.07, 123.55, 116.50, 110.54, 43.20 and 14.50 ppm. ESI mass m/z = 386 $[M + H]^+$.

### *2.3. Antimicrobial Evaluation*

Different organisms were selected to cover the Gram-positive and Gram-negative bacteria; namely Staphylococcus aureus ATCC 29213 and *Bacillus subtilis* ATCC 3366 as Gram-positive, *Escherichia coli* ATCC 25922, *Klebsiella pneumoniae* ATCC 13883, *Salmonella typhi* ATCC 25566 and *Pseudomonas aeruginosa* ATCC 27853 as Gram-negative microorganisms. Antifungal effect was investigated using *Candida albicans* NRRL Y-477. Disc diffusion method was used as the method of the microbiological study, where the standard antibiotic discs were Ampicillin 10 μg/disc, erythromycin 10 μg/disc and anti-fungal fluconazole. The negative control was DMSO (100%). The result was collected to estimate the existence of antibacterial efficiencies of the prepared compounds depending on the criterion procedure by Bauer et al [34]. Bacterial culture, at first, which is adjusted to 0.5 McFarland standards, was applied to Mueller Hinton agar plates in-house. After that, the plates were dried for 15 min. and then applied for the sensitivity test. The discs were impregnated with the tested compounds on the MH agar surface together with the control antibiotic and the negative control and incubated at 37 °C for 24 h. After incubation, the plates were examined for any inhibition zone. To ensure reliability of the result, the test was repeated three times.

### 2.4. *Antiproliferative Assay*

A549 and Hela cells (Produced from ATCC, Rockville, MD, USA) were cultured in DMEM (Dulbecco's Modified Eagle Medium), supplemented with 10% FBS, 100 mg/mL of streptomycin and 100 U/mL of penicillin at 37 °C in a 5%-$CO_2$ humidified atmosphere. Cell cytotoxicity assay was done by MTT method [35]; 3-(4,5-dimethylthiazol-2-yl)-2,5-diphenyltetrazolium bromide method. Concisely, cells were seeded in 96-well tissue culture plates and incubated for 48 h at 37 °C, 5% $CO_2$. The culture medium was removed and replaced with fresh medium containing the synthesized compounds in different concentrations to the wells and incubated for another 24 h. Thereafter, the MTT solution (1 mg/mL) was added and incubated for 3 h. To dissolve the reduced MTT crystals, one hundred microliters of ethanol was added to each well, then, the optical density of each well was measured at 492/630 nm with enzyme immunoassay instrument.

### 2.5. *In Silico ADME Prediction Analysis*

In this experiment, in silico ADME analysis was done in QikProp module of Schrodinger Maestro [36] where the important physiological descriptors like predicted $IC_{50}$ for blocking hERG K+ channel in vitro, predicted octanol or water partition coefficient [log P(o/w)], number of hydrogen bond acceptors (HBA), number of hydrogen bond donors (HBD), predicted aqueous solubility (logS), MDCK cell permeability (MDCK), blood–brain partition coefficient (logBB), percentage human oral absorption rate, etc. were analyzed.

## 3. Results and Discussion

### 3.1. *Synthesis of* **6a–j**

The synthetic pathway of the compounds (**6**) is illustrated in Scheme 1. *N*-benzylisatin (**3**) was prepared using our previously reported method [31] by heating same equivalent of isatin (**1**) and benzyl bromide (**2**) at 80 °C in dimethylformamide (DMF) for 12 h. *N*-benzylisatin monohydrazone (**4**) also was prepared using previously reported hydrazone formation method [27,32,33,37] using 1.2 equivalent of hydrazine hydrate and *N*-benzylisatin (**3**) at refluxing condition for 3–4 h in alcoholic solution (MeOH/EtOH). Designed *N*-benzylisatin-aryl hydrazones (**6a–j**) were synthesized by reacting compound **4** with substituted aryl aldehydes (**5a–j**) in absolute ethanol in presence of catalytic amount of glacial acetic acid [27]. The choice of the substituents and its positions on the aromatic ring of benzaldehyde is to offer various electron withdrawal and donation on the aromatic ring which will, in turn, give different degrees of electron density on the biologically active parts of isatin molecule.

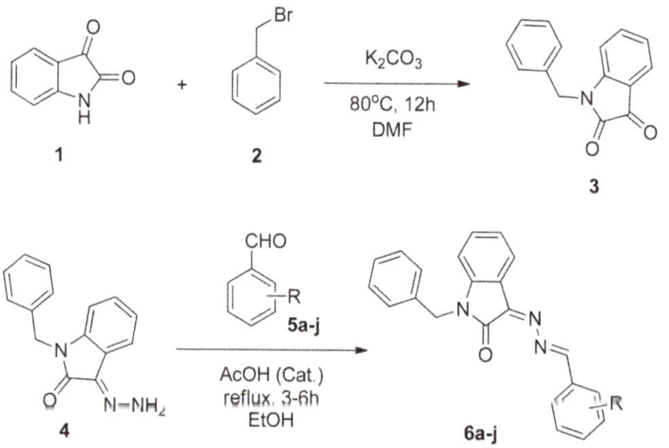


| 6 | R |
|---|---|
| a | 4-N(CH3)2 |
| b | 4-Cl |
| c | 4-OH |
| d | 2-Br |
| e | 3-Br |
| f | 4-Br |
| g | 2-CH3 |
| h | 3-CH3 |
| i | 4-CH3 |
| j | 4-SCH3 |

**Scheme 1.** Synthesis of *N*-benzylisatin-aryl hydrazones (**6a–j**).

## *3.2. Biological Evaluation*

### 3.2.1. Antimicrobial Evaluation of **6a–j**

Synthesized *N*-benzylisatin-aryl hydrazones (**6a–j**) were evaluated for their antibacterial activity against two Gram-positive Staphylococcus aureus ATCC 29,213 and *Bacillus subtilis* ATCC 3366, four Gram-negative, *Escherichia coli* ATCC 25922, *Klebsiella pneumoniae* ATCC 13883, *Salmonella typhi* ATCC 25,566 and *Pseudomonas aeruginosa* ATCC 27,853 bacterial strains. As expected, the result of the antibacterial evaluation revealed no effect of these compounds (**6a–j**) on selected strains of bacteria in concentration up to 100 µg/disc (where no significant inhibition zones were detected). In case of antifungal activity, no inhibition zone was detected against *Candida albicans* NRRL Y-477. Excitingly, our hypothesis matched with the antimicrobial data of the evaluated compounds **6a–j**, which are all having low molecular weight (353–418 Da), conforming to O'Shea and Moser's findings. In addition, the antimicrobial data of **6a–j** also meet the rule of five (RO5) with logP value of 9.36–11.8 (please see Section 3.4), which is far more than to have anti-microbial activity. In case of antifungal activity, there were no inhibition zones detected against *Candida albicans* NRRL Y-477 either.

### 3.2.2. Antiproliferative Activity of **6a–j**

The antiproliferative activity of compounds **6a–j** are listed in Table 1 as values of $IC_{50}$. All the tested compounds displayed excellent antiproliferative activity against non-small cell lung cancer cell lines 'A549' and human cervical cancer cells lines 'HeLa' tested. Among them, compound **6c** displayed the highest antiproliferative activity against the cell lines 'A549' (4.35 µM) and 'Hela' (4.09 µM). Whereas positive control gefitinib showed less antiproliferative activity against the cell lines 'A549' (15.23 µM) and 'Hela' (7.35 µM). Besides compound **6c**, another three bromo derivatives of *N*-benzylisatin-aryl hydrazones (**6d**, **6f** and **6g**) showed excellent antiproliferative activity, which is comparable to the control gefitinib. Detail explanations and SARs study are given below.

**Table 1.** Antiproliferative activity of *N*-benzylisatin-aryl hydrazones (**6a–j**).

| **Compound** | 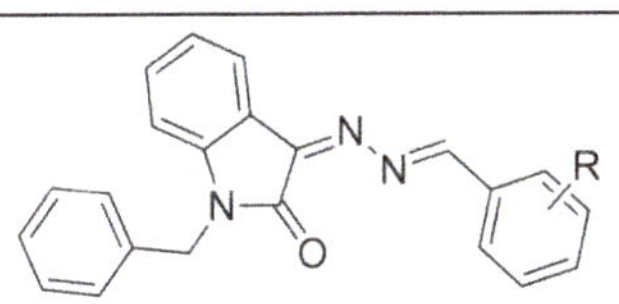 **(R = )** | **$IC_{50}$ (µM)** | |
|---|---|---|---|
| | | **A549** | **Hela** |
| **4** | – | 14 ± 2.15 | 46 ± 1.30 |
| **6a** | 4-N($CH_3$)$_2$ | 23.7 ± 0.69 | 25.1 ± 0.95 |
| **6b** | 4-Cl | 22 ± 1.11 | 14.8 ± 0.94 |
| **6c** | 4-OH | 4.35 ± 0.05 | 4 ± 0.09 |
| **6d** | 2-Br | 7.3 ± 0.52 | 6.7 ± 0.17 |
| **6e** | 3-Br | 41 ± 1.20 | 9.1 ± 0.35 |
| **6f** | 4-Br | 7.30 ± 0.03 | 7.4 ± 0.20 |
| **6g** | 2-$CH_3$ | 7.5 ± 0.11 | 7.7 ± 0.21 |
| **6h** | 3-$CH_3$ | 8.4 ± 0.25 | 13.4 ± 0.35 |
| **6i** | 4-$CH_3$ | 17 ± 0.70 | 18.5 ± 0.65 |
| **6j** | 4-$SCH_3$ | 14.6 ± 0.53 | 9 ± 0.86 |
| **Gefitinib** | – | 15.2 ± 0.48 | 7.3 ± 0.42 |

Data represent means ± SD obtained from three different experiments. Cell lines used as follows, non-small cell lung cancer cell lines 'A549' and human cervical cancer cells lines 'HeLa'. Gefitinib was used as positive controls.

### 3.3. *Structure Activity Relationships (SARs) of* ***6a–j***

The compound, namely (E)-1-benzyl-3-hydrazonoindolin-2-one (**4**) showed moderate antiproliferative activity (14.15 μM) against the non-small cell lung cancer cell lines 'A549' similar to that of gefitinib (15.23 μM), but six-fold less active against human cervical cancer cells lines 'HeLa' (46.03 μM). Compound **4** reacted with 4-substituted aryl aldehydes forming compound **6a** having 4-(*N*,*N*-)-dimethyl group (**6a**), chloro group (**6b**) and methyl group (**6i**) decreased the antiproliferative activity than the gefitinib against both cell lines tested. While substituted by methylthio group (**6j**) shows similar antiproliferative activity comparing gefitinib. Interestingly, as depicted in Table 2, entry **6c**, substituted by hydroxy group at 4-position showed higher antiproliferative activity which is around four fold than the gefitinib against non-small cell lung cancer cell lines 'A549' (4.35 μM) and two fold higher than the HeLa cell lines (4.09 μM). In case of bromo substituents at the 2- and 4-positions of the aryl ring, the compounds **6d** and **6f**, respectively, the antiproliferative activity dramatically increased more than two-fold for A549, and similar for HeLa cell lines in comparison to the gefitinib. While addition at the 3-position by a bromo group (**6e**) and by a methyl group (**6h**) of the aryl ring showed similar or less antiproliferative activity comparing to gefitinib. Interestingly, the methyl group at 2-position (**6g**) increased the activity by two-fold compared to gefitinib, against A549 and similarly against HeLa cell lines. In conclusion, ortho and para substitution by bromo and hydroxy group proved to be the most active potential agents for non-small cell lung cancer cell lines 'A549' among the compound tested.

**Table 2.** Analysis of drug likeness properties and pharmacokinetic properties by QikProp for **6a–j**.

| No. | MW [b] | HBD [c] | HBA [d] | logPo/w [e] | logS [f] | logP | HERG [g] | Caco-2 [h] | BBB [i] | MDCK [j] | HOA(%) [k] |
|---|---|---|---|---|---|---|---|---|---|---|---|
| **6a** | 382 | 0 | 6.5 | 4.77 | −5.76 | 10.2 | −7.35 | 3189 | −0.36 | 1733.1 | 100 |
| **6b** | 374 | 0 | 5.5 | 4.86 | −5.60 | 9.4 | −7.16 | 3720 | −0.02 | 5054.6 | 100 |
| **6c** | 355 | 1 | 6.25 | 3.82 | −5.07 | 11.8 | −7.13 | 1102 | −0.83 | 549.7 | 100 |
| **6d** | 418 | 0 | 5.5 | 4.87 | −5.58 | 9.5 | −7.19 | 3656 | −0.04 | 4745.7 | 100 |
| **6e** | 418 | 0 | 5.5 | 4.93 | −5.74 | 9.4 | −7.20 | 3637 | −0.02 | 5296.9 | 100 |
| **6f** | 418 | 0 | 5.5 | 4.93 | −5.72 | 9.4 | −7.19 | 3719 | −0.01 | 5433.8 | 100 |
| **6 g** | 353 | 0 | 5.5 | 4.61 | −5.30 | 9.4 | −7.14 | 3754 | −0.19 | 2067.3 | 100 |
| **6 h** | 353 | 0 | 5.5 | 4.67 | −5.46 | 9.4 | −7.17 | 3638 | −0.21 | 1998.3 | 100 |
| **6i** | 353 | 0 | 5.5 | 4.67 | −5.44 | 9.3 | −7.16 | 3734 | −0.20 | 2054.9 | 100 |
| **6j** | 385 | 0 | 6.0 | 5.01 | −5.93 | 9.7 | −7.38 | 3479 | −0.21 | 3246.6 | 100 |
| **G** [l] | **447** | **1** | **7.7** | **4.37** | **−5.23** | **10.8** | **−7.12** | **1053** | **0.31** | **2313.1** | **100** |

[b] Molecular weight in Daltons (acceptable range: <500); [c] Hydrogen bond donor (acceptable range: ≤5); [d] Hydrogen bond acceptor (acceptable range: ≤10); [e] Predicted octanol/water partition coefficient (acceptable range: −2–6.5); [f] Predicted aqueous solubility, S in mol/dm$^{-3}$ (acceptable range: −6.5–0.5); [g] Predicted $IC_{50}$ value for blockage of hERG K+ channels (concern: below −5); [h] Caco−2 value, permeability to Caco−2 (human colorectal carcinoma) cells in vitro; [I] Blood−brain barrier permeability (acceptable range: ~−0.4); j Predicted apparent MDCK cell permeability in nm/sec, QPPMDCK= >500 is great, <25 is poor; k Predicted human oral absorption on 0% to 100% scale (<25% is poor and >80% is high); l G = Gefitinib.

### 3.4. *In Silico Drug Likeness Property Analysis*

In modern drug discovery approaches, evaluation of absorption, distribution, metabolism and excretion (ADME) of drug candidates impose significant value to the rational drug design. In vitro and in vivo ADME prediction has now become faster and more accurate with the computational chemistry tools which has been developed recently and aids the pharmaceutical industries to screen many compounds within a very short time [38]. In this experiment, we analyzed the predicted ADME values of the designed compounds (**6a–j**) and are summarized in Table 2. Since high molecular weight compounds are always less effective in terms of intestinal absorption, our synthesized compounds' (**6a–j**) molecular weight (353–418Da) of this study are less than the established drug gefitinib (447Da), which supports the established parameters [9,39]. Subsequently, the studied compounds showed recommended values for hydrogen bond donor (<5) and acceptor (<10). Gefitinib showed HBD value of 1 and HBA value of 7.7 whereas the studied compounds showed HBD values of 0 (except **6c** having HBD 1) and HBA values of maximum 6.5 (compound **6a**), 6.25 (compound **6c**), 6.0 (compound **6j**) and other compounds **6b**, **6d–6i** having HBA values of 5.5, which indicates superior values to the gefitinib (HBA values is 7.7). In 2002, Jorgensen and Duffy established a parameter to check the bioavailability of the drug compound which can be determined by octanol/water partition coefficient and solubility scoring where the recommended values are −2–6.5 and −6.5–0.5 mol/dm$^{-3}$, respectively [40]. Compounds of this study showed scoring within the reference values including gefitinib. Octanol/water partition coefficient and solubility score in between 3.82–5.01 and −5.93––5.07, respectively. The hERG K+ channel blockers are potentially toxic for the heart and thus the recommended range for predicted log $IC_{50}$ values for blockage of hERG K+ channels (loghERG) is > −5 [41]. Intriguingly, all the compounds of this study showed higher value for loghERG score (>−7.13) than gefitinib (−7.12) which proved their less toxicity than gefitinib. The Caco-2 cell, considered as the reliable in vitro models to estimate oral drug absorption and transdermal delivery [42], was high for all compounds except gefitinib and compound **6c**, which signifies the improved and well oral drug absorption and transdermal delivery efficiency of the studied compounds than gefitinib. The blood–brain barrier separates CNS from blood and a successful compound must pass through the blood stream [43] which depends on several factors such as molecular weight which must be below 480. Since our synthesized compounds have less molecular weight than the recommended values therefore its showed significant result. Madin–Darby canine kidney (MDCK) cell permeability is considered as the measurement of the blood–brain barrier permeability where greater than 500 is of great value and less than 25 indicates very poor result according to Jorgensen's rule of 3 [44]. Compounds **6b**, **6d**, **6e**, **6f** and **6j** gave much higher MDCK value than gefitinib while others showed around similar values except **6c**. The synthesized compounds also gave a predicted human oral absorption rate of 100%. Taken together, all the designed compounds of this study showed higher predicted ADME values than gefitinib which can be tested further by in vitro and in vivo experiments to establish successful drug candidates.

## 4. Conclusions

A series of novel *N*–benzylisatin–aryl–hydrazones were designed and synthesized with good to moderate yields for their antimicrobial and antiproliferative activity evaluation, for the development of potent anticancer therapeutics with no or minimal side effects. Six bacterial strains and a fungal strain were used for the antimicrobial screening of the synthesized compounds and no inhibitory effect was found on different microbes in concentration up to 100-µg/disc. On the other hand, four compounds showed two–to–four-fold antiproliferative activity than the FDA approved first–line treatments for lung cancer drug gefitinib. For example, $IC_{50}$ of compound **6c** is 4.35 µM, whereas gefitinib shows 15.23-µM concentration against non–small cell lung cancer cell lines 'A549.' In case of HeLa cell lines, antiproliferative activity of compound **6c** also showed two-fold higher than the gefitinib. The strong inhibitory effect of **6c**, among the tested compounds, on the growth of cancerous cells accompanied by their complete safety on the growth of microbial cells, indicate a high level of target selectivity and a unique mechanism of action. Higher predicted ADME values were obtained than the known gefitinib. In our observation, this *N*–benzylisatin–aryl–hydrazones can be a potential agent for anticancer therapeutics. Studies on the mechanism of action of antiproliferative activity of these derivative are ongoing and the results will be explored in due courses.

**Supplementary Materials:** The following are available online at http://www.mdpi.com/2076-3417/10/11/3669/s1, Figure of the $^{13}$CNMR spectra of **6a–j**.

**Author Contributions:** Conceptualization, A.F.M.M.R. and H.S.A.-S.; data curation, H.A.A., I.S.I., A.Z.M., A.A. and M.A.; formal analysis, I.S.I., A.Z.M. and M.A.; funding acquisition, H.S.A.; investigation, H.S.A. and A.F.M.M.R.; methodology, H.A.A., I.S.I., A.Z.M., A.A. and M.A.; resources, H.S.A.–S. and A.F.M.M.R.; supervision, H.S.A.–S. and A.F.M.M.R; validation, H.S.A.-S.; writing—original draft, H.S.A.-S. and A.F.M.M.R.; writing—review and editing, H.S.A.–S. and A.F.M.M.R.; All authors have read and agreed to the published version of the manuscript.

**Funding:** This research was funded by the King Saud University, Research Center of the Female Campus for Scientific and Medical Studies.

**Acknowledgments:** This research project was supported by a grant from the research center of the Female Campus for Scientific and Medical Studies, King Saud University, Riyadh, Saudi Arabia.

**Conflicts of Interest:** The authors declare no conflict of interest.

## References

1. Crespo-Ortiz, M.P.; Wei, M.Q. Antitumor activity of artemisinin and its derivatives: From a well-known antimalarial agent to a potential anticancer drug. *J. Biomed. Biotechnol.* **2012**. [CrossRef] [PubMed]
2. National Academies of Sciences, Engineering, and Medicine. *The Drug Development Paradigm in Oncology: Proceedings of a Workshop*; The National Academies Press: Washington, DC, USA, 2018. [CrossRef]
3. Burotto, M.; Manasanch, E.E.; Wilkerson, J.; Fojo, T. Gefitinib and erlotinib in metastatic non-small cell lung cancer: A meta-analysis of toxicity and efficacy of randomized clinical trials. *Oncologist* **2015**, *20*, 400–410. [CrossRef] [PubMed]
4. Hsiue, E.H.-C.; Lee, J.-H.; Lin, C.-C.; Yang, J.C.-H. Safety of gefitinib in non-small cell lung cancer treatment. *Expert Opin. Drug Saf.* **2016**, *15*, 993–1000. [CrossRef] [PubMed]
5. Ding, P.N.; Lord, S.J.; Gebski, V.; Links, M.; Bray, V.; Gralla, R.J.; Yang, J.C.-H.; Lee, C.K. Risk of treatment-related toxicities from EGFR tyrosine kinase inhibitors: A meta-analysis of clinical trials of gefitinib, erlotinib, and afatinib in advanced EGFR-mutated non-small cell lung cancer. *J. Thorac. Oncol.* **2016**, *12*, 633–643. [CrossRef] [PubMed]
6. Katakami, N.; Atagi, S.; Goto, K.; Hida, T.; Horai, T.; Inoue, A.; Ichinose, Y.; Koboyashi, K.; Takeda, K.; Kiura, K.; et al. LUX-Lung 4: A phase II trial of afatinib in patients with advanced non-small-cell lung cancer who progressed during prior treatment with erlotinib, gefitinib, or both. *J. Clin. Oncol.* **2013**, *31*, 3335–3341. [CrossRef]
7. Roy, S.; Hagen, K.D.; Maheswari, P.U.; Lutz, M.; Spek, A.L.; Reedijk, J.; van Wezel, G.P. Phenanthroline derivatives with improved selectivity as DNA-targeting anticancer or antimicrobial drugs. *ChemMedChem* **2008**, *3*, 1427–1434. [CrossRef]
8. O'Shea, R.; Moser, H.E. Physicochemical properties of antibacterial compounds: Implications for drug discovery. *J. Med. Chem.* **2008**, *51*, 2871–2878. [CrossRef]
9. Lipinski, C.A.; Lombardo, F.; Dominy, B.W.; Feeney, P.J. Experimental and computational approaches to estimate solubility and permeability in drug discovery and development settings. *Adv. Drug Deliv. Rev.* **2001**, *46*, 3–26. [CrossRef]
10. Vine, K.L.; Matesic, L.; Locke, J.M.; Skropeta, D. Recent highlights in the development of isatin-based anticancer agents. *Adv. Anticancer Agents Med. Chem.* **2013**, 2, 254–312. [CrossRef]
11. Hajare, R.; Kulkarni, S.; Thakar, M.; Paranjape, R. Isatin anti-HIV agent: A review. *World J. Pharm. Pharm. Sci.* **2016**, *5*, 569–575. [CrossRef]
12. Pakravan, P.; Kashanian, S.; Khodaei, M.M.; Harding, F.J. Biochemical and pharmacological characterization of isatin and its derivatives: From structure to activity. *Pharmacol. Rep.* **2013**, *65*, 313–335. [CrossRef]
13. Li, C.; Zhang, F. Single step incorporation of isatin to enaminone: A recyclable catalyst towards assembly of diverse four ring fused pyrrolo[2,3,4-kl]acridin-1-ones. *RSC Adv.* **2016**, *6*, 75359–75364. [CrossRef]
14. Loloiu, G.; Loloiu, T.; Maior, O. Chemistry of isatin. Synthesis of 2,3-oxo-1H-pyrrolo[2,3-b]dibenzo-p-dioxins. *Rev. Chim. (Buchar.)* **1998**, *49*, 861–864.
15. Popp, F.D. Chemistry of isatin. *Adv. Heterocycl. Chem.* **1975**, *18*, 1–58.
16. Rajopadhye, M. The chemistry of isatin and isatin derivatives: Synthesis of novel ring systems. *Diss. Abstr. Int. B* **1988**, *48*, 2325–2326.
17. Sumpter, W.C. The chemistry of isatin. *Chem. Rev.* **1944**, *34*, 393–434. [CrossRef]
18. Da Silva, J.F.M.; Garden, S.J.; Pinto, A.C. The chemistry of isatins: A review from 1975 to 1999. *J. Braz. Chem. Soc.* **2001**, *12*, 273–324. [CrossRef]
19. Lashgari, N.; Mohammadi Ziarani, G. *Synthesis of Heterocyclic Compounds Based on Isatin Through 1,3-Dipolar Cycloaddition Reactionsp*; ARKIVOC: Gainesville, FL, USA, 2012.
20. Medvedev, A.; Igosheva, N.; Crumeyrolle-Arias, M.; Glover, V. Isatin: Role in stress and anxiety. *Stress* **2005**, *8*, 175–183. [CrossRef]

21. Vine, K.L.; Matesic, L.; Locke, J.M.; Ranson, M.; Skropeta, D. Cytotoxic and anticancer activities of isatin and its derivatives: A comprehensive review from 2000–2008. *Anti-Cancer Agents Med. Chem.* **2009**, *9*, 397–414. [CrossRef]
22. Matesic, L.; Locke, J.M.; Bremner, J.B.; Pyne, S.G.; Skropeta, D.; Ranson, M.; Vine, K.L. N-Phenethyl and N-naphthylmethyl isatins and analogues as in vitro cytotoxic agents. *Bioorg. Med. Chem.* **2008**, *16*, 3118–3124. [CrossRef]
23. Han, K.; Zhou, Y.; Liu, F.; Guo, Q.; Wang, P.; Yang, Y.; Song, B.; Liu, W.; Yao, Q.; Teng, Y.; et al. Design, synthesis and in vitro cytotoxicity evaluation of 5-(2-carboxyethenyl)isatin derivatives as anticancer agents. *Bioorg. Med. Chem. Lett.* **2014**, *24*, 591–594. [CrossRef] [PubMed]
24. Ibrahim, H.S.; Abou-seri, S.M.; Ismail, N.S.M.; Elaasser, M.M.; Aly, M.H.; Abdel-Aziz, H.A. Bis-isatin hydrazones with novel linkers: Synthesis and biological evaluation as cytotoxic agents. *Eur. J. Med. Chem.* **2016**, *108*, 415–422. [CrossRef]
25. Pervez, H.; Ahmad, M.; Zaib, S.; Yaqub, M.; Naseer, M.M.; Iqbal, J. Synthesis, cytotoxic and urease inhibitory activities of some novel isatin-derived bis-Schiff bases and their copper(II) complexes. *MedChemComm* **2016**, *7*, 914–923. [CrossRef]
26. Ganguly, S.; Debnath, B. Molecular docking studies and ADME prediction of novel isatin analogs with potent anti-EGFR activity. *Med. Chem. (Los Angeles, CA, USA)* **2014**, *4*, 558–568. [CrossRef]
27. Rahman, A.F.M.M.; Park, S.-E.; Kadi, A.A.; Kwon, Y. Fluorescein hydrazones as novel nonintercalative topoisomerase catalytic inhibitors with low DNA toxicity. *J. Med. Chem.* **2014**, *57*, 9139–9151. [CrossRef]
28. Dweedar, H.E.; Mahrous, H.; Ibrahim, H.S.; Abdel-Aziz, H.A. Analogue-based design, synthesis and biological evaluation of 3-substituted-(methylenehydrazono)indolin-2-ones as anticancer agents. *Eur. J. Med. Chem.* **2014**, *78*, 275–280. [CrossRef] [PubMed]
29. Eldehna, W.M.; Altoukhy, A.; Mahrous, H.; Abdel-Aziz, H.A. Design, synthesis and QSAR study of certain isatin-pyridine hybrids as potential anti-proliferative agents. *Eur. J. Med. Chem.* **2015**, *90*, 684–694. [CrossRef]
30. Islam, M.S.; Al-Majid, A.M.; El-Senduny, F.F.; Badria, F.A.; Rahman, A.F.M.M.; Barakat, A.; Elshaier, Y.A.M.M. Synthesis, anticancer activity, and molecular modeling of new halogenated spiro[pyrrolidine-thiazolo-oxindoles] derivatives. *Appl. Sci.* **2020**, *10*, 2170. [CrossRef]
31. Kadi, A.A.; Al-Shakliah, N.S.; Rahman, A.F.M.M. Synthesis and fragmentation behavior study of n-alkyl/benzyl Isatin derivatives present in small/complex molecules: Precursor for the preparation of biological active heterocycles. *Mass Spectrom. Lett.* **2015**, *6*, 65–70. [CrossRef]
32. Coffey, K.E.; Moreira, R.; Abbas, F.Z.; Murphy, G.K. Synthesis of 3,3-dichloroindolin-2-ones from isatin-3-hydrazones and (dichloroiodo)benzene. *Org. Biomol. Chem.* **2015**, *13*, 682–685. [CrossRef]
33. Popp, F.D. Potential anticonvulsants. IX. Some isatin hydrazones and related compounds. *J. Heterocycl. Chem.* **1984**, *21*, 1641–1645. [CrossRef]
34. Bauer, A.W.; Kirby, W.M.; Sherris, J.C.; Turck, M. Antibiotic susceptibility testing by a standardized single disk method. *Am. J. Clin. Pathol.* **1966**, *45*, 493–496. [CrossRef]
35. Stewart, M.J.; Watson, I.D. Standard units for expressing drug concentrations in biological fluids. *Br. J. Clin. Pharmacol.* **1983**, *16*, 3–7. [CrossRef] [PubMed]
36. Arifuzzaman, M.; Mitra, S.; Jahan, S.I.; Jakaria, M.; Abeda, T.; Absar, N.; Dash, R. A Computational workflow for the identification of the potent inhibitor of type II secretion system traffic ATPase of Pseudomonas aeruginosa. *Comput. Biol. Chem.* **2018**, *76*, 191–201. [CrossRef] [PubMed]
37. Murukan, B.; Kumari, B.S.; Mohanan, K. Synthesis, spectral, electrochemical and antibacterial studies of copper(II) complexes with isatin derived bishydrazone and different co-ligands. *J. Coord. Chem.* **2007**, *60*, 1607–1617. [CrossRef]
38. Ekins, S.; Waller, C.L.; Swaan, P.W.; Cruciani, G.; Wrighton, S.A.; Wikel, J.H. Progress in predicting human ADME parameters in silico. *J. Pharmacol. Toxicol. Methods* **2000**, *44*, 251–272. [CrossRef]
39. Ghose, A.K.; Viswanadhan, V.N.; Wendoloski, J.J. A knowledge-based approach in designing combinatorial or medicinal chemistry libraries for drug discovery. 1. A qualitative and quantitative characterization of known drug databases. *J. Comb. Chem.* **1999**, *1*, 55–68. [CrossRef]
40. Jorgensen, W.L.; Duffy, E.M. Prediction of drug solubility from structure. *Adv. Drug Deliv. Rev.* **2002**, *54*, 355–366. [CrossRef]
41. Chemi, G.; Gemma, S.; Campiani, G.; Brogi, S.; Butini, S.; Brindisi, M. Computational tool for fast in silico evaluation of hERG K+ channel affinity. *Front. Chem.* **2017**, *5*. [CrossRef] [PubMed]

42. Kulkarni, A.; Han, Y.; Hopfinger, A.J. Predicting Caco-2 cell permeation coefficients of organic molecules using membrane-interaction QSAR. *J. Chem. Inf. Comput. Sci.* **2002**, *42*, 331–342. [CrossRef]
43. Clark, D.E. In silico prediction of blood-brain barrier permeation. *Drug Discov. Today* **2003**, *8*, 927–933. [CrossRef]
44. Lionta, E.; Spyrou, G.; Vassilatis, D.K.; Cournia, Z. Structure-based virtual screening for drug discovery: Principles, applications and recent advances. *Curr. Top. Med. Chem.* **2014**, *14*, 1923–1938. [CrossRef] [PubMed]

applied sciences

MDPI

Article

# In Vitro and In Silico Screening of 2,4,5-Trisubstituted Imidazole Derivatives as Potential Xanthine Oxidase and Acetylcholinesterase Inhibitors, Antioxidant, and Antiproliferative Agents

**Eduardo Noriega-Iribe [1], Laura Díaz-Rubio [1,*], Arturo Estolano-Cobián [1], Victor Wagner Barajas-Carrillo [1], José M. Padrón [2], Ricardo Salazar-Aranda [3], Raúl Díaz-Molina [4], Victor García-González [4], Rocio Alejandra Chávez-Santoscoy [1,5], Daniel Chávez [6] and Iván Córdova-Guerrero [1,*]**

1 Facultad de Ciencias Químicas e Ingeniería, Universidad Autónoma de Baja California, Tijuana 22390, Mexico; eduardo.noriega@uabc.edu.mx (E.N.-I.); arturo.estolano@uabc.edu.mx (A.E.-C.); wbarajas@uabc.edu.mx (V.W.B.-C.); ale.santoscoy@gmail.com (R.A.C.-S.)
2 BioLab, Instituto Universitario de Bio-Orgánica "Antonio González" (IUBO-AG), Universidad de La Laguna, c/Astrofísico Francisco Sánchez 2, 38206 La Laguna, Spain; jmpadron@ull.es
3 Departamento de Química Analítica, Facultad de Medicina, Universidad Autónoma de Nuevo León, San Nicolás de los Garza, Monterrey 64460, Mexico; salazar121212@yahoo.com.mx
4 Departamento de Bioquímica, Facultad de Medicina Mexicali, Universidad Autónoma de Baja California, Mexicali 21000, Mexico; rauldiaz@uabc.edu.mx (R.D.-M.); vgarcia62@uabc.edu.mx (V.G.-G.)
5 Tecnológico de Monterrey, Escuela de Ingeniería y Ciencias, Av. Eugenio Garza Sada 2501 Sur, Monterrey 64849, Mexico
6 Centro de Graduados e Investigación en Química, Tecnológico Nacional de México/Instituto Tecnológico de Tijuana, Tijuana 22510, Mexico; dchavez@tectijuana.mx
* Correspondence: ldiaz26@uabc.edu.mx (L.D.-R.); icordova@uabc.edu.mx (I.C.-G.); Tel.: +52-664-120-7741 (I.C.-G.)

Received: 18 February 2020; Accepted: 8 April 2020; Published: 22 April 2020

**Abstract:** The employment of privileged scaffolds in medicinal chemistry supplies scientists with a solid start in the search for new and improved therapeutic molecules. One of these scaffolds is the imidazole ring, from which several derivatives have shown a wide array of biological activities. A series of 2,4,5-triphenyl imidazole derivatives were synthesized, characterized, and evaluated in vitro as antioxidant molecules using 1,1-diphenyl-2-picrylhydrazyl ($DPPH^{\cdot}$) and 2-2′-azino-bis-(3-ethylbenzothiazoline-6-sulfonate) ($ABTS^{\cdot+}$) assays, acetylcholinesterase (AChE) and xanthine oxidase (XO) inhibitors as well as antiproliferative agents. Additional in silico studies such as docking and determination of their absorption, distribution, metabolism, and excretion (ADME) properties were calculated. Compounds **3** and **10** were the most active antioxidants in both the DPPH and ABTS assays ($EC_{50}$ of 0.141 and 0.174 mg/mL, and 0.168 and 0.162 mg/mL, respectively). In the enzymatic inhibition, compound **1** showed the best activity, inhibiting 25.8% of AChE at a concentration of 150 μg/mL, and compound **3** was the most active XO inhibitor with an $IC_{50}$ of 85.8 μg/mL. Overall, against the six different evaluated cancerous cell lines, molecules **2**, **10**, and **11** were the most antiproliferative compounds. In silico predictions through docking point out **11**, and ADME analysis to **11** and **12**, as good candidates for being lead compounds for further derivations.

**Keywords:** imidazole; antiproliferative; antioxidant activities; docking; DPPH; ABTS; acetylcholinesterase; xanthine oxidase

## 1. Introduction

Imidazole (1,3-diaza-2,4-cyclopentadiene) is a heterocyclic aromatic compound that can be found in many biological molecules such as histidine, histamine, or in natural nucleotides. It is a highly versatile pharmacophore; therefore, there are several reports of a wide range of biological activities in molecules containing an imidazole motif such as antifungal, antituberculosis, antibiotic, cytotoxic, anti-inflammatory, antioxidant, and analgesic, amongst many others [1–4].

Imidazole derivatives, being di-, tri-, and tetra-substituted, have shown antioxidant activity through different antioxidant methodologies [5–7]. This is a useful property to counteract oxidative stress, a condition when reactive oxygen species (ROS) overcome the natural cellular antioxidant defense system. As the aging process, along with several chronic and degenerative human diseases, have been linked to oxidative stress such as cardiovascular, neurodegenerative, and cancerous ones [8,9], compounds with antioxidant properties are of high interest for researchers.

One of the neurodegenerative diseases in which oxidative stress has been regarded as one of the underlying causes is Alzheimer's disease (AD) [10], being that this disease is the most frequent cause of dementia in elderly people [11]. As the cholinergic deficit is heavily related to the disease progression, inhibitors of the enzyme acetylcholinesterase (AChEI) are potential drugs for the treatment of AD patients [12]. Imidazole bearing molecules have been also evaluated as AChEI with interesting results [13].

Xanthine oxidase (XO) is a key enzyme in purine metabolism, and is involved in uric acid production as the final metabolite. High production of uric acid can lead to gout; therefore, inhibition of this enzyme has been targeted as a therapeutic approach, with imidazole having been employed for a long time as a scaffold for XO inhibitors [14]. As the activity of XO produces both uric acid and reactive oxygen species, a XO inhibitor with antioxidant properties could show a good therapeutic profile, inhibiting the enzyme and controlling the oxidative damage to tissues near it [14,15].

The literature has shown numerous imidazole derivatives with tri-substitutions, of both alkyl and aryl types, with the aryl types frequently heterocyclic in nature. In a broad sense, in recent years, trisubstituted imidazoles have been synthesized many times, providing new synthetic methodologic alternatives, or in the search of particular biological properties [4]. Alternatively, this article proposes a group of trisubstitutions, where only small variations are introduced in one of them, to conduct a more finely-detailed structure–activity relationship (SAR) of the biological assays performed.

Based on the broad literature for the biological activities of imidazole derivatives and the above-mentioned SAR strategy, in this work, we present the synthesis of 2,4,5-triphenylimidazoles with substitutions in their A ring to perform an initial screening of their activities as antiproliferative, antioxidant, AChE, and XO inhibitor compounds, in order to find new leaders with these biological profiles. To complement the in vitro evaluations, molecular docking and in silico analysis of their ADME properties was made to select the best candidates and set the path for studies on new drug families.

## 2. Materials and Methods

### *2.1. General Information*

All reagents for the synthetic methodology and solvents went through purification before being used. Melting point measurements were made on a SMP11 melting point apparatus (Stuart). Different models of UV–Vis spectrophotometers were employed for the UV–Vis spectra, a Genesys 20 model was used for the antioxidant assays, a Microplate reader Multiskan™ FC was used for the acetylcholinesterase assay (both from Thermo Scientific), and a Microplate reader model PowerWave™ XS (from BioTek) was used for the antiproliferative assay and expressed in nanometers (nm). Fourier-transform infrared spectroscopy (FTIR) was performed on a Spectrum One (Perkin-Elmer) and a Nicolet is 5 spectrophotometer (from Thermo Scientific). Nuclear magnetic resonance (NMR) spectra were obtained on a Bruker spectrometer; model Avance DPX of 400 MHz. The chemical shifts (represented by $\delta$) are shown using tetramethylsilane (TMS) with $\delta$: 0.00 as the internal standard. Gas chromatography mass spectrometry (GCMS) results were obtained on a TRACE 1310 and an ISQ LT models (GC and MS, respectively) from Thermo Scientific. The purification of the synthesized molecules was realized through column chromatography, employing Sigma-Aldrich Silica Gel 60 Å (230–400 mesh). To confirm the achieved purity, compounds were verified by thin-layer chromatography (TLC) employing silica plates backed on aluminum (from Merck), revealing the plates using an UV light at 254 nm.

### 2.2. Synthesis of Triphenyl Imidazole Derivatives

A mixture of ammonium acetate (5 Eq) and acetic acid (10 mL) were refluxed; after five minutes of constant dripping, 1 Eq of the appropriate aldehyde (**1–13**) was added; finally after another five minutes, 1 Eq of benzil was added. Reflux was continued until completion of the reaction (verified by Thin-layer chromatography). To stop the reaction, ammonium hydroxide was added up to a pH of 9, the formed precipitate was filtered, washed using cold water, and dried. To purify the product, column chromatography or recrystallization was employed. Confirmation of all structures were achieved by mass and NMR spectra, as discussed below:

2,4,5-triphenylimidazole (**1**): White powder (yield 95%). $C_{21}H_{16}N_2$. Mp = >250 °C. IR (KBr, $cm^{-1}$) = 3037(C–H aromatic), 1599 (C–C), 1323 (C–N) $cm^{-1}$. $^1$H NMR (400 MHz, $CDCl_3$) δ: 7.97 (d, $J$ = 7.6 Hz, 2H), 7.51 (d, $J$ = 7.2 Hz, 4H), 7.47 (t, $J$ = 7.2 Hz, 2H), 7.40–7.27 (m, 7H). $^{13}$C NMR (100 MHz, $CDCl_3$) δ: 146.29, 132.63, 131.93, 129.04, 128.75, 128.56, 128.21, 127.92, 127.27. GC-MS ($m/z$) = 296 $[M]^+$ (97), 281 (18), 207 (63), 165 (100), 147 (20), 73 (46).

2-(4,5-diphenyl-1$H$-imidazol-2-yl) phenol (**2**): White powder (yield 99%). $C_{21}H_{16}N_2O$. Mp = 210–212 °C. IR (KBr, $cm^{-1}$) = 3205 (O–H), 1601 (C–C aromatic), 1326 (C–N) $cm^{-1}$. $^1$H NMR (400 MHz, $CDCl_3$) δ: 7.70 (dd, $J_1$ = 6.4, $J_2$ = 1.0 Hz, 1H), 7.54 (dd, $J_1$ = 8.0, $J_2$ = 1.6 Hz, 4H), 7.37 (m, 6H), 7.25 (m, 1H), 7.11 (d, $J$ = 8.0 Hz, 1H), 6.86 (dd, $J_1$ = 8.0, $J_2$ = 1.0 Hz, 1H). $^{13}$C NMR (100 MHz, $CDCl_3$) δ: 157.03, 145.17, 130.90, 128.81, 128.38, 128.23, 127.88, 124.30, 119.13, 117.87, 111.81. GC-MS ($m/z$) = 312.3 $[M]^+$ (100), 283.1 (8), 209.1 (4), 165.2 (65), 77.2 (15).

4-(4,5-diphenyl-1$H$-imidazol-2-yl) phenol (**3**): White powder (yield 94%). $C_{21}H_{16}N_2O$. Mp = 248–250 °C. IR (KBr, $cm^{-1}$) = 3423 (O–H), 3056 (C–H aromatic), 1609 (C–C aromatic), 1280 (C–N) $cm^{-1}$. $^1$H NMR (400 MHz, $CDCl_3$) δ: 6.80 (d, 2H), 7.85 (d, $J$ = 8.0 Hz, 2H), 7.46–7.19 (m, 10H). $^{13}$C NMR (100 MHz, $CDCl_3$) δ: 157.27, 146.25, 127.64, 126.56, 121.36, 114.98. GC-MS ($m/z$) = 312 $[M]^+$ (93), 281 (14), 207 (55), 165 (100), 73 (39).

2-(4-methoxyphenyl)-4,5-diphenyl-1$H$-imidazole (**4**): White powder (yield 91%). $C_{22}H_{18}N_2O$. Mp = 230–232 °C. IR (KBr, $cm^{-1}$) = 2958 (C–H aromatic), 1492 (C–C aromatic), 1251 (C–O), 1027 (C–N) $cm^{-1}$. $^1$H NMR (400 MHz, $CDCl_3$) δ: 7.86 (d, $J$= 8.8 Hz, 2H), 7.50 (d, $J$ = 7.2 Hz, 4H), 7.32–7.23 (m, 6H), 6.94 (d, $J$ = 8.8 Hz, 2H), 3.80 (s, 3H). $^{13}$C NMR (100 MHz, $CDCl_3$) δ: 159.83, 146.56, 128.23, 127.89, 127.00, 122.70, 113.97, 55.11. GC-MS ($m/z$) = 326.2 $[M]^+$ (100), 311.2 (32), 283.1 (12), 165.2 (24), 77.1 (9).

2-(3-methoxyphenyl)-4,5-diphenyl-1$H$-imidazole (**5**): White powder (yield 90%). $C_{22}H_{18}N_2O$. Mp = >250 °C. IR (KBr, $cm^{-1}$) = 2998 (C–H aromatic), 2961 (C–H, aliphatic), 1485 (C–C aromatic), 1243 (C–O), 1201 (C-N) $cm^{-1}$. $^1$H NMR (400 MHz, $CDCl_3$) δ: 12.68 (s, 1H), 7.69 (m, 2H), 7.56 (d, $J$ = 7.6 Hz, 2H), 7.51 (d, $J$ = 7.2 Hz, 2H), 7.46 (t, $J$ = 7.2 Hz, 2H), 7.39 (t, $J$ = 7.6 Hz, 2H), 7.31 (t, $J$ = 7.2 Hz, 2H), 7.23 (t, $J$ = 6.8 Hz, 1H), 6.95 (d, $J$ = 7.2 Hz, 1H), 3.83 (s, 3H). $^{13}$C NMR (100 MHz, $CDCl_3$) δ: 160.02, 145.81, 137.52, 135.60, 132.10, 131.53, 130.25, 129.11, 128.95, 128.63, 128.26, 127.52, 126.96, 118.08, 114.67, 110.64, 55.67. GC-MS ($m/z$) = 326 $[M]^+$ (100), 282 (10), 207 (10), 165 (74), 89 (28), 77 (21), 44 (61).

2-(2-methoxyphenyl)-4,5-diphenyl-1$H$-imidazole (**6**): Pale yellow powder (yield 93%). $C_{22}H_{18}N_2O$. Mp = 200–202 °C. IR (KBr, $cm^{-1}$) = 1601 (C–C aromatic), 1480 (C–C aromatic), 1240 (C–O), 766 (C–H) $cm^{-1}$. $^1$H NMR (400 MHz, $CDCl_3$) δ: 10.48 (s, 1H), 8.48 (dd, 1H, $J_1$ = 8 Hz, $J_2$ = 1.2 Hz), 7.66–7.49 (m, 4H), 7.33–7.23 (m, 7H), 7.11 (t, $J$ = 7.6 Hz, 1H), 7.00 (d, $J$ = 8.4 Hz, 1H), 3.99 (s, 3H). $^{13}$C NMR (100 MHz, $CDCl_3$) δ: 155.65, 143.98, 129.46, 128.54, 127.72, 121.58, 118.05, 111.12, 55.80. GC-MS ($m/z$) = 326.3 $[M]^+$ (100), 308.2 (80), 295.1 (39), 221.2 (39), 165.2 (57), 77.2 (16).

4-(4,5-diphenyl-1*H*-imidazol-2-yl)-2-methoxyphenol **(7)**: White powder (yield 92%). $C_{22}H_{18}N_2O_2$. Mp = 246–248 °C. IR (KBr, $cm^{-1}$) = 3510 (O–H), 2996 (C–H), 1601 (C–C aromatic), 1496 (C–C aromatic), 1274 (C–O), 695 (C–H) $cm^{-1}$. $^1$H NMR (400 MHz, $CDCl_3$) δ: 7.57 (d, *J* = 1.6 Hz, 1H), 7.49 (d, *J* = 6.8 Hz, 4H), 7.36–7.23 (m, 7H), 6.88 (d, *J* = 8.4 Hz, 1H), 3.88 (s, 3H). $^{13}$C NMR (100 MHz, $CDCl_3$) δ: 147.23, 146.75, 146.63, 132.47, 128.26, 127.91, 127.11, 121.78, 118.60, 114.76, 109.08, 55.71. GC-MS (*m*/*z*) = 342 $[M]^+$ (4), 341 (15), 311 (14), 295 (5), 165 (30), 105 (100), 77 (56), 44 (29).

2-(3,4-dimethoxyphenyl)-4,5-diphenyl-1*H*-imidazole **(8)**: White powder (yield 84%). $C_{23}H_{20}N_2O_2$. Mp = 220–222 °C. IR (KBr, $cm^{-1}$) = 2959 (C–H), 1591 (C–C aromatic), 1495 (C–C aromatic), 1253 (C–O), 762 (C–H) $cm^{-1}$.$^1$H NMR (400 MHz, $CDCl_3$) δ: 7.60 (s, 1H), 7.49 (m, 5H), 7.32 (m, 6H), 6.88 (d, *J* = 8.4 Hz, 1H), 3.89 (s, 3H), 3.86 (s, 3H). $^{13}$C NMR (100 MHz, $CDCl_3$) δ: 149.28, 148.90, 146.51, 132.66, 128.23, 127.90, 127.05, 122.95, 118.12, 111.00, 109.04, 55.68. GC-MS (*m*/*z*) = 356 $[M]^+$ (1), 342 (1), 281 (3), 207 (16), 193 (12), 176 (100), 165 (14), 69 (65).

2-(2-chlorophenyl)-4,5-diphenyl-1*H*-imidazole **(9)**: Pale yellow powder (yield 93%). $C_{21}H_{15}ClN_2$. Mp = 190–192 °C. IR (KBr, $cm^{-1}$) = 2924 (C–H), 1602 (C–C aromatic), 1479 (C–C aromatic), 763 (C–H), 696 (C–Cl) $cm^{-1}$. $^1$H NMR (400 MHz, $CDCl_3$) δ: 10.25 (s, 1H), 8.44 (dd, 1H, $J_1$ = 7.6 Hz, $J_2$ = 1.2 Hz), 7.66 (m, 2H), 7.47–7.25 (m, 11H). $^{13}$C NMR (100 MHz, $CDCl_3$) δ: 143.20, 137.91, 134.57, 130.88, 130.48, 129.59, 129.57, 129.04, 129.02, 128.36, 128.06, 127.96, 127.79, 127.67, 127.52, 127.09. GC-MS (*m*/*z*) = 330 $[M]^+$ (18), 281 (18), 207 (65), 176 (71), 165 (61), 89 (35), 44 (100).

4-(4,5-diphenyl-1*H*-imidazol-2-yl)-N,N-dimethylaniline **(10)**: Brown yellow powder (yield 83%). $C_{23}H_{21}N_3$. Mp = 234–236 °C. IR (KBr, $cm^{-1}$) = 3000 (C–H), 1618 (C–C aromatic), 1497 (C–C aromatic), 1200 (C–N), 765 (C–H), 696 (C–Cl) $cm^{-1}$. $^1$H NMR (400 MHz, $CDCl_3$) δ: 9.70 (s, 1H), 7.77 (d, *J* = 8.8 Hz, 2H), 7.51 (d, *J* = 7.2 Hz, 4H), 7.30–7.21 (m, 6H), 6.70 (d, *J* = 8.8 Hz, 2H), 2.96 (s, 6H). $^{13}$C NMR (100 MHz, $CDCl_3$) δ: 150.74, 147.01, 132.98, 129.94, 129.06, 128.48, 127.85, 127.15, 126.56, 112.14, 40.32. GC-MS (*m*/*z*) = 339 $[M]^+$ (6), 325 (4), 313 (17), 269 (23), 178 (53), 165 (100), 89 (42), 77 (38).

2-(4-nitrophenyl)-4,5-diphenyl-1*H*-imidazole **(11)**: Yellow powder (yield 90%). $C_{21}H_{15}N_3O_2$. Mp = 230–232 °C. IR (KBr, $cm^{-1}$) = 2923 (C–H), 1600 (C–C aromatic), 1519 (N–O), 1486 (C–C aromatic), 1339 (N–O), 765 (C–H) $cm^{-1}$. $^1$H NMR (400 MHz, $CDCl_3$) δ: 8.19 (d, *J* = 8.8 Hz, 2H), 8.06 (d, *J* = 8.8 Hz, 2H), 7.53–7.45 (m, 4H), 7.33 (m, 6H). $^{13}$C NMR (100 MHz, $CDCl_3$) δ: 147.03, 143.89, 135.83, 128.53, 128.04, 125.70, 124.14. GC-MS (*m*/*z*) = 341 $[M]^+$ (1), 330 (10), 281 (4), 220 (10), 176 (100), 165 (24), 89 (21), 69 (60), 45 (40).

2-(2-nitrophenyl)-4,5-diphenyl-1*H*-imidazole **(12)**: Red powder (yield 77%). $C_{21}H_{15}N_3O_2$. Mp = 210–212 °C. IR (ATR diamond, $cm^{-1}$) = 2926 (C–H), 1598 (C–C aromatic), 1517 (N–O), 1485 (C–C aromatic), 1331 (N–O), 759 (C–H) $cm^{-1}$. $^1$H NMR (400 MHz, $CDCl_3$) δ: 8.34 (d, *J* = 8.0 Hz, 2H), 8.11 (d, *J* = 8.0 Hz, 2H), 7.69–7.30 (m, 10 H). $^{13}$C NMR (100 MHz, $CDCl_3$) δ: 147.44, 143.40, 139.83, 135.48, 133.99, 130.27, 129.39, 129.10, 128.57, 128.47, 127.97, 127.71, 127.47, 125.46, 124.43. GC-MS (*m*/*z*) = 341 $[M]^+$ (48), 311 (10), 237 (5), 165 (28), 135 (31), 104 (100), 89 (60), 79 (29).

2-(anthracen-9-yl)-4,5-diphenyl-1*H*-imidazole **(13)**: Pale yellow powder (yield 64%). $C_{29}H_{20}N_2$. Mp = 204–206 °C. IR (ATR diamond, $cm^{-1}$) = 3074–3020 (C–H), 1609 (C–C aromatic), 1447 (C–C aromatic) $cm^{-1}$. $^1$H NMR (400 MHz, $CDCl_3$) δ: 8.47 (s, 1H), 8.00 (d, *J* = 8.0 Hz, 2H), 7.92 (d, *J* = 8.0 Hz, 2H), 7.64–7.27 (m, 14H). $^{13}$C NMR (100 MHz, $CDCl_3$) δ: 143.60, 131.49, 131.10, 128.91, 128.61, 128.42, 127.83, 127.40, 126.55, 125.83, 125.31, 124.58. GC-MS (*m*/*z*) = 396.2 $[M]^+$ (100), 323.1 (3), 291.1 (4), 203 (10), 165 (16), 105 (8), 77 (4).

### 2.3. *In Vitro Antioxidant Activity Assay*

#### 2.3.1. 1,1-Diphenyl-2-Picrylhydrazyl (DPPH) Radical-Scavenging Assay

For the determination of the radical-scavenging activity, we used our implementation of the Salazar-Aranda et al. [16] method. A set of serial dilutions in methanol were prepared for each sample. Then, 0.5 mL aliquots of each dilution were mixed with a solution of 1,1-diphenyl-2-picrylhydrazyl (DPPH) in methanol (0.5 mL, 76 μM). The resulting mixtures were kept in the dark at room temperature for 30 min. The absorbance of each sample was measured at 517 nm ($A_{517}$) and methanol was used as the blank. To calculate the radical-scavenging activity as DPPH decoloration percentage, the formula below was used:

$$DPPH\ (\%) = [1 - (B/A)] \times 100$$

where A represents the absorbance value of the DPPH solution (used as control) and B is the absorbance of the DPPH solution with the sample. Results were expressed as $EC_{50}$, which represents the required concentration to diminish the absorbance of DPPH by 50%. Quercetin was employed as the reference compound.

#### 2.3.2. ABTS Radical-Scavenging Assay

For the determination of the ABTS radical cation ($ABTS^+$) scavenging activity, we used our implementation of the Re et al. and Kuskoski et al. [17,18] method. $ABTS^+$ was produced by reacting an ABTS stock solution (7 mM in water) with 2.45 mM potassium persulfate. The resulting mixture was kept at room temperature in the dark for 16–18 h before its use. Methanol was used to dilute the $ABTS^+$ solution (150 µL) to give an absorbance of 0.7 ± 0.02 at 754 nm. This value was taken as the initial absorbance (A1). For each sample, aliquots were prepared mixing 980 µL of the $ABTS^+$ methanolic solution with 20 µL of the samples at diverse concentrations. Each mixture was stirred, incubated at room temperature for 7 min, and its absorbance was read (754 nm). This value was considered as the final absorbance (A2). To calculate the radical-scavenging activity as a percentage of ABTS decoloration, the employed formula was:

$$\%\ \text{of inhibition} = [(A_1 - A_2)/A_1] \times 100$$

All determinations were performed in triplicate. Results were expressed as $EC_{50}$, which represents the required concentration to diminish the absorbance of ABTS by 50%. Quercetin was used as the reference compound.

### 2.4. *In Vitro Acetylcholinesterase Inhibitory Assay*

The determination of acetylcholinesterase activity was done using our implementation of the methodology reported by Adewusi et al. [19]. Employing a 96-well plate, 75 µL of Trizma-HCl buffer (50 mM, pH 8) was added along with 75 µL of the synthesized compound diluted, obtaining a 150 µg/mL concentration (0.15% for the dimethyl sulfoxide DMSO) at the end. Subsequently, 25 µL of a buffer solution of 15 mM acetylthiocholine chloride (ATCl) was added to each well with 125 µL of a 3 mM buffer solution of Ellman's reagent (DTNB) (Ellman's reagent), giving both of them concentrations of 1.5 mM at the end. Employing a microplate reader every 45 s, the absorbance was measured at a wavelength of 405 nm, for three consecutive times. After these lectures, to each well 25 µL of an enzyme buffer solution with a concentration 2 U/mL of acetylcholinesterase was supplied, enriched with 0.1 mg/mL bovine serum albumin, obtaining an enzyme 0.2 U/mL final concentration. Five consecutive lectures were taken every 45 s. Of each plate, six wells served as the control for the acetylcholinesterase 100% activity, having no tested compound on them. Galantamine was used as the positive control. A correction for the substrate's spontaneous hydrolysis was made by subtracting the absorbance from before the addition of the enzyme from the enzyme containing wells. Using the equation:

$$\text{Inhibition}\ \% = 1 - (A_{\text{sample}}/A_{\text{control}}) \times 100$$

we obtained the percentage of acetylcholinesterase inhibition, where the absorbances were the 0 and 225 s differences of the sample evaluated and for the enzyme 100% activity control previously described. All experiments were performed in triplicate.

### 2.5. In Vitro Xanthine Oxidase Inhibitory Assay

The XO inhibition activity was evaluated using our implementation of the protocol reported by Almada-Taylor et al. [20]. To a volume of 0.33 mL of a xanthine 150 mM solution, phosphate buffer 120 mM with a pH of 7.8 was added (0.4 mL) and mixed with 0.25 mL of a solution of the compound to be analyzed. The reaction was started with the addition of a 0.5 U/mL solution of xanthine oxidase enzyme (0.02 mL). This was allowed to incubate for 3 min at 24 °C, followed by absorbance lecture at 295 nm ($A_{295}$) for the measurement of the formation of uric acid. As a reference, allopurinol was employed, and the control was an absorbance lecture without an inhibitor. Employing the formula:

$$\% \text{ of Xanthine Oxidase inhibition} = [1 - (A_S / A_C)] \times 100,$$

the percentage of xanthine oxidase inhibition activity was determined. $A_S$ indicates the initial velocity of reaction of the sample, and $A_C$ indicates that for the control. All determinations were made in duplicate, and repeated at least three times. Using interpolation from a linear regression analysis, the required concentration to diminish the XO activity by 50% ($IC_{50}$) was calculated.

### 2.6. Cell Lines and Culture Conditions

The in vitro antiproliferative activity of the investigated compounds was evaluated against six human solid tumor cell lines: A549 (non-small cell lung), HBL-100 (breast), HeLa (cervix), and SW1573 (non-small cell lung) as drug sensitive lines; and T-47D (breast) and WiDr (colon) as drug resistant lines. These cell lines were a kind gift from Prof. G. J. Peters (VU Medical Center, Amsterdam, The Netherlands). Cells were maintained in 25 $cm^2$ culture flasks in Roswell Park Memorial Institute (RPMI) 1640 media enriched with 5% FCS (Fetal Calf Serum) and 2 mM L-glutamine in a 37 °C, 5% $CO_2$, and 95% humidified air incubator.

### 2.7. In vitro Antiproliferative Assay

Cells were trypsinized, resuspended in medium containing 5% FCS and antibiotics (100 U/mL of penicillin G and 0.1 mg/mL of streptomycin), counted (Moxi Z automated cell counter), and diluted to reach the appropriate cell densities (2500 cells/well A549, HBL-100, HeLa and SW1573, and 5000 cell/well for T-47D and WiDr) for inoculation onto 96-well plates. Twenty-four hours later, compounds were added at concentrations in the range 0.01–100 μM. Cisplatin and etoposide were used as the positive control and DMSO (0.25% v/v) was used as the negative control. Drug incubation times were 48 h. Then, cells were fixed using 25 μL ice-cold trichloroacetic acid (TCA) solution (50% w/v) for 60 min at 4 °C, after which time the plates were rinsed with water. Next, 25 μL of a sulforhodamine B (SRB) solution (0.4% w/v in 1% acetic acid) was added for 15 min. Unbound SRB was rinsed with 1% acetic acid. The remaining dye was dissolved with 150 μL of Tris solution (10 mM, pH 10.5). The optical density of each well was determined at 530 and 620 nm using a microplate reader. The anti-proliferative activity, expressed as 50% growth inhibition ($GI_{50}$), was calculated according to NCI formulas [21].

### 2.8. Molecular Docking

The molecular models of the synthesized compounds were obtained inserting their SMILES strings in University of California, San Francisco (UCSF) Chimera 1.11.2 [22]. Energy minimization of the structures was done using Chimera default conditions with Molecular Modelling Toolkit (MMTK) and Antechamber parameters [23]. AutoDock Tools 1.5.6 [24] was employed to define the rotatable bonds and atomic charges for each ligand. Download of the crystallographic structures of the receptors EGFR (PDB ID: 4HJO) and HER2 (PDB ID;3PP0) was done through Protein Data Bank (https://www.rcsb.org/) [25]. Each receptor was prepared with AutoDock Tools, removing the co-crystalized ligand along with the molecules of water included in the model, adding hydrogens and calculating the Gasteiger charges. AutoDock 4.2 [26] was employed for the docking analysis by using a grid box of 72 × 72 × 72 Å with x = 24.5, y = 9, z = −1 as the center coordinates for EGFR and x = 17.5, y = 17.5, z = 27 for HER2, with a grid point spacing of 0.375 Å. A Lamarckian genetic algorithm was used with a population size of 150, maximum number of evaluations $2.5 \times 10^6$, maximum number of generations 27000, rate of gene mutation 0.02, and rate of crossover 0.8, generating 10 docked conformations for each analyzed compound.

### 2.9. *In Silico Drug-Likeness Prediction*

To determine the pharmacokinetics and physicochemical properties related to drug-likeness of the synthesized compounds, the SwissADME web server was employed [27].

## 3. Results and Discussion

### 3.1. *Synthesis of Triphenyl Imidazole Derivatives*

The 2,4,5-trisubstituted imidazole derivatives **1–13** were prepared from a 1,2-diketone (benzil), ammonium acetate and the corresponding aldehydes, following the known Radziszewski reaction and the methodology proposed by Puratchikody et al. with some modifications (Scheme 1) [28], with reaction yields of 64–99%. All compounds were characterized by IR and mass spectroscopy, $^1$H- and $^{13}$C-NMR. In the $^1$H NMR spectra of compounds **1–13**, the corresponding signals for the aromatic protons of the rings of position four and five of the imidazole heterocycle were observed, with typical displacements between 7.19–7.69 ppm. For the aromatic system of position two, all of the protons' expected shifts were observed, as were their coupling constants. In the $^{13}$C-NMR spectra, the carbons that formed the imidazole ring were observed at shifts of 159.83–143.20 ppm for carbon two, while those of position four and five were seen at 128.54–127.64 ppm. NMR spectra of the selected derivatives can be observed in Figures S1–S13 in the Supplementary Materials.

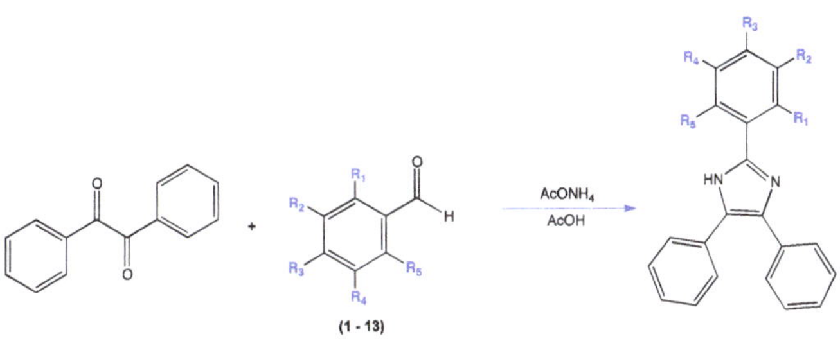


| | $R_1$ | $R_2$ | $R_3$ | $R_4$ | $R_5$ |
|---|---|---|---|---|---|
| **1** | –H | –H | –H | –H | –H |
| **2** | –H | –H | –H | –H | –OH |
| **3** | –H | –H | –OH | –H | –H |
| **4** | –H | –H | $–OCH_3$ | –H | –H |
| **5** | –H | $–OCH_3$ | –H | –H | –H |
| **6** | $–OCH_3$ | –H | –H | –H | –H |
| **7** | –H | $–OCH_3$ | –OH | –H | –H |
| **8** | –H | $–OCH_3$ | $–OCH_3$ | –H | –H |
| **9** | –Cl | –H | –H | –H | –H |
| **10** | –H | –H | $–N(CH_3)_2$ | –H | –H |
| **11** | –H | –H | $–NO_2$ | –H | –H |
| **12** | $–NO_2$ | –H | –H | –H | –H |
| **13** | | | 9-anthracene | | |

**Scheme 1.** General reaction scheme for the synthesis of 2,4,5-triphenyl-1*H*-imidazole derivatives.

### 3.2. Antioxidant Activity

Both in the DPPH and ABTS assays, imidazole presented $EC_{50}$ of >15 and >10 mg/mL, respectively (Table 1), which compared to most of the results shown by its derivatives, suggests that the 2,4,5-triphenyl substitution in the imidazole heterocyclic is relevant for the antioxidant activity of these compounds, where the effect of their substitutions on their A ring is further developed below.

The DPPH (2,2-diphenyl-1-picrylhydrazyl) radical scavenging method is widely used to evaluate antioxidant activities in a relatively short period of time compared to other methods. The results of this assay are shown in Table 1, comparing the synthesized products with the standard quercetin, where the most active synthesized imidazole derivatives were **3**, **10**, **7**, and **2** with values of $EC_{50}$ of 0.141, 0.174, 0.341, and 1.389 mg/mL, respectively. These results show that the presence of electron donating groups such as hydroxy and *p*-dimethylamino on an aromatic ring bonded to imidazole are essential in the antioxidant activity. The consulted literature indicates that this could be due to the free pair of electrons in nitrogen or in the oxygen of the hydroxy group, which can react with free radicals, being favored due to their aromatic ring stabilization [29]. The rest of the compounds presented low activity in this assay, mainly because of their lack of acidic hydrogen in the aromatic system of position two (A ring); instead, compounds **4**, **5**, **6** and **8** bear methoxy groups, there is a chlorine atom in **9** ($EC_{50}$ of 5.62 mg/mL), an electron attractor effect of the $NO_2$ group in products **11** and **12**, and an anthracene group in **13**.

It is interesting to point out the difference in antioxidant activity between isomers **2** and **3**, where it is shown from the last one that there was a higher oxidative inhibitory potential in both employed techniques (DPPH and ABTS). It is well known that the antioxidant mechanisms of phenolic compounds are hydrogen atom transfer and single electron transfer, in order to inhibit free radicals, which are the expected mechanisms for the phenolic hydroxyls present in these isomers. These different results could be due to the fact that even though both compounds can transfer their hydrogens because of their high acidity, hydroxyl in **2** is in an *ortho* position, favoring the formation of a hydrogen bond along with a nitrogen of the imidazole nucleus, and forming a 6-membered stable ring. It is referenced that these hydrogen bond interactions can diminish the hydrogen dissociation and therefore the antioxidant ability of these groups [30].

In 2015, Hemalatha et al. [31] evaluated the antioxidant activity with a DPPH assay of compounds **2**, **3**, and **10**, reporting $IC_{50}$ values of 0.003, 0.0037, and 0.0031 mg/mL, respectively, while the $IC_{50}$ values for the same compounds in our analysis were 1.389, 0.141 and 0.174 mg/mL, respectively. Even though there were notorious differences between both results, establishing a direct comparation was complicated due to differences in the methodologies employed for the assay, as in [31], a higher concentration of the DPPH radical was employed, and the incubation times for the reactions were not stated.

In a similar way to the DPPH methodology, the ABTS radical-scavenging assay showed that compounds **10**, **3**, **2**, and **7** with $EC_{50}$ values of 0.162, 0.168, 0.188 and 0.199 mg/mL, respectively, were the most active products, however, compound **13** showed moderate activity, while products **1**, **4**, **5**, **6**, **8**, **11**, and **12** presented low activity, as can be seen in Table 1. With these compounds, once again, it is important to emphasize that the participation of hydroxyl and dimethylamino groups play an important role as free radical scavengers. Several reports have discussed the possible mechanisms involved in $ABTS^+$ quenching, suggesting the mixed hydrogen atom transfer/single electron transfer reaction mechanisms [32], and some groups have these properties of chemical reactivity such as *N,N*-dimethylaniline derivatives, which can generate efficient and stable radicals [33].

### 3.3. Acetylcholinesterase Inhibitory Assay

In this assay, galantamine was more active than the products evaluated. Nevertheless, as an initial screening a structure activity relationship is attempted to obtain valuable information for future research.

Among the synthesized products, compound **1** showed the best activity with 25.8% of inhibition (Figure 1). AChE inhibitors bond with the enzyme in a well-known gorge, which in its bottom presents a Trp residue (Trp84 for *Tetronarce californica* AChE, the enzyme used for the in vitro assay). This residue is of crucial importance for ligand interaction by means of a π–cation interaction [34,35]; however, it can also have purely hydrophobic interactions. In the case of galantamine and donepezil [34,36] this residue presents classical π–π stacking with a galantamine double bond, and with the benzyl ring in donepezil. In a similar way, compound **1** could adopt a similar position against AChE, presenting a π–π interaction with Trp84 through its A ring, which has no substitutions that could affect the π electron cloud in the ring, thus explaining the result shown. Hydrophobic and π–π interactions tend to be the most observed ones between AChE and the scaffolds of different inhibitors [37–39].

The next compounds with high inhibition percentages were compounds **11** and **12**, which presented a nitro functionality in their *p*- and *o*- positions. The nitrogen atom in this group is positively charged; in this manner, these compounds could have π–cation interactions with Trp84, or even with Phe330, which is another residue that commonly has this interaction. This could explain why **11** and **12** followed compound **1** with the best results.

Some tendencies seen in the results when comparing **2** (which has an *o*-OH substitution) against **6** (which presents an *o*-OMe one), we can see that the inhibition activity diminishes; the same pattern was observed with **7** and its methoxy version **8**, although the comparison between **3** and **4** appeared as the exception of this behavior. Compound **9** had only 5.9% inhibition activity; as π–π interactions with AChE are important, the chloride presence in **9** could alter the electron cloud from the A ring, disturbing the π–π interactions that can be made.

**Table 1.** Antioxidant activity ($EC_{50}$) of synthesized compounds **1–13**.

| Compounds | Antioxidant Activity ($EC_{50}$, mg/mL) | |
|---|---|---|
| | DPPH | ABTS |
| **1** | 3.25 ± 0.137 | 34.312 ± 0.245 |
| **2** | 1.389 ± 0.631 | 0.188 ± 0.011 |
| **3** | 0.141 ± 0.094 | 0.168 ± 0.046 |
| **4** | 16.74 ± 0.003 | 1.644 ± 0.584 |
| **5** | 16.89 ± 0.636 | 37.223 ± 2.629 |
| **6** | 7.12 ± 1.916 | 15.643 ± 0.324 |
| **7** | 0.341 ± 0.101 | 0.199 ± 0.001 |
| **8** | 12.23 ± 3.042 | 1.964 ± 0.37 |
| **9** | 5.62 ± 1.752 | ND |
| **10** | 0.174 ± 0.041 | 0.162 ± 0.006 |
| **11** | ND | 8.025 ± 0.771 |
| **12** | ND | 42.158 ± 2.697 |
| **13** | 4.00 ± 0.135 | 0.449 ± 0.03 |
| Imidazole | >15 | >10 |
| * Quercetin | 0.052 ± 0.037 | 0.075 ± 0.002 |

* Served as the reference compound. Values are mean ± SD, DPPH n = 2, ABTS n = 3. ND = Non-detected in the evaluated concentrations. $EC_{50}$ = Concentration required to decrease the absorbance by 50%.

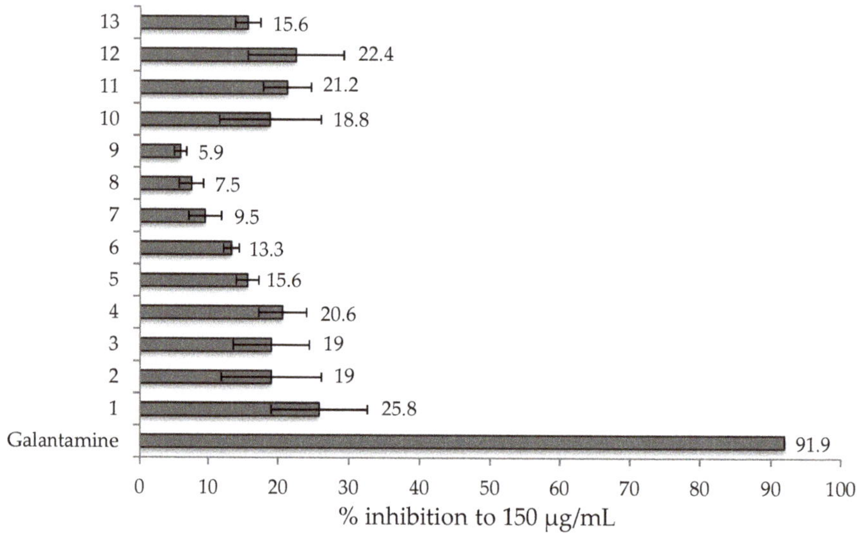


**Figure 1.** Percentage of acetylcholinesterase inhibition of triphenyl imidazole compounds **1–13** (150 μg/mL). Galantamine served as the reference compound.

### *3.4. Xanthine Oxidase Assay*

Although not being as active as the positive control allopurinol, some tendencies in the structure activity relationship of the synthesized compounds can be noticed, as seen in Figure 2.

Comparing compounds **2–6** where hydroxy and methoxy substitutions are present, the *p*- substitution can be inferred as a significant requirement for this products, as only *p*-OH and *p*-OMe products showed activity. This was also the case for compounds **7** and **8**, with hydroxy and methoxy groups as substitutions, while having a *para* substitution besides a *meta* one, allowed them to show activity.

It appears that not only the *p*- position is of importance, but also that the functionality in these synthesized compounds must be of -OH or -OMe type, bearing an oxygen as a heteroatom bonded to the aromatic ring. Products **10** and **11** also have substitutions in this position, but with nitrogen as the heteroatom (an amine and nitro group, respectively) and in their case, the *para* position with a nitrogenated group showed no activity. For the synthesized products, the interaction with xanthine oxidase, instead of being similar to the one for allopurinol, which interacts with one of its aromatic nitrogen to bond with molybdenum in the catalytic site of the enzyme [40], could be similar to the topiroxostat one. This inhibitor interacts with the xanthine oxidase molybdenum with its oxygen in a covalent bond [41]. While compound **10** has its nitrogen in a tertiary amine, and **11** in a nitro group, it could be more difficult for them to bond with the Mo center of the enzyme, favoring in our products the presence of oxygen over nitrogen.

Product **3**, having a *p*-OH group and no other substitution that diminishes its activity, resulted in the most active compound from the synthesized ones. Between the hydroxy and methoxy substitutions, it appeared as the first one favored the inhibition activity over xanthine oxidase. Compound **3** with a *p*-OH substitution showed an $IC_{50}$ of 85.8 μg/mL, while **4**, which has a *p*-OMe, showed almost double the $IC_{50}$; again, between **7** and **8**, we could see that the methoxy version was less potent than the hydroxy one. This can be related with the observation made for different products with alcohol groups in their structure such as polyphenols, which can form hydrogen bonds with XO via their hydroxyl groups [14,42].

However, the exception to the structure activity relationship discussed was **12**, having an *o*-$NO_2$, which lacked a *para* position and oxygen heteroatom functionality. This compound was achieved as one of the few products with xanthine oxidase inhibition, although it showed the second lowest activity. Further *ortho* nitrogen containing products must be synthesized to expand this analysis.

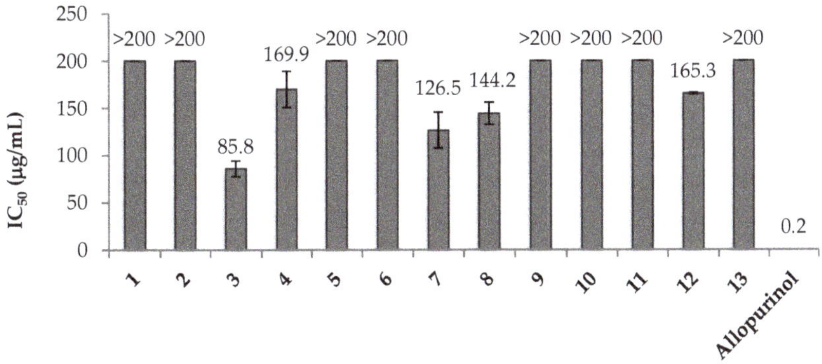


**Figure 2.** Xantine oxidase inhibitory activity of synthesized compounds **1–13**. Allopurinol served as the reference compound. Bars are mean ± SD, n = 3.

### *3.5. Antiproliferative Assay*

The antiproliferative activity evaluation of the synthesized triphenyl imidazole derivatives was made with the sulforhodamine B (SRB) assay. The tumoral cell lines employed were adherent epithelial cells from different anatomic origins. All results were expressed as growth inhibition 50 ($GI_{50}$), as the concentration needed to inhibit the 50% of cell population, and calculated and expressed as micromolar (µM). As positive controls, different antitumor drugs were employed such as cisplatin, etoposide, and camptothecin, and imidazole was used as the structural reference of the synthesized compounds.

The obtained results from the evaluation of the 13 synthesized compounds with the SRB assay against the tumor cell lines (Table 2) showed no selectivity by any specific line. Based on the results in Table 2 and the $GI_{50}$ range (Figure 3), the most active compounds of the series were **10** and **11**. As initial highlights, imidazole had no inhibitory activity in the compounds employed as controls; on the other hand, one of the most resistant cell lines against the synthesized compounds and drugs was A549, which corresponds to lung adenocarcinoma, and this is in agreement with the literature, as it has been documented that lung type cancers are usually chemotherapy resistant, even to one of the most used antitumor drugs, taxol [43].

From the 13 synthesized compounds, lower activity was shown from derivative **1**, this being the triphenyl imidazole bearing no substitutions, as against five of the six evaluated cell lines, it showed no significant activity, and a low one against SW1573 (89 µM). Likewise, between the molecules with one methoxy substitution **4** (*p*-OMe), **5** (*m*-OMe) and **6** (*o*-OMe), which are position isomers, only **5** showed a low activity against SW1573 with an $GI_{50}$ of 76 µM, while **4** and **6** presented no significant activity against all of the evaluated cell lines.

Following these general low active compounds, derivatives **9** (*o*-Cl) and **12** (*o*-$NO_2$) were partially active, as they presented different degrees of activity, but against only a couple of cell lines. Compound **9** showed good to moderate activity only against two cell lines, which were HeLa with a $GI_{50}$ of 7.7 µM and SW1573 with 17 µM; compound **12** presented activity against the same cell lines with $GI_{50}$ of 6.1 and 66 µM respectively.

The following molecules with better results were the compounds **3** (*p*-OH), **7** (*m*-OMe, *p*-OH), **8** (*m*-OMe, *p*-OMe), and **13** (anthracene), as these molecules presented activity against all evaluated cell lines, the only exception being **13** against lines T-47D and WiDr; however, **13** also showed one of the best particular results, this being a $GI_{50}$ of 4.2 µM against SW1573. In the case of **3**, activity was shown against all analyzed cell lines, with HeLa being the most sensitive with a $GI_{50}$ of 13 μM, followed by SW1573 and HBL-100 with 15 and 16 μM, respectively, while in the rest of the lines, the results were between 19 and 22 μM.

**Table 2.** Antiproliferative activity of compounds **1–13** against six human solid tumor cell lines [a].

| Compound | Substituent | Cell Lines ($GI_{50}$, μM) | | | | | |
|---|---|---|---|---|---|---|---|
| | | A549 | HBL-100 | HeLa | SW1573 | T-47D | WiDr |
| 1 [b] | -H | >100 | >100 | >100 | 89 | >100 | >100 |
| 2 | *o*-OH | 11 ± 5.5 | **7.0 ± 2.0** | **4.3 ± 0.6** | **3.6 ± 0.2** | 18 ± 0.3 | 19 ± 0.5 |
| 3 | *p*-OH | 19 ± 4.0 | 16 ± 0.4 | 13 ± 2.5 | 15 ± 2.3 | 20 ± 1.5 | 22 ± 0.7 |
| 4 | *p*-OMe | >100 | >100 | >100 | >100 | >100 | >100 |
| 5 [b] | *m*-OMe | >100 | >100 | >100 | 76 | >100 | >100 |
| 6 | *o*-OMe | >100 | >100 | >100 | >100 | >100 | >100 |
| 7 | *m*-OMe, *p*-OH | 17 ± 1.3 | 17 ± 0.7 | 15 ± 1.7 | 15 ± 0.5 | 20 ± 2.0 | 17 ± 1.2 |
| 8 | *m*-OMe, *p*-OMe | 26 ± 1.0 | 15 ± 2.0 | 10 ± 0.4 | 15 ± 3.1 | 16 ± 1.2 | 13 ± 6.8 |
| 9 [b] | *o*-Cl | >100 | >100 | **7.7** | 17 | >100 | >100 |
| 10 | *p*-$N(CH_3)_2$ | **3.8** [b] | **5.9 ± 3.1** | **4.5 ± 1.2** | **4.4 ± 2.5** | **5.3 ± 1.9** | **4.5 ± 0.8** |
| 11 | *p*-$NO_2$ | **6.3 ± 3.4** | **3.3 ± 1.6** | **3.0 ± 0.9** | **2.9 ± 0.4** | **5.5 ± 0.2** | **4.6 ± 0.2** |
| 12 [b] | *o*-$NO_2$ | >100 | >100 | **6.1** | 66 | >100 | >100 |
| 13 | 9-anthracene | 45 ± 14 | 23 ± 10 | 12 ± 1.5 | **4.2 ± 1.4** | >100 | >100 |
| Imidazole | - | >100 | >100 | >100 | >100 | >100 | >100 |
| CDDP | - | **4.9 ± 0.3** | **1.9 ± 0.2** | **1.9 ± 0.4** | **2.7 ± 0.4** | 17 ± 2.3 | 23 ± 4.3 |
| VP-16 | - | **1.5 ± 0.3** | **1.2 ± 0.3** | **2.4 ± 0.9** | 15 ± 1.5 | 18 ± 4.4 | 24 ± 2.6 |

[a] $GI_{50}$ values are given in μM. Standard deviation was calculated from two to four independent experiments. Cisplatin (CDDP) and etoposide (VP-16) were used as reference antiproliferative drugs. Values in bold represent the best anti-proliferative data against tumor cell lines with $GI_{50}$ values < 10 μM. [b] Only one experiment was performed.

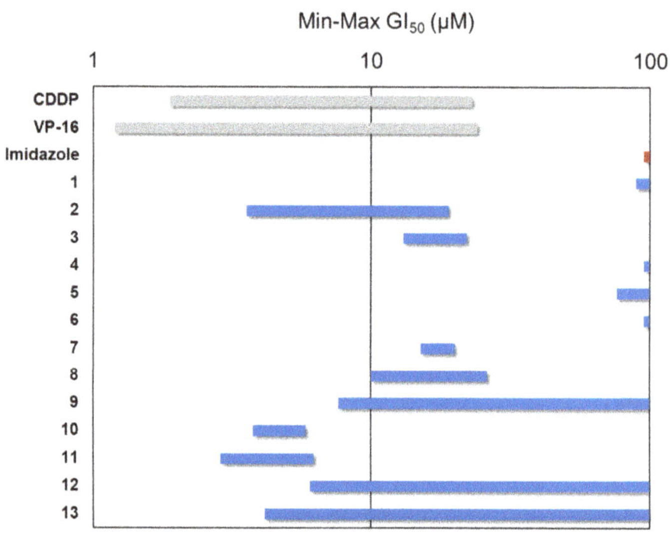


**Figure 3.** $GI_{50}$ range plot of the tested compounds.

With molecules **7** and **8**, very similar $GI_{50}$ values could be seen between them. Comparing them against monosubstituted compounds **4**, **5**, and **6** (which have a methoxy group in different positions), **7** and **8** showed that di-substitution enhanced the antiproliferative activity against these cell lines, these being substitutions of the hydroxy and methoxy type. In the literature, compound **8** was reported to show antiproliferative activity against the breast cancer cell line MDA-MB-231 with a $GI_{50}$ of 21 μM [44]. This value is consistent with the results obtained in our study.

The best set of synthesized molecules were **2** (*o*-OH), **10** (*p*-$N(CH_3)_2$) and **11** (*p*-$NO_2$) as they showed the overall lowest $GI_{50}$ values. In particular, **2** presented significant activity against lines SW1573 and HeLa with values of 3.6 and 4.3 μM; **10** showed significant activity against the majority of cell lines, where the outstanding results were against A549 with $GI_{50}$ of 3.8 μM, and against SW1573 with 4.4 μM. The most active compound against all of the analyzed cell lines was **11**, as its $GI_{50}$ was between 2.9 to 6.3 μM, with SW1573 being the most sensitive line.

These two most active compounds had a substitution in their A ring in the *para* position, where the second most active was a tertiary amine nitrogen, and the most active had a nitro group. One of the most sensitive cell lines to the synthesized compounds and the evaluated drugs was SW1573, which is from alveolar carcinoma. This is despite the line belonging to a lung cancer lineage which, as mentioned before with A549, are pharmacotherapy resistant carcinomas. However, these two lung related cell lines showed that they were sensitive to compound **11**, with $GI_{50}$ values of 6.3 for A549 and 2.9 μM for SW1573, which were the second lowest and the lowest values, respectively, for these lines.

In 2017, Dake's research team [45] reported the synthesis and evaluation of triphenyl imidazole derivatives with substitutions in their A ring against the A549 line, where their compound 6f showed an $IC_{50}$ of 15 μM. This molecule has *m*-I, *m*-OMe, and *p*-OH substitutions, where the iodine is structurally similar to **7**. The presence of this heteroatom improved activity by a 2 μM difference compared to not having it (17 μM for molecule **7**).

The *p*-$NO_2$ substitution in compound **11** bears an important role in antiproliferative activity, which could be due to the nitroaromatic structure. Nitroaromatic compounds have gained interest as chemotherapeutic agents against cancer because molecules with nitro groups in their metabolism can go through bio-reduction, which generates reactive species that cause damage to cell components by oxidative stress; additional reductions are favored in hypoxic conditions, which generates highly cytotoxic species [46]. Even though molecule **12** is an isomer of **11**, in comparison, it showed much lower activity. This could be due to **12** having the nitro group in the *ortho* position, where it could interact with the hydrogen in the imidazole ring, diminishing the generation of the reactive species needed for the antiproliferative activity.

### *3.6. Molecular Docking*

Encouraged by previous reports from our group where docking techniques were applied with good results [47,48], in the present work, docking was employed to search possible imidazole receptors.

Many solid tumors are characterized by aberrant signal transduction through different receptors belonging to the ErbB family of receptor tyrosine kinases, where the EGFR and HER2 receptors belong; therefore, one therapeutic approach in oncotherapy is the inhibition of one or both of these receptors [49,50]. The ErbB receptors and their ligands are overexpressed in the majority of solid neoplasms; EGFR and ErbB-3 are found on average in 50% to 70% of lung, colon, and breast carcinomas [51]. HER2 is mainly related with breast cancer (is expressed in 30% of primary breast carcinomas [51]), but is also related with ovary, colon, lung, uterine cervix, and esophagus cancers, amongst others [52]. As co-expression of different ErbB receptors occurs commonly, 87% of EGFR positive tumors also express HER2 [51]. Due to all of the above, EGFR and HER2 receptors have been selected in the literature [53] to relate in vitro anti-cancer activity to in silico docking calculations. In this last reference, the results from the docking of imidazole derivatives against EGFR and HER2 showed a general good agreement with their cytotoxic results. They evaluated two imidazoles that are reported in the presented work, **11** and **12**, with generally closely related results; having the same docking algorithm and protocol, differences may arise due to different ligand preparation as this step can influence the final result [54]. In the present work, the proposed docking protocol was employed for an initial screening for both EGFR and HER2 as potential cancer-mediated receptors for the synthesized imidazole derivatives.

The binding energies results from the docking analysis are shown in Table 3, which includes imidazole as a negative control and lapatinib, an EGFR and HER2 inhibitor [55], as the positive control. All synthesized compounds showed better results than the imidazole, suggesting the derivatization improved their affinity for these receptors. Although lapatinib showed the best result against both enzymes compared to our compounds, it was closely followed by some products. From the synthesized compounds, **11**, **12**, **5**, **9**, and **7** presented the best results interacting with both EGFR and HER2, as they were in the first five places with lower binding energies. After that, there were variations in the order in which the synthesized products interacted with the selected receptors. Comparing the results for the docking in each receptor, against the in vitro results for each of the evaluated cell lines, there was little agreement between them. This can be explained in several ways, one could be the use of a specific docking algorithm, while each one presents differences in the way results are achieved. The employment of different algorithms with the present work dataset of ligand structures and $GI_{50}$ values could be further explored to find the most suitable algorithm for the synthesized ligands. On other hand, it could be possible that the biological receptors where the compounds interact are different to those in EGFR and HER2, explaining the little correlation shown. Additionally, it has been reported that docking results could be significantly improved with post-docking energy refining through semi-empirical methods such as PM7 [56].

The compound that represented good agreement between its in vitro and in silico results was **11**, bearing a *p*-nitro substitution (Figure 4). It showed −9.11 and −9.19 kcal/mol binding energy with EGFR and HER2, respectively, having the second-best affinity with both receptors. On the other hand, it was the first or second most active compound against the six evaluated cell lines. This suggests that **11** could be one potential lead compound for further derivatization in the search for new active antiproliferative agents.

**Table 3.** Docking scores of the synthesized triphenyl imidazoles with their controls.

| Compound | Binding Energy (kcal/mol) | |
|---|---|---|
| | EGFR | HER2 |
| **1** | −8.32 | −8.59 |
| **2** | −7.92 | −8.92 |
| **3** | −8.24 | −8.1 |
| **4** | −7.89 | −8.4 |
| **5** | −8.87 | −9.13 |
| **6** | −8.2 | −8.58 |
| **7** | −8.68 | −8.99 |
| **8** | −7.63 | −8.99 |
| **9** | −8.49 | −9.1 |
| **10** | −8.28 | −8.98 |
| **11** | −9.11 | −9.19 |
| **12** | −9.88 | −9.31 |
| **13** | −8.37 | −8.23 |
| **Imidazole** | −2.89 | −3.21 |
| **Lapatinib** | −10.48 | −9.88 |

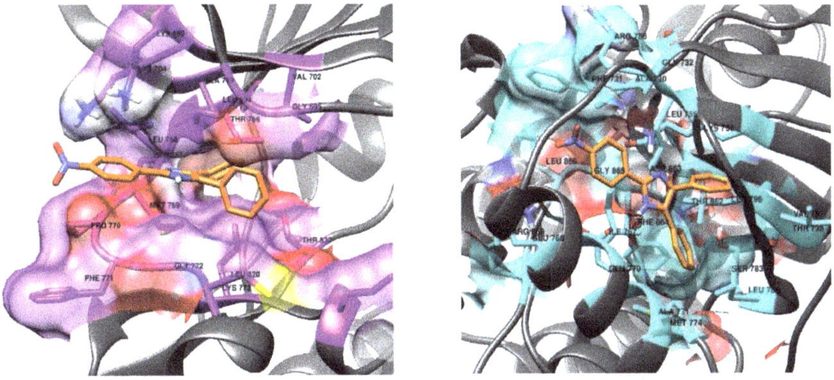

(**a**) (**b**)

**Figure 4.** Compound **11** (orange) depicted with both of the analyzed receptors. Highlighted are the residues at distances <5.0 Å from **11**, with some surfaces active to show the cavity in which the docking analysis put the best docked pose. (**a**) EGFR, with the near residues in purple. (**b**) HER2, with the near residues in cyan.

### *3.7. In Silico Drug-Likeness Prediction*

As can be seen from Table 4, the calculations from the SwissADME website allow for the analysis of which synthesized compounds have better pharmacokinetics and drug-like properties.

All of them had a TPSA between the limits suggested for good bioavailability (20–130 Å$^2$). The vast majority are inhibitors to cytochrome enzymes, which could affect the metabolism and present drug–drug interactions [27], **13** being the least CYP inhibitor, followed by **11**, **12**, and **9**. Although their water solubility was moderate, all of them are predicted to have a high gastrointestinal (GI) absorption (although this can be partially limited for **1–10**, being P-gp substrates). The exception to this is compound **13,** which is poorly soluble and has low GI absorption. The great majority seem to be able to be permeate the blood–brain barrier (BBB), although this was not the case for compounds **9**, **11**, **12**, and **13**. However, as all the BBB permeant compounds are also P-gp substrates, they would be pumped out from the brain and we would expect no interactions with the central nervous system due to this. Due to these enlisted data, we could expect the synthesized compounds to be, in general, suitable for oral administration.

Lipinski's rule of five [57] can be applied as a first filter, which accounts for the physicochemical properties related to the drug-likeness of a molecule. The molecular weight, number of H-bond donors and acceptors, and lipophilicity are in general accordance to the Lipinski rule. Only compounds **9** and **13** presented a violation, in both cases related to their very high lipophilicity.

Considering the predicted pharmacokinetics and drug-likeness, compounds **11** and **12** with nitro substitution can be considered as promising lead compounds for further studies, which can be additionally supported by the fact they were amongst the most active in vitro compounds, both as AChE inhibitors, **12** as an XO inhibitor, and **11** as part of the antiproliferative imidazoles in cancer cell lines.

**Table 4.** Pharmacokinetics and drug-likeness calculations made by SwissADME for the synthesized compounds **1–13**.

| Descriptors | Compound | | | | | | | | | | | | |
|---|---|---|---|---|---|---|---|---|---|---|---|---|---|
| | **1** | **2** | **3** | **4** | **5** | **6** | **7** | **8** | **9** | **10** | **11** | **12** | **13** |
| MW (g/mol) | 296.37 | 312.36 | 312.36 | 326.39 | 326.39 | 326.39 | 342.39 | 356.42 | 330.81 | 339.43 | 341.36 | 341.36 | 396.48 |
| #H–bond acceptors | 1 | 2 | 2 | 2 | 2 | 2 | 3 | 3 | 1 | 1 | 3 | 3 | 1 |
| #H–bond donors | 1 | 2 | 2 | 1 | 1 | 1 | 2 | 1 | 1 | 1 | 1 | 1 | 1 |
| TPSA (Å$^2$) | 28.68 | 48.91 | 48.91 | 37.91 | 37.91 | 37.91 | 58.14 | 47.14 | 28.68 | 31.92 | 74.5 | 74.5 | 28.68 |
| Consensus Log P | 4.59 | 4.26 | 4.17 | 4.56 | 4.55 | 4.52 | 4.27 | 4.54 | 5.15 | 4.6 | 4 | 3.95 | 6.36 |
| ESOL Log S | –5.4 | –5.24 | –5.24 | –5.44 | –5.44 | –5.44 | –5.29 | –5.49 | –5.97 | –5.59 | –5.42 | –5.42 | –7.6 |
| ESOL Class | MS | MS | MS | MS | MS | MS | MS | MS | MS | MS | MS | MS | PS |
| GI absorption | High | High | High | High | High | High | High | High | High | High | High | High | Low |
| BBB permeant | Yes | Yes | Yes | Yes | Yes | Yes | Yes | Yes | No | Yes | No | No | No |
| P–gp substrate | Yes | Yes | Yes | Yes | Yes | Yes | Yes | Yes | Yes | Yes | No | No | No |
| CYP1A2 inhibitor | Yes | Yes | Yes | Yes | Yes | Yes | Yes | Yes | Yes | Yes | Yes | Yes | No |
| CYP2C19 inhibitor | Yes | Yes | Yes | Yes | Yes | Yes | Yes | Yes | Yes | Yes | Yes | Yes | Yes |
| CYP2C9 inhibitor | No | No | No | No | No | No | No | Yes | No | No | No | No | No |
| CYP2D6 inhibitor | Yes | Yes | Yes | Yes | Yes | Yes | Yes | Yes | No | Yes | Yes | Yes | No |
| CYP3A4 inhibitor | Yes | Yes | Yes | Yes | Yes | Yes | Yes | Yes | Yes | Yes | No | No | No |
| Lipinski #violations | 0 | 0 | 0 | 0 | 0 | 0 | 0 | 0 | 1 | 0 | 0 | 0 | 1 |

MW = Molecular weight; TPSA = Topological polar surface area; Log P = Logarithm of the partition coefficient; ESOL Log S = ESOL model logarithm of molar solubility in water; ESOL class = Solubility class in Log S scale; MS = Moderately soluble; PS = Poorly soluble; GI = Gastrointestinal; BBB = Blood–brain barrier; P-gp = Permeability glycoprotein; CYP = Cytochrome.

## 4. Conclusions

A series of 13 derivatives of 2,4,5-trisubstituted imidazole were synthesized and their structures were characterized and confirmed through a series of spectroscopic and spectrometric techniques. Their antioxidant activities were analyzed with DPPH radical-scavenging and ABTS radical cation scavenging assays. In DPPH, the most active compounds were **3** and **10** ($EC_{50}$ of 0.141 and 0.174 mg/mL, respectively), bearing a *p*-OH and *p*-dimethylamino substitution in their A ring; in ABTS, the most active compounds were again **10** and **3** with an $EC_{50}$ of 0.162 and 0.168 mg/mL, respectively. This suggests the important role of heteroatoms with a free pair of electrons and acid phenolic hydrogens, so future derivatives should maintain these characteristics for improved antioxidant activity.

In the enzymatic assays, though not as active as the controls, **1** showed the best activity in AChE inhibition with 25.8% of inhibition, followed by the nitro containing compounds **12** (22.4%) and **11** (21.2%). The most active XO inhibitor was **3**, with an $IC_{50}$ of 85.8 μg/mL and a *p*-OH substitution. Present results point out that aromatic and positively charged groups are important for AChE inhibition activity, as the literature suggests. For XO inhibition, an oxygen in the *para* position appears to improve triphenyl imidazole derivatives activities, though an unexpected result for compound **12** suggests that future derivatives with nitrogen in the *ortho* position should be further explored.

The antiproliferative activity was evaluated against six cell lines from different anatomic origins, and the synthesized compounds showed from moderate to very good activities. Amongst the most active compounds were **2** (*o*-OH), **10** (*p*-$N(CH_3)_2$), and **11** (*p*-$NO_2$), where the last was outstanding as it was the first or second most active against all of the evaluated cell lines. Further expansion of this family of derivatives could maintain a nitrogen in the *para* position of the A ring, as it appears this favors their antiproliferative activity, with additional structure modulations.

In the in silico analysis, the docking against the EGFR and HER2 receptors had the agreement of **11** being amongst the two better binding affinities results. The ADME predictions of the 13 synthesized compounds showed that they are overall suitable for oral administration, with **11** and **12** having better pharmacokinetics and drug-likeness properties, which combined with their in vitro results point them as good candidates for being lead compounds in further derivations in the search of new drugs, especially as AChE inhibitors or as antiproliferative agents.

**Supplementary Materials:** The following are available online at http://www.mdpi.com/2076-3417/10/8/2889/s1, Figure S1: 1H-NMR Compound 2, Figure S2: 1H-NMR extension Compound 2, Figure S3: 13C-NMR Compound 2, Figure S4: 1H-NMR Compound 4, Figure S5: 13C-NMR Compound 4, Figure S6: 1H-NMR Compound 7, Figure S7: 13C-NMR Compound 7, Figure S8: 1H-NMR Compound 8, Figure S9: 13C-NMR Compound 8, Figure S10: 1H-NMR Compound 9, Figure S11: 13C-NMR Compound 9, Figure S12: 1H-NMR Compound 11, Figure S13: 13C-NMR Compound 11.

**Author Contributions:** Conceptualization, I.C.-G.; Methodology, E.N.-I., J.M.P., R.S.-A., D.C., L.D.-R., and V.W.B.-C.; Software, A.E.-C. and V.G.-G.; Formal analysis, E.N.-I. and R.A.C.-S.; Investigation, I.C.-G. and R.D.-M.; Writing—original draft preparation, E.N.-I. and I.C.-G.; Writing—review and editing, A.E.-C., R.A.C.-S., R.D.-M., and L.D.-R.; Supervision, I.C.-G. and L.D.-R.; Funding acquisition, V.G.-G. and R.D.-M. All authors have read and agreed to the published version of the manuscript.

**Funding:** This research received no external funding.

**Acknowledgments:** We gratefully acknowledge the Facultad de Ciencias Químicas e Ingeniería, Universidad Autónoma de Baja California for the financing given for the realization of this project. E.N.-I. acknowledges CONACYT for the doctoral fellowship. J.M.P. thanks the Spanish Government for financial support through project PGC2018-094503-B-C22 (MCIU/AEI/FEDER, UE).

**Conflicts of Interest:** The authors declare no conflict of interest.

## References

1. Shingalapur, R.V.; Hosamani, K.M.; Keri, R.S. Synthesis and evaluation of in vitro anti-microbial and anti-tubercular activity of 2-styryl benzimidazoles. *Eur. J. Med. Chem.* **2009**, *44*, 4244–4248. [CrossRef] [PubMed]
2. Achar, K.C.; Hosamani, K.M.; Seetharamareddy, H.R. In-vivo analgesic and anti-inflammatory activities of newly synthesized benzimidazole derivatives. *Eur. J. Med. Chem.* **2010**, *45*, 2048–2054. [CrossRef] [PubMed]

3. De Luca, L. Naturally occurring and synthetic imidazoles: Their chemistry and their biological activities. *Curr. Med. Chem.* **2006**, *13*, 1–23. [CrossRef] [PubMed]
4. Heravi, M.M.; Daraie, M.; Zadsirjan, V. Current advances in the synthesis and biological potencies of tri- and tetra-substituted 1H-imidazoles. *Mol. Divers.* **2015**, *19*, 577–623. [CrossRef] [PubMed]
5. Ghorbani-Vaghei, R.; Izadkhah, V.; Mahmoodi, J.; Karamian, R.; Ahmadi Khoei, M. The synthesis of imidazoles and evaluation of their antioxidant and antifungal activities. *Monatsh. Chem. Chem. Mon.* **2018**, *149*, 1447–1452. [CrossRef]
6. Sorrenti, V.; Salerno, L.; Di Giacomo, C.; Acquaviva, R.; Siracusa, M.A.; Vanella, A. Imidazole derivatives as antioxidants and selective inhibitors of nNOS. *Nitric Oxide* **2006**, *14*, 45–50. [CrossRef] [PubMed]
7. Abdel-Wahab, B.F.; Awad, G.E.A.; Badria, F.A. Synthesis, antimicrobial, antioxidant, anti-hemolytic and cytotoxic evaluation of new imidazole-based heterocycles. *Eur. J. Med. Chem.* **2011**, *46*, 1505–1511. [CrossRef]
8. Giustarini, D.; Dalle-Donne, I.; Tsikas, D.; Rossi, R. Oxidative stress and human diseases: Origin, link, measurement, mechanisms, and biomarkers. *Crit. Rev. Clin. Lab. Sci.* **2009**, *46*, 241–281. [CrossRef]
9. Halliwell, B. Oxidative stress and neurodegeneration: Where are we now? *J. Neurochem.* **2006**, *97*, 1634–1658. [CrossRef]
10. Angoa Pérez, M.; Rivas Arancibia, S. Estrés oxidativo y neurodegeneración: ¿causa o consecuencia? *Arch. Neurocien.* **2007**, *12*, 45–54.
11. Tran, T.D.; Nguyen, T.C.V.; Nguyen, N.S.; Nguyen, D.M.; Nguyen, T.T.H.; Le, M.T.; Thai, K.M. Synthesis of novel chalcones as acetylcholinesterase inhibitors. *Appl. Sci.* **2016**, *6*, 198. [CrossRef]
12. Ferreira-Vieira, T.H.; Guimaraes, I.M.; Silva, F.R.; Ribeiro, F.M. Alzheimer's Disease: Targeting the cholinergic system. *Curr. Neuropharmacol.* **2016**, *14*, 101–115. [CrossRef]
13. Da Costa, J.S.; Bizarro Lopes, J.P.; Russowsky, D.; Petzhold, C.L.; de Amorim Borges, A.C.; Ceschi, M.A.; Konrath, E.; Batassini, C.; Santana Lunardi, P.; Saraiva Gonçalves, C.A. Synthesis of tacrine-lophine hybrids *via* one-pot four component reaction and biological evaluation as acetyl- and butyrylcholinesterase inhibitors. *Eur. J. Med. Chem.* **2013**, *62*, 556–563. [CrossRef] [PubMed]
14. Šmelcerović, A.; Tomović, K.; Šmelcerović, Ž.; Petronijević, Ž.; Kocić, G.; Tomašič, T.; Jakopin, Ž.; Anderluh, M. Xanthine oxidase inhibitors beyond allopurinol and febuxostat; an overview and selection of potential leads based on *in silico* calculated physico-chemical properties, predicted pharmacokinetics and toxicity. *Eur. J. Med. Chem.* **2017**, *135*, 491–516. [CrossRef] [PubMed]
15. Zhang, T.; Lv, Y.; Lei, Y.; Liu, D.; Feng, Y.; Zhao, J.; Chen, S.; Meng, F.; Wang, S. Design, synthesis and biological evaluation of 1-hydroxy-2-phenyl-4-pyridyl-1*H*-imidazole derivatives as xanthine oxidase inhibitors. *Eur. J. Med. Chem.* **2018**, *146*, 668–677. [CrossRef] [PubMed]
16. Salazar-Aranda, R.; Granados-Guzmán, G.; Pérez-Meseguer, J.; González, G.M.; Waksman de Torres, N. Activity of Polyphenolic Compounds against *Candida glabrata*. *Molecules* **2015**, *20*, 17903–17912. [CrossRef] [PubMed]
17. Re, R.; Pellegrini, N.; Proteggente, A.; Pannala, A.; Yang, M.; Rice-Evans, C. Antioxidant activity applying an improved ABTS radical cation decolorization assay. *Free Radic. Biol. Med.* **1999**, *26*, 1231–1237. [CrossRef]
18. Kuskoski, E.M.; Asuero, A.G.; Troncoso, A.M.; Mancini-Filho, J.; Fett, R. Aplicación de diversos métodos químicos para determinar actividad antioxidante en pulpa de frutos. *Food Sci. Technol.* **2005**, *25*, 726–732. [CrossRef]
19. Adewusi, E.A.; Moodley, N.; Steenkamp, V. Antioxidant and acetylcholinesterase inhibitory activity of selected southern African medicinal plants. *S. Afr. J. Bot.* **2011**, *77*, 638–644. [CrossRef]
20. Almada-Taylor, G.; Díaz-Rubio, L.; Salazar-Aranda, R.; Waksman de Torres, N.; Uranga-Solis, C.; Delgadillo-Rodríguez, J.; Ramos, M.A.; Padrón, J.M.; Hernández-Martínez, R.; Córdova-Guerrero, I. Biological Activities of Extracts from Aerial Parts of *Salvia pachyphylla* Epling Ex Munz. *Plants* **2018**, *7*, 105. [CrossRef] [PubMed]
21. Monks, A.; Scudiero, D.; Skehan, P.; Shoemaker, R.; Paull, K.; Vistica, D.; Hose, C.; Langley, J.; Cronise, P.; Vaigro-Wolff, A.; et al. Feasibility of a high-flux anticancer drug screen using a diverse panel of cultured human tumor cell lines. *J. Nat. Cancer Inst.* **1991**, *83*, 757–766. [CrossRef]
22. Pettersen, E.F.; Goddard, T.D.; Huang, C.C.; Couch, G.S.; Greenblatt, D.M.; Meng, E.C.; Ferrin, T.E. UCSF Chimera—A visualization system for exploratory research and analysis. *J. Comput. Chem.* **2004**, *25*, 1605–1612. [CrossRef]

23. Wang, J.; Wang, W.; Kollman, P.A.; Case, D.A. Automatic atom type and bond type perception in molecular mechanical calculations. *J. Mol. Graph. Model.* **2006**, *25*, 247–260. [CrossRef]
24. Sanner, M.F. Python: A programming language for software integration and development. *J. Mol. Graph. Model.* **1999**, *17*, 57–61. [CrossRef]
25. RSCB PDB Protein Data Bank. Available online: https://www.rcsb.org/ (accessed on 1 August 2019).
26. Morris, G.M.; Huey, R.; Lindstrom, W.; Sanner, M.F.; Belew, R.K.; Goodsell, D.S.; Olson, A.J. Autodock4 and AutoDockTools4: Automated docking with selective receptor flexibility. *J. Comput. Chem.* **2009**, *16*, 2785–2791. [CrossRef] [PubMed]
27. Daina, A.; Michielin, O.; Zoete, V. SwissADME: A free web tool to evaluate pharmacokinetics, drug-likeness and medicinal chemistry friendliness of small molecules. *Sci. Rep.* **2017**, *7*, 42707. [CrossRef] [PubMed]
28. Puratchikody, A.; Doble, M. Antinociceptive and antiinflammatory activities and QSAR studies on 2-substituted-4,5-diphenyl-1H-imidazoles. *Bioorg. Med. Chem.* **2007**, *15*, 1083–1090. [CrossRef] [PubMed]
29. Singh, P.; Kumar, R.; Tiwari, S.; Khanna, R.S.; Tewari, A.K.; Khanna, H.D. Docking, synthesis and evaluation of antioxidant activity of 2,4,5-triaryl imidazole. *Clin. Med. Biochem. Open Access* **2015**, *1*, 105. [CrossRef]
30. Foti, M.C. Antioxidant properties of phenols. *J. Pharm. Pharmacol.* **2007**, *59*, 1673–1685. [CrossRef]
31. Hemalatha, S.; Naveena, P.A. Synthesis and biological evaluation of substituted diphenyl imidazole derivatives. *World J. Pharm. Res.* **2015**, *4*, 1321–1333.
32. Zheng, L.; Zhao, M.; Xiao, C.; Zhao, Q.; Su, G. Practical problems when using ABTS assay to assess the radical-scavenging activity of peptides: Importance of controlling reaction pH and time. *Food Chem.* **2016**, *192*, 288–294. [CrossRef] [PubMed]
33. Mateo, J.L.; Bosch, P.; Lozano, A.E. Reactivity of Radicals derived from dimethylanilines in acrylic photopolymerization. *Macromolecules* **1994**, *27*, 7794–7799. [CrossRef]
34. Kryger, G.; Silman, I.; Sussman, J.L. Structure of acetylcholinesterase complexed with E2020 (Aricept®): Implications for the design of new anti-Alzheimer drugs. *Structure* **1999**, *7*, 297–307. [CrossRef]
35. Harel, M.; Schalk, I.; Ehret-Sabatier, L.; Bouet, F.; Goeldner, M.; Hirth, C.; Axelsen, P.H.; Silman, I.; Sussman, J.L. Quaternary ligand binding to aromatic residues in the active-site gorge of acetylcholinesterase. *Proc. Natl. Acad. Sci. USA* **1993**, *90*, 9031–9035. [CrossRef]
36. Bartolucci, C.; Perola, E.; Pilger, C.; Fels, G.; Lamba, D. Three-dimensional structure of a complex of galanthamine (Nivalin®) with acetylcholinesterase from Torpedo californica: Implications for the design of new anti-Alzheimer drugs. *Proteins* **2001**, *42*, 182–191. [CrossRef]
37. Zhao, X.-J.; Gong, D.-M.; Jiang, Y.-R.; Guo, D.; Zhu, Y.; Deng, Y.-C. Multipotent AChE and BACE-1 inhibitors for the treatment of Alzheimer's disease: Design, synthesis and bio-analysis of 7-amino-1,4-dihydro-2*H*-isoquilin-3-one derivates. *Eur. J. Med. Chem.* **2017**, *138*, 738–747. [CrossRef] [PubMed]
38. Mughal, E.U.; Sadiq, A.; Murtaza, S.; Rafique, H.; Zafar, M.N.; Riaz, T.; Khan, B.A.; Hameed, A.; Khan, K.M. Synthesis, structure-activity relationship and molecular docking of 3-oxoaurones and 3-thioaurones as acetylcholinesterase and butyrylcholinesterase inhibitors. *Bioorg. Med. Chem.* **2017**, *25*, 100–106. [CrossRef]
39. De la Torre, P.; Astudillo Saavedra, L.; Caballero, J.; Quiroga, J.; Alzate-Morales, J.H.; Gutiérrez Cabrera, M.; Trilleras, J. A novel class of selective acetylcholinesterase inhibitors: Synthesis and evaluation of (*E*)-2-(benzo[*d*]thiazol-2-yl)-3-heteroarylacrylonitriles. *Molecules* **2012**, *17*, 12072–12085. [CrossRef]
40. Okamoto, K.; Eger, B.T.; Nishino, T.; Pai, E.F.; Nishino, T. Mechanism of inhibition of xanthine oxidoreductase by allopurinol: Crystal structure of reduced bovine milk xanthine oxidoreductase bound with oxipurinol. *Nucleosides Nucleotides Nucleic Acids* **2008**, *27*, 888–893. [CrossRef]
41. Matsumoto, K.; Okamoto, K.; Ashizawa, N.; Nishino, T. FXY-051: A novel and potent hybrid-type inhibitor of xanthine oxidoreductase. *J. Pharmacol. Exp. Ther.* **2011**, *336*, 95–103. [CrossRef]
42. Lin, S.; Zhang, G.; Liao, Y.; Pan, J. Inhibition of chrysin on xanthine oxidase activity and its inhibition mechanism. *Int. J. Biol. Macromol.* **2015**, *81*, 274–282. [CrossRef] [PubMed]
43. Huang, H.; Tong, T.T.; Yau, L.F.; Chen, C.Y.; Mi, J.N.; Wang, J.R.; Jiang, Z.H. LC-MS based sphingolipidomic study on A549 human lung adenocarcinoma cell line and its taxol-resistant strain. *BMC Cancer* **2018**, *18*, 799. [CrossRef] [PubMed]
44. Mohan, C.D.; Srinivasa, V.; Rangappa, S.; Mervin, L.; Mohan, S.; Paricharak, S.; Baday, S.; Li, F.; Shanmugam, M.K.; Chinnathambi, A.; et al. Trisubstituted-imidazoles induce apoptosis in human breast cancer cells by targeting the oncogenic PI3K/Akt/mTOR signaling pathway. *PLoS ONE* **2016**, *11*, e0153155. [CrossRef] [PubMed]

45. Dake, S.A.; Kharat, K.R.; Yadav, A.R.; Kendrekar, P.S.; Pawar, R.P. In-vitro antiproliferative activity study of 2,4,5-triphenyl-1H-imidazole derivatives. *J. Org. Inorg. Chem.* **2017**, *3*, 5. [CrossRef]
46. Lopes, M.S.; de Andrade Sena, C.F.; Silva, B.L.; de Souza, C.M.; Ramos, J.P.; Cassali, G.D.; de Souza-Fagundes, E.M.; Alves, R.J.; de Oliveira, M.C.; de Oliveira, R.B. Synthesis of nitroaromatic compounds as potential anticancer agents. *Anticancer Agents Med. Chem.* **2015**, *15*, 206–216. [CrossRef]
47. Galindo-Hernández, O.; Córdova-Guerrero, I.; Díaz-Rubio, L.J.; Pulido-Capiz, A.; Díaz-Villanueva, J.F.; Castañeda-Sánchez, C.Y.; Serafín-Higuera, N.; García-González, V. Protein translation associated to PERK arm is a new target for regulation of metainflammation: A connection with hepatocyte cholesterol. *J. Cell. Biochem.* **2019**, *120*, 4158–4171. [CrossRef]
48. Díaz-Rubio, L.; Hernández-Martínez, R.; Estolano-Cobián, A.; Chávez-Velasco, D.; Salazar-Aranda, R.; Waksman de Torres, N.; Rivero, I.A.; García-González, V.; Ramos, M.A.; Córdova-Guerrero, I. Synthesis, biological evaluation and docking studies of chalcone and flavone analogs as antioxidants and acetylcholinesterase inhibitors. *Appl. Sci.* **2019**, *9*, 410. [CrossRef]
49. Park, J.H.; Liu, Y.; Lemmon, M.A.; Radhakrishnan, R. Erlotinib binds both inactive and active conformations of the EGFR tyrosine kinase domain. *Biochem. J.* **2012**, *448*, 417–423. [CrossRef]
50. Aertgeerts, K.; Skene, R.; Yano, J.; Sang, B.-C.; Zou, H.; Snell, G.; Jennings, A.; Iwamoto, K.; Habuka, N.; Hirokawa, A.; et al. Structural analysis of the mechanism of inhibition and allosteric activation of the kinase domain of HER2 protein. *J. Biol. Chem.* **2011**, *286*, 18756–18765. [CrossRef]
51. Normanno, N.; De Luca, A.; Bianco, C.; Strizzi, L.; Mancino, M.; Maiello, M.R.; Carotenuto, A.; De Feo, G.; Caponigro, F.; Salomon, D.S. Epidermal growth factor receptor (EGFR) signaling in cancer. *Gene* **2006**, *366*, 2–16. [CrossRef]
52. Iqbal, N.; Iqbal, N. Human epidermal growth factor receptor 2 (HER2) in cancers: Overexpression and therapeutic implications. *Mol. Biol. Int.* **2014**, *2014*, 852748. [CrossRef] [PubMed]
53. Guda, R.; Kumar, G.; Korra, R.; Balaji, S.; Dayakar, G.; Palabindela, R.; Myadaraveni, P.; Yellu, N.R.; Kasula, M. EGFR, HER2 target based molecular docking analysis, in vitro screening of 2,4,5-trisubstituted imidazole derivatives as potential anti-oxidant and cytotoxic agents. *J. Photochem. Photobiol. B Biol.* **2017**, *176*, 69–80. [CrossRef] [PubMed]
54. Butler, K.T.; Luque, F.J.; Barril, X. Toward accurate relative energy predictions of the bioactive conformation of drugs. *J. Comput. Chem.* **2009**, *30*, 601–610. [CrossRef] [PubMed]
55. Montemurro, F.; Valabrega, G.; Aglietta, M. Lapatinib: A dual inhibitor of EGFR and HER2 tyrosine kinase activity. *Expert Opin. Biol. Ther.* **2007**, *7*, 257–268. [CrossRef] [PubMed]
56. Sulimov, A.V.; Kutov, D.C.; Katkova, E.V.; Sulimov, V.B. Combined docking with classical force field and quantum chemical semiempirical method PM7. *Adv. Bioinform.* **2017**, *2017*, 7167691. [CrossRef] [PubMed]
57. Lipinski, C.A.; Lombardo, F.; Dominy, B.W.; Feeney, P.J. Experimental and computational approaches to estimate solubility and permeability in drug discovery and development settings. *Adv. Drug Deliv. Rev.* **2001**, *46*, 3–26. [CrossRef]

*applied sciences*

MDPI

*Article*

# Synthesis, Anticancer Activity, and Molecular Modeling of New Halogenated Spiro[pyrrolidine-thiazolo-oxindoles] Derivatives

**Mohammad Shahidul Islam [1], Abdullah Mohammed Al-Majid [1], Fardous F. El-Senduny [2], Farid A. Badria [3], A. F. M. Motiur Rahman [4], Assem Barakat [1,5,*] and Yaseen A. M. M. Elshaier [6,*]**

1 Department of Chemistry, College of Science, King Saud University, P. O. Box 2455, Riyadh 11451, Saudi Arabia; mislam@ksu.edu.sa (M.S.I.); amajid@ksu.edu.sa (A.M.A.-M.)
2 Department of Chemistry, Faculty of Science, Mansura University, Mansura 35516, Egypt; biobotany@gmail.com
3 Department of Pharmacognosy, Faculty of Pharmacy, Mansoura University, Mansoura 35516, Egypt; faridbadria@gmail.com
4 Department of Pharmaceutical Chemistry, College of Pharmacy, King Saud University, Riyadh 11451, Saudi Arabia; afmrahman@ksu.edu.sa
5 Department of Chemistry, Faculty of Science, Alexandria University, P.O. Box 426, Ibrahimia, Alexandria 21321, Egypt
6 Department of Organic and Medicinal chemistry, Faculty of Pharmacy, University of Sadat City, Sadat City, Menoufiya 32958, Egypt
* Correspondence: ambarakat@ksu.edu.sa (A.B.); yaseen.elshaier@fop.usc.edu.eg (Y.A.M.M.E.); Tel.: +966-11467-5901 (A.B.); Fax: +966-11467-5992 (A.B.)

Received: 26 February 2020; Accepted: 12 March 2020; Published: 23 March 2020

**Abstract:** A one-pot, single-step, and an atom-economical process towards the synthesis of highly functionalized spirooxindoles analogues was efficiently conducted to produce a satisfactory chemical yields (70–93%) with excellent relative diastereo-, and regio-selectivity. An in vitro antiproliferative assay was carried out on different cancer cell lines to evaluate the biological activity of the synthesized tetrahydro-1′*H*-spiro[indoline-3,5′-pyrrolo[1,2-*c*]thiazol]-2-one **5a–n.** The prepared hybrids were then tested *in vitro* for their antiproliferative effects against three cancer cell lines, namely, HepG2 (liver cancer), MCF-7 (breast cancer), and HCT-116 (colon cancer). The spirooxindole analogue **5g** exhibited a broad activity against HepG2, MCF-7, and HCT-116 cell lines of liver, breast, and colorectal cancers when compared to *cis*platin. Modeling studies including shape similarity, lipophilicity scores, and physicochemical parameters were calculated. The results of this study indicated that spirooxindole analogue **5g** retained a good physiochemical parameters with acceptable lipophilicity scores.

**Keywords:** spirooxindole; 1,3-dipolar cycloaddition; eco-friendly chemistry; ROCS; shape alignment; lipophilicity; anticancer activity

## 1. Introduction

The design of highly complex spiro-heterocycles with multifunctional and potential pharmaceutical efficacy has attracted considerable attention from synthetic and medicinal chemists [1]. One of the most privileged *aza*-heterocyclic scaffolds is spiro[pyrrolidine-oxindole] [2], which is present in natural products and useful as a building block for the synthesis of significant biologically active compounds. This class of *aza*-heterocyclic compounds has gained great interest, owing to several reports of its pharmaceutical potency, including anticancer [3], antitumor [4], 5-HT3 receptor antagonist [5], acetylcholinesterase-inhibitory [6], antibacterial [7], antibiotic [8], and MDM2–p53 inhibitor [9] effects; selective cyclooxygenase COX-1 with TNF-α and IL-6 inhibitors [10]; and potential hypoglycemic dual inhibitory activity against $\alpha$-amylase and $\alpha$-glucosidase [11] (Figure 1). To date, prolonged efforts have been exerted to expand divergent complexity and to develop efficient synthetic routes for these valuable privileged *aza*-heterocyclic scaffolds, which would remarkably enhance their bioactivity [1,12]. In particular, [3+2] cycloaddition is one of the most efficient synthetic approaches to produce these valuable scaffolds with stereoselective method and high yield [13]. To extend our previous research, we explored the effect of halogen substitution on the isatin ring.

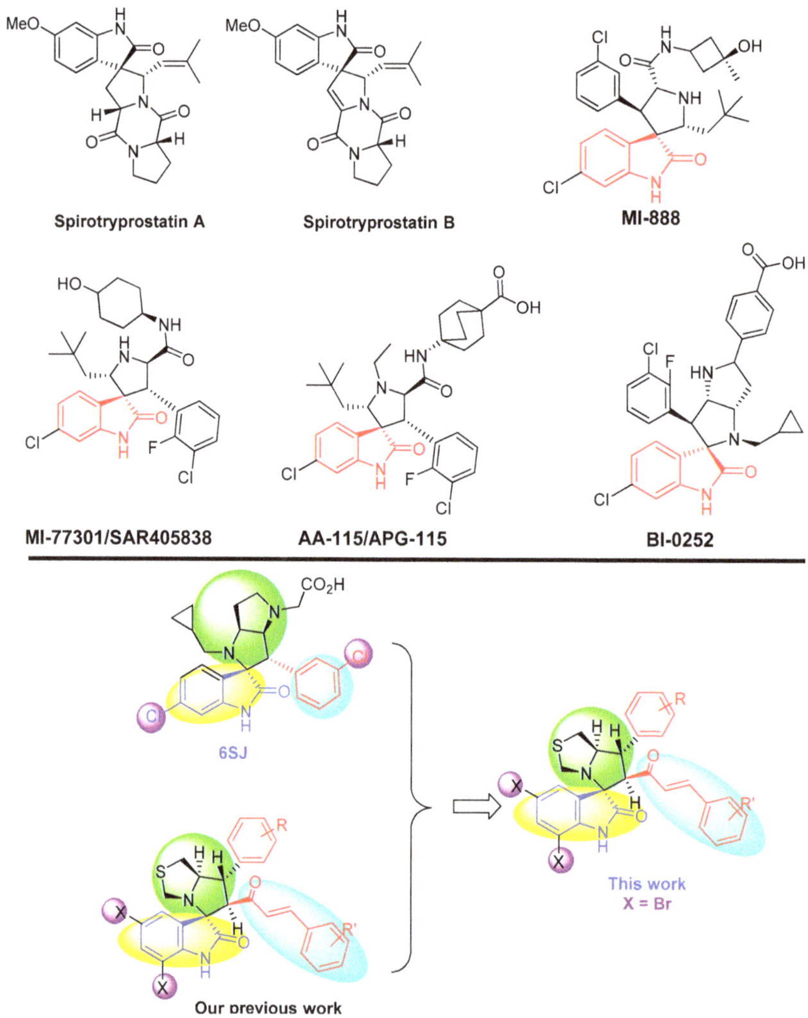


**Figure 1.** Natural (Spirotryprostatin A and B) and other synthetic spirooxindole scaffolds with high biological importance and structure-activity relationship.

Our previous studies [9] revealed that the presence of dihalide substitution on acyl moiety substantially increased the anticancer activity of the resulting product(s). Moreover, it was reported [9] that chlorinated indole moiety retained better activity, as illustrated in Figure 1a. Subsequently, this study was designed to introduce two bromo atoms on the indole ring, presumably to enhance the activity of the examined spirooxindole compounds shown in Figure 1. The *aza*-heterocyclic compounds were prepared via a multicomponent eco-friendly strategy using oxindole as a core structure. The resulting hybrids were biologically evaluated using an *in vitro* antiproliferative assay against three different cell lines for liver, breast, and colorectal cancer. In addition, molecular properties and lipophilicity studies were conducted to get insight about "drug properties consideration" and to discover the compounds' structure-property relationship (SPR).

## 2. Results and Discussion

### 2.1. Synthesis of **5a–n**

The requisite spirooxindoles analogous were prepared by a multicomponent reaction (Scheme 1). The advantages of this efficient method were low-cost and readily available synthons for the synthesis of highly divergent compounds with high-importance applications. Fourteen analogues were prepared through the reactions of *bis*-benzylidine **1a–n,** which had been prepared according to our previous publication [9] with thioproline and 5,7-dibromoisatin, to afford the requisite target compounds. The chemical features of the requisite compounds were assigned based on HNMR, CNMR, IR, and CHN analysis.

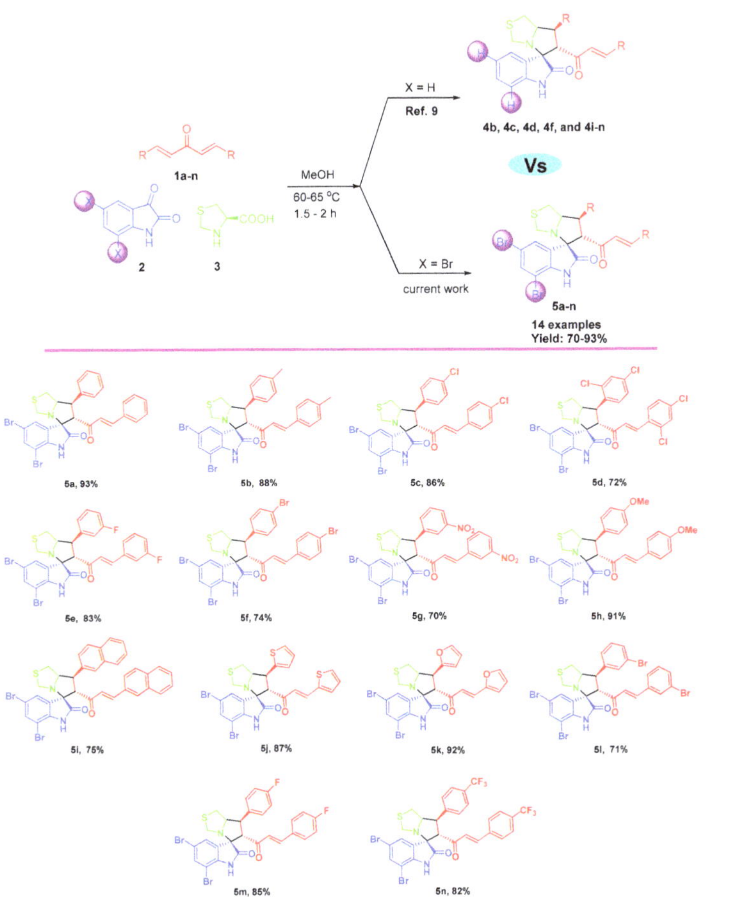


**Scheme 1.** Synthesis of tetrahydro-1′*H*-spiro[indoline-3,5′-pyrrolo [1,2-*c*]thiazol]-2-one **5a–n**.

According to Scheme 2 and based on our previous study [9], the reaction proceeds *via* one pot reaction, in which initially 5,7-dibromoisatin **2** reacted with thioproline **3** affording the azomethine ylide after the removal of carbon dioxide from the intermediate. Subsequently, the azomethine ylide reacted with the *bis*-benzylidine **1a–n** to provide the target compounds in a regioselective and diastereoselective manner. The reaction proceeded *via* path A regio-selectively to afford the regioisomer products **5a–n**, while the second regio-isomers **5a–n′** did not occur (path B). There are possible diastereoselective products that could be formed, but in this case only diastereoselective compounds **5a–n** occurred *via* the path C not D.

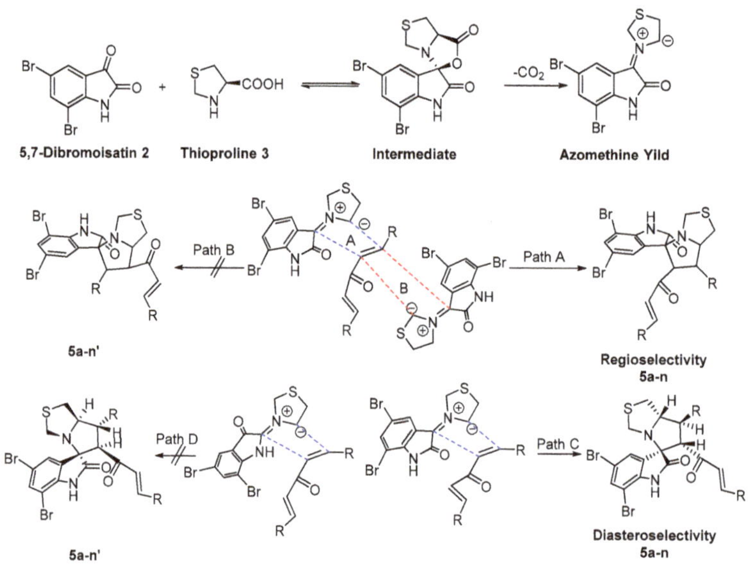


**Scheme 2.** Plausible reaction mechanism of the synthesized compounds.

## 2.2. Biological Activity

The compounds were subjected to an initial evaluation for potential cytotoxic activity against different cancer cell lines, namely, HepG2, MCF-7, and HCT-116 cells, at 50 $\mu$M. Cell viability was measured using MTT assay. Among the screened 14 compounds, 3 (**5h**, **5i**, and **5j**) did not show any cytotoxic activity against HepG2 cells. The concentration of the active compounds that killed 50% of the cells ($IC_{50}$) was evaluated against HepG2 cells. Compound **5g** ($IC_{50}$ = 5.00 ± 0.66 $\mu$M) was the most potent active compound, showing more potent activity than that of the standard chemotherapeutic drug *cis*platin ($IC_{50}$ = 9.00 ± 0.76 $\mu$M) (Table 1). Moderate anticancer activity against HepG2 cells was observed for compounds **5a** and **5m** ($IC_{50}$ = 10.00 ± 0.47 and 17.00 ± 0.68 $\mu$M, respectively).

The same three inactive compounds (**5h**, **5i**, and **5j**) did not show activity against MCF-7 or HCT-116 cells (Table 1). The other 11 tested compounds ($IC_{50} \leq 9.00$ $\mu$M) showed superior activity to that of cisplatin ($IC_{50}$ = 9.00 ± 0.29 $\mu$M) against MCF-7 cells (Table 1); only compounds (**5c**, **5f**, **5g**, and **5l**) ($IC_{50} < 3.00$ $\mu$M) were more potent than cisplatin ($IC_{50}$ = 3.00 ± 0.24 $\mu$M) against colon cancer cells (Table 1). The present study showed that compound **5g** retained broad anticancer activity against the three tested cell lines of liver, breast, and colorectal cancers; HepG2, MCF-7, and HCT-116 cells, respectively.

**Table 1.** Results of anticancer activity against HepG2, MCF-7, and HCT-116 cells.

| Compound | R | Cancer Type/Cell Line | | |
|---|---|---|---|---|
| | | Liver HepG2 ($IC_{50}$ [a], $\mu$M) | Breast MCF-7 ($IC_{50}$, $\mu$M) | Colon HCT-116 ($IC_{50}$, $\mu$M) |
| 4b | 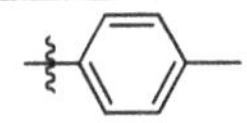 | 3.57 ± 0.50 | NT [c] | 8.00 ± 1.20 |
| 4c | 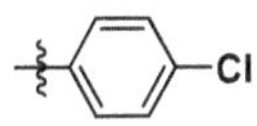 | 2.00 ± 0.50 | NT | 3.00 ± 0.50 |
| 4d | 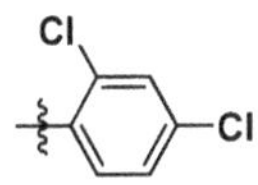 | 0.85 ± 0.20 | NT | 2.00 ± 0.60 |
| 4f | 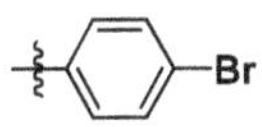 | 0.80 ± 0.10 | NT | 3.00 ± 0.50 |
| 4i | 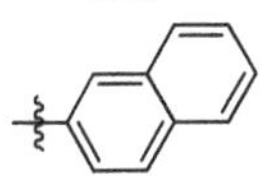 | 2.40 ± 1.00 | NT | 8.00 ± 0.30 |
| 4j | 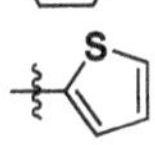 | >50.00 | NT | 14.50 ± 1.50 |
| 4k | 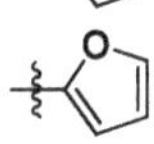 | >50.00 | NT | 19.00 ± 2.00 |
| 4l | 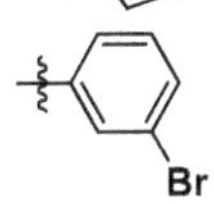 | 0.90 ± 0.10 | NT | 1.57 ± 0.30 |
| 4m | 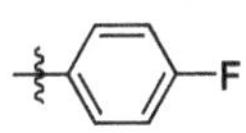 | 2.40 ± 0.40 | NT | 5.00 ± 0.30 |
| 4n | 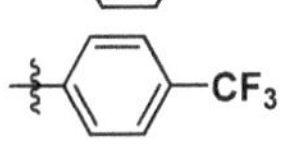 | 0.90 ± 0.20 | NT | 2.90 ± 0.40 |
| 5a | 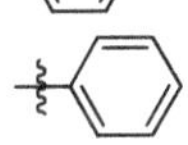 | 10.00 ± 0.47 | 6.00 ± 0.13 | 4.50 ± 0.05 |
| 5b | 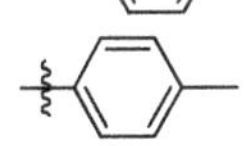 | 30.00 ± 0.38 | 5.50 ± 0.47 | 5.00 ± 0.30 |
| 5c | 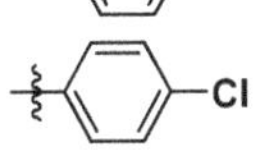 | 25.00 ± 0.09 | 3.00 ±1.26 | 2.90 ± 0.25 |
| 5d | 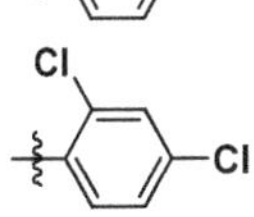 | >50.00 ± 0.28 | 9.00 ± 0.05 | 8.50 ± 0.10 |
| 5e | 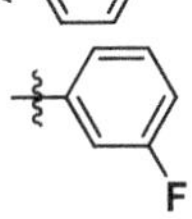 | 22.00 ± 1.02 | 3.00 ± 0.32 | 5.00 ± 0.12 |
| 5f | 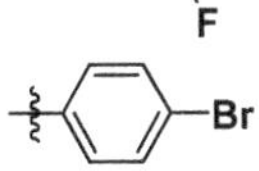 | 50.00 ± 0.38 | 2.50 ± 1.66 | 2.20 ± 0.15 |
| 5g | 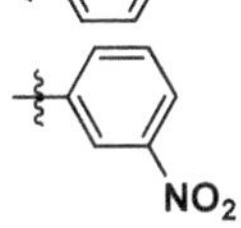 | 5.00 ± 0.66 | 4.00 ± 0.29 | 2.80 ± 0.20 |
| 5h | 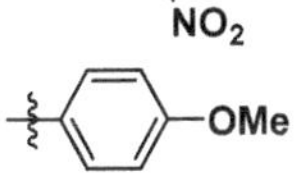 | NA [b] | NA | NA |

**Table 1.** *Cont.*

| Compound | R | Cancer Type/Cell Line | | |
|---|---|---|---|---|
| | | Liver HepG2 ($IC_{50}$ [a], $\mu M$) | Breast MCF-7 ($IC_{50}$, $\mu M$) | Colon HCT-116 ($IC_{50}$, $\mu M$) |
| 5i | 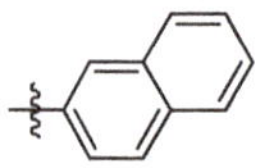 | NA | NA | NA |
| 5j | 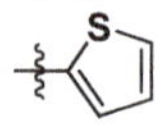 | NA | NA | NA |
| 5k | 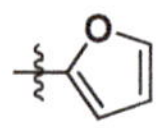 | 40.00 ± 0.57 | 8.00 ± 0.20 | 13.00 ± 0.72 |
| 5l | 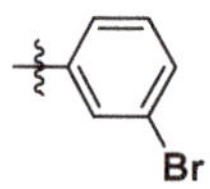 | 35.00 ± 0.45 | 3.00 ± 0.04 | 2.80 ± 0.19 |
| 5m | 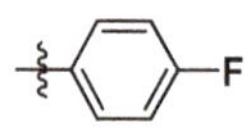 | 17.00 ± 0.678 | 4.50 ± 0.08 | 4.00 ± 0.52 |
| 5n | 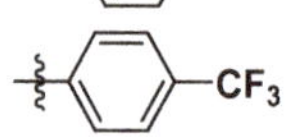 | 30.00 ± 0.79 | 5.00 ± 0.16 | 3.70 ± 1.04 |
| *cis*platin | | 9.00 ± 0.76 | 9.00 ± 0.29 | 3.00 ± 0.24 |

[a] $IC_{50}$ ($\mu M$) was evaluated using MTT assay and ± is the standard deviation from three independent experiments. [b] NA: means that the tested compound did not show anticancer activity at 50 $\mu M$. [c] NT: did not tested against the MCF-7 cells.

### 2.3. Effect of the Dibromo on the Anticancer Activity

The structure-activity relationship between the previously reported spirooxindole analogues **4b**, **4c**, **4d**, **4f**, and **4i-n** [9] and the diboromo-substituted spiroxindoles **5b**, **5c**, **5d**, **5f**, and **5i-n** is described. In fact, the $IC_{50}$ values of Table 1 clearly show that the replacement of the H atoms of the previously reported compounds **4b**, **4c**, **4d**, **4f**, and **4i-n** with that of its analogues with the Br produced a significant decrease in the inhibitory growth effect on the HEPG2 cell line. On the other hand, compounds **5b**, **5c**, **5f**, **5k,** and **5m** (dibromo-substituted) showed better activity against HCT-116 cells than their dibromo-unsubstituted indole counterparts. Compounds **5d**, **5l**, and **5n** showed less activity than the compounds **4d**, **4l**, and **4n,** respectively. Compounds **5i** and **5j** were not active and compounds **4i** and **4j** presented some activity (Table 1).

### 2.4. Shape Alignment by Rapid Overlay Chemical Structure (ROCS) Analysis

Shape and electrostatic potential are two fundamental molecular descriptors for computational drug discovery, because in protein ligand binding, the shape of a ligand has to conform in large degree to the shape of a protein binding site. The electrostatic potentials presented in the binding site have to complement the electrostatic potential of the protein. Accordingly, it is very important to model and understand protein ligand bindings correctly. The 3D shape structure exhibits good neighborhood behavior, in which high similarity in shape reflects high similarity in biology. Shape similarity can have different applications, such as virtual screening, lead-hopping, molecular alignment, pose generation, and predictions.

ROCS is a tool used in shape similarity studies. ROCS requires a query, which is an active molecule with some biological activities in at least one 3D conformation. It also requires a database of the molecules of the compounds of interest. Consistent with these standards, our compounds (database set) showed similarity to standard compound **BI-0225** (Figure 2). Compound **5g** showed high similarity to **BI-0225** in terms of its oxindole moiety and oxoindole ring.

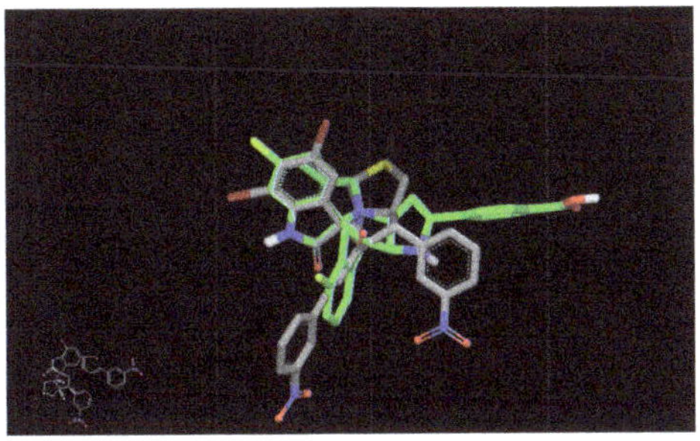

**Figure 2.** Shape similarity of **5g** with **BI-0252** as analyzed by Rapid Overlay Shape Chemical Structure (ROCS) and visualized by VIDA application.

*2.5. Predicted Pharmacokinetics and Pharmacodynamics Parameters*

Absorption, distribution, metabolism, excretion, and toxicity (ADMET) prediction for drug candidates is mandatory in the drug design process, as these parameters contribute to determining the failure of approximately 60% of all drugs in the development and approval phases. It is well-known that ADMET prediction is performed at the last stage of the drug development process with high cost and effort. At present, ADMET is determined at the beginning of drug discovery stages in order to eliminate molecules with poor ADMET properties from the drug discovery pipeline with the aim to save research costs. In this regard, computational tools were used to predict ADMET properties in this study [14].

The Caco-2 cell, percentage of human intestinal absorption (HIA), and skin permeability models have all been suggested as reliable in vitro models to estimate oral drug absorption and transdermal delivery [15]. Drug penetration to the blood brain barrier (BBB) provides insight into drugs that act on the central nervous system and on plasma protein binding (PPB). Compared to the other compounds, **5g** showed the lowest BBB penetration value (0.017) and a low value in the Caco-2 cell model (18.80). All compounds showed high PPB and HIA values, as well as very low skin permeability values in the range of −1.80 to −2.79 (Table 2).

**Table 2.** Predicted pharmacokinetic and pharmacodynamic parameters of the most active compounds.

| # | Lipinski's Rule | | | | PreADMET [16] Prediction | | | | | |
|---|---|---|---|---|---|---|---|---|---|---|
| | MW | LogP [17] | HBD | HBA | BBB | PPB | HIA | Caco-2 Value | Skin Permeability | Drug-Likeness Model Score [17] |
| **5a** | 607.98 | 5.02 | 1 | 4 | 0.39 | 100.00 | 97.99 | 39.51 | −2.45 | 0.36 |
| **5b** | 636.01 | 5.82 | 1 | 4 | 0.95 | 94.44 | 98.03 | 41.48 | −2.28 | 0.60 |
| **5c** | 679.25 | 6.44 | 1 | 4 | 1.64 | 100.00 | 98.04 | 44.04 | −2.49 | 0.83 |
| **5e** | 646.34 | 5.56 | 1 | 4 | 0.51 | 100.00 | 97.99 | 40.35 | −2.75 | 0.63 |
| **5f** | 768.15 | 6.72 | 1 | 4 | 1.96 | 100.00 | 97.96 | 46.69 | −2.15 | 0.71 |
| **5g** | 700.35 | 4.47 | 1 | 8 | 0.017 | 100.00 | 98.94 | 18.80 | −2.46 | 0.44 |
| **5l** | 768.15 | 6.72 | 1 | 4 | 1.69 | 100.00 | 97.96 | 46.60 | −2.15 | 0.63 |
| **5m** | 646.34 | 5.56 | 1 | 4 | 0.49 | 100.00 | 97.99 | 40.35 | −2.79 | 0.72 |
| **5n** | 746.34 | 7.49 | 1 | 4 | 3.33 | 100.00 | 98.03 | 48.45 | −1.80 | 0.47 |

HBD, hydrogen bond donor; HBA, hydrogen bond acceptor; BBB, blood brain barrier; PPB, plasma protein binding; HIA, percentage human intestinal absorption; Caco-2 value, permeability to Caco-2 (human colorectal carcinoma) cells in vitro.

### *2.6. Ligand Efficiency (LE) and Lipophilic Efficiency (LipE)*

In the current study, for optimization assessment, LE was calculated [18]. The parameter LE has a crucial role in "lead optimization for drug-like candidate" properties [19]. Compounds with the highest activity were selected for evaluation against sensitive cancer cell lines (breast and colon cancer cells).

LE was calculated using the following equation [20]:

$$LE = (pIC_{50} \times 1.37)/NHA$$

$IC_{50}$ = half-maximal inhibitory concentration (in terms of molar concentration); NHA = non-hydrogen atom.

The compounds had an LE value in the range of 0.19–0.26 except for compound **5n** (Table 3). All compounds exhibited higher LE values in breast cancer cells than in colon cancer cells, especially compounds **5c**, **5e**, and **5l** (LE = 0.26), all of which were structural isomers.

The recommended LE value should be in the range of 0.3. The acceptable LE value should be higher than 0.3.

### *2.7. Lipophilic Efficiency (LipE) or Ligand Lipophilic Efficiency (LEE)*

Lip E or LLE is an avenue to determine compound affinity with respect to its lipophilicity.

Nowadays, the lipophilic efficiency (LipE) index (LEE), which includes lipophilicity and potency, is becoming more and more popular in drug design. It allows for the normalization of observed potency with changes in the lipophilicity, and it is considered an effective and practical tool for keeping lipophilicity under control to avoid any "molecular obesity".

LipE or LLE is calculated as the difference between the potency and log P as illustrated in the following equation:

$$Lip\ E = pIC50 - cLog\ P$$

According to data revealed in Table 3, compound **5g** showed best value in comparison to other derivatives between both cell lines.

**Table 3.** Summary of ligand efficiency scores for the target compounds.

| Compounds | R | NHA | cLog P | Breast Cancer Cells | | | Colon Cancer Cells | | |
|---|---|---|---|---|---|---|---|---|---|
| | | | | $pIC_{50}$ | LE | LipE (LEE) | $pIC_{50}$ | LE | LipE (LEE) |
| 5a | 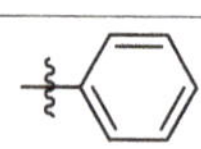 | 35 | 5.02 | 5.22 | 0.2 | 0.20 | 5.34 | 0.2 | 0.32 |
| 5b | 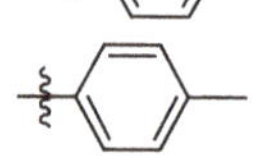 | 37 | 5.82 | 5.26 | 0.19 | -0.56 | 5.3 | 0.2 | -0.52 |
| 5c | 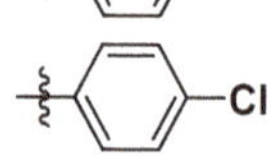 | 37 | 6.44 | 5.52 | 0.26 | -0.92 | 5.53 | 0.26 | -0.91 |
| 5e | 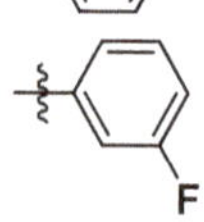 | 37 | 5.56 | 5.52 | 0.26 | -0.04 | 5.3 | 0.2 | -0.26 |
| 5f | 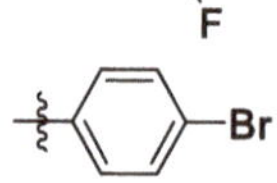 | 37 | 6.72 | 5.6 | 0.2 | -1.12 | 5.66 | 0.21 | -1.06 |

**Table 3.** *Cont.*

| Compounds | R | NHA | cLog P | Breast Cancer Cells | | | Colon Cancer Cells | | |
|---|---|---|---|---|---|---|---|---|---|
| | | | | $pIC_{50}$ | LE | LipE (LEE) | $pIC_{50}$ | LE | LipE (LEE) |
| 5g | 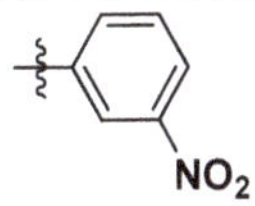 | 41 | 4.47 | 5.39 | 0.22 | 0.92 | 5.55 | 0.19 | 1.08 |
| 5l | 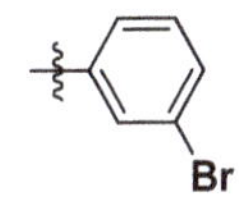 | 37 | 6.72 | 5.52 | 0.26 | -1.2 | 5.55 | 0.2 | -0.17 |
| 5m | 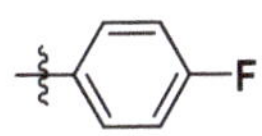 | 37 | 5.56 | 5.34 | 0.2 | -0.22 | 5.39 | 0.2 | -0.17 |
| 5n | 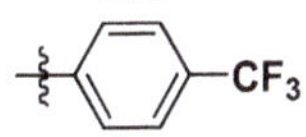 | 43 | 7.49 | 5.3 | 0.17 | -2.19 | 5.43 | 0.17 | -2.06 |

### 2.8. Structure-Activity Relationship

The activity of the 2,4-dichloro derivative (compound **5c** or **4c** ) was better than those of the 4- chloro analogues (compound **5d** or **4d** respectively), emphasizing the geometrical role of aryl moieties in the activities of the compounds. This result was consistent with that of our previous studies, which indicated the effect of such substitution patterns and showed that the 2,4-dichloro substitution was favorable to the activities of the compounds [19,20].

Hetero aryl (compound **5k**, **5j**), bulky (compound **5i**), or EDG (**5h**) reduced the activity indicated the site which was adopted by aryl groups in the side chain. The substitution site on both phenyl rings was important in the activity and physicochemical parameters of the compounds. This was clearly observed in compounds **5c**, **5e**, and **5l** compared to compounds **5f** and **5m**. Strong EWG (**5g**) exhibited the best activity.

## 3. Materials and Methods

General information of the equipment used in the synthesis and the characterization of the compounds can be found in the supplementary materials. Additionally, the anticancer activity along with shape alignment and ROCS can be found in the supplementary materials.

## 4. Conclusions

In summary, we synthesized a series of new spirooxindole analogues based on di-substituted isatin. The anticancer activity of the compounds against three different cancer cell lines was explored. Among the analogues, the compound spirooxindole analogue **5g** had an inhibitory growth potency in HCT116 similar to that of cisplatin, but it is ca. 1.8 (in HepP2) or 2.25 (in MCF7) times more potent than the reference drug, and also showed good physicochemical parameters and lipophilicity value. Further investigation of the mechanism of action of compound **5g** is required.

**Supplementary Materials:** The following are available online at http://www.mdpi.com/2076-3417/10/6/2170/s1, characterization of the synthesized compounds; protocol for anticancer activity; ROCS protocol, Figures of the NMR spectrum.

**Author Contributions:** Conceptualization, A.B.; Data curation, F.F.E.-S., and F.A.B.; Formal analysis, F.F.E.-S., F.A.B., and A.F.M.M.R.; Funding acquisition, A.M.A.-M.; Investigation, M.S.I.; Methodology, M.S.I.; Resources, A.M.A.-M.; Supervision, Y.A.M.M.E.; Validation, A.B.; Writing—original draft, A.B. and Y.A.M.M.E.; Writing—review and editing, F.A.B., A.B., and Y.A.M.M.E. All authors have read and agreed to the published version of the manuscript.

**Funding:** This research was funded by King Saud University, research group (No. RGP-044).

**Acknowledgments:** The authors would like to extend their sincere appreciation to the Deanship of Scientific Research at King Saud University for providing funding to the research group (No. RGP-044). Thanks to RSSU for their technical support. Yaseen A. M. M. Elshaier acknowledges the OpenEye scientific software for providing academic license.

**Conflicts of Interest:** The authors declare no conflict of interest.

## References

1. Santos, M.M. Recent advances in the synthesis of biologically active spirooxindoles. *Tetrahedron* **2014**, *52*, 9735–9757. [CrossRef]
2. Pavlovska, T.L.; Redkin, R.G.; Lipson, V.V.; Atamanuk, D.V. Molecular diversity of spirooxindoles. Synthesis and biological activity. *Mol. Divers.* **2016**, *20*, 299–344. [CrossRef]
3. Arun, Y.; Bhaskar, G.; Balachandran, C.; Ignacimuthu, S.; Perumal, P. Facile one-pot synthesis of novel dispirooxindole-pyrrolidine derivatives and their antimicrobial and anticancer activity against a549 human lung adenocarcinoma cancer cell line. *Bioorg. Med. Chem. Lett.* **2013**, *23*, 1839–1845. [CrossRef] [PubMed]
4. Girgis, A.S. Regioselective synthesis of dispiro [1H-indene-2, 3′-pyrrolidine-2′, 3 ″-[3H] indole]-1, 2 ″(1 ″h)-diones of potential anti-tumor properties. *Eur. J. Med. Chem.* **2009**, *44*, 91–100. [CrossRef]
5. Islam, M.S.; Park, S.; Song, C.; Kadi, A.A.; Kwon, Y.; Rahman, A.M. Fluorescein hydrazones: A series of novel non-intercalative topoisomerase IIα catalytic inhibitors induce G1 arrest and apoptosis in breast and colon cancer cells. *Eur. J. Med. Chem.* **2017**, *125*, 49–67. [CrossRef]
6. Ali, M.A.; Ismail, R.; Choon, T.S.; Yoon, Y.K.; Wei, A.C.; Pandian, S.; Kumar, R.S.; Osman, H.; Manogaran, E. Substituted spiro [2.3′] oxindolespiro [3.2 ″]-5, 6-dimethoxy-indane-1 ″-one-pyrrolidine analogue as inhibitors of acetylcholinesterase. *Bioorg. Med. Chem. Lett.* **2010**, *20*, 7064–7066. [CrossRef]
7. Sharma, P.; Kumar, A.; Sahu, V.; Upadhyay, S.; Singh, J. Synthesis of bioactive spiro-2-[3′-(2′-phenyl)-3h-indolyl]-1-aryl-3-phenylaziridines and SAR studies on their antimicrobial behavior. *Med. Chem. Res.* **2009**, *18*, 383–395. [CrossRef]
8. Uchida, R.; Imasato, R.; Shiomi, K.; Tomoda, H.; Ōmura, S. Yaequinolones j1 and j2, novel insecticidal antibiotics from penicillium sp. Fki-2140. *Org. Lett.* **2005**, *7*, 5701–5704. [CrossRef]
9. Barakat, A.; Islam, M.S.; Ghawas, H.M.; Al-Majid, A.M.; El-Senduny, F.F.; Badria, F.A.; Elshaier, Y.A.; Ghabbour, H.A. Design and synthesis of new substituted spirooxindoles as potential inhibitors of the MDM2–p53 interaction. *Bioorg. Chem.* **2019**, *86*, 598–608. [CrossRef]
10. Altowyan, M.S.; Barakat, A.; Al-Majid, A.M.; Al-Ghulikah, H.A. Spiroindolone analogues bearing benzofuran moiety as a selective cyclooxygenase COX-1 with TNF-α and IL-6 inhibitors. *Saudi J. Biol. Sci.* **2020**. [CrossRef]
11. Altowyan, M.S.; Barakat, A.; Al-Majid, A.M.; Al-Ghulikah, H. Spiroindolone analogues as potential hypoglycemic with dual inhibitory activity on α-amylase and α-glucosidase. *Molecules* **2019**, *24*, 2342. [CrossRef] [PubMed]
12. Singh, G.S.; Desta, Z.Y. Isatins as privileged molecules in design and synthesis of spiro-fused cyclic frameworks. *Chem. Rev.* **2012**, *112*, 6104–6155. [CrossRef] [PubMed]
13. Bariwal, J.; Voskressensky, L.G.; Van der Eycken, E.V. Recent advances in spirocyclization of indole derivatives. *Chem. Soc. Rev.* **2018**, *47*, 3831–3848. [CrossRef] [PubMed]
14. Kuntz, I.; Chen, K.; Sharp, K.; Kollman, P. The maximal affinity of ligands. *Proc. Natl. Acad. Sci. USA* **1999**, *96*, 9997–10002. [CrossRef] [PubMed]
15. Kulkarni, A.; Han, Y.; Hopfinger, A.J. Predicting caco-2 cell permeation coefficients of organic molecules using membrane-interaction qsar analysis. *J. Chem. Inf. Comput. Sci.* **2002**, *42*, 331–342. [CrossRef] [PubMed]
16. Available online: https//preadmet.Bmdrc.Kr/druglikeness (accessed on 12 November 2019).
17. Available online: http://molsoft.Com/mprop/ (accessed on 12 November 2019).
18. Jabeen, I.; Pleban, K.; Rinner, U.; Chiba, P.; Ecker, G.F. Structure-activity relationships, ligand efficiency, and lipophilic efficiency profiles of benzophenone-type inhibitors of the multidrug transporter p-glycoprotein. *J. Med. Chem.* **2012**, *55*, 3261–3273. [CrossRef] [PubMed]
19. Islam, M.S.; Ghawas, H.M.; El-Senduny, F.F.; Al-Majid, A.M.; Elshaier, Y.A.; Badria, F.A.; Barakat, A. Synthesis of new thiazolo-pyrrolidine (spirooxindole) tethered to 3-acylindole as anticancer agents. *Bioorg. Chem.* **2019**, *82*, 423–430. [CrossRef] [PubMed]

20. Arnott, J.A.; Kumar, R.; Planey, S.L. Lipophilicity indices for drug development. *J. Appl. Biopharm. Pharmacokinet.* **2013**, *1*, 31–36.

**Sample Availability:** Samples of the compounds **5a–n** are available from the authors.

*applied sciences*

MDPI

*Article*

# Sol–Gel Treatment of Textiles for the Entrapping of an Antioxidant/Anti-Inflammatory Molecule: Functional Coating Morphological Characterization and Drug Release Evaluation

**Francesco Puoci [1], Carmela Saturnino [2], Valentina Trovato [3], Domenico Iacopetta [1], Elpida Piperopoulos [4], Claudia Triolo [5], Maria Grazia Bonomo [2], Dario Drommi [6], Ortensia Ilaria Parisi [1], Candida Milone [4], Maria Stefania Sinicropi [1,*], Giuseppe Rosace [3,*] and Maria Rosaria Plutino [7,*]**

[1] Department of Pharmacy, Health and Nutritional Sciences, University of Calabria, Via Pietro Bucci, 87036 Arcavacata di Rende (CS), Italy; francesco.puoci@unical.it (F.P.); domenico.iacopetta@unical.it (D.I.); ortensiailaria.parisi@unical.it (O.I.P.)

[2] Department of Science, University of Basilicata, Viale dell'Ateneo Lucano 10, 85100 Potenza, Italy; carmela.saturnino@unibas.it (C.S.); mariagrazia.bonomo@unibas.it (M.G.B.)

[3] Department of Engineering and Applied Sciences, University of Bergamo, Viale Marconi 5, 24044 Dalmine (BG), Italy; valentina.trovato@unibg.it

[4] Department of Engineering, University of Messina, Contrada di Dio, S.Agata, 98166 Messina (ME), Italy; epiperopoulos@unime.it (E.P.); candida.milone@unime.it (C.M.)

[5] Department of Mathematics and Computer science, Physical and Earth Sciences, University of Messina, Viale F. Stagno d'Alcontres 31, Vill. S. Agata, 98166 Messina (ME), Italy; claudia.triolo@unime.it

[6] Department of ChiBioFarAm, University of Messina, Viale F. Stagno d'Alcontres 31, Vill. S. Agata, 98166 Messina (ME), Italy; ddrommi@unime.it

[7] Institute for the Study of Nanostructured Materials, ISMN—CNR, Palermo, c/o Department of ChiBioFarAm, University of Messina, Viale F. Stagno d'Alcontres 31, Vill. S. Agata, 98166 Messina (ME), Italy

[*] Correspondence: s.sinicropi@unical.it (M.S.S.); giuseppe.rosace@unibg.it (G.R.); mariarosaria.plutino@cnr.it (M.R.P.); Tel.: +39-0984-493200 (M.S.S.); +39-035-2052021 (G.R.); +39-090-6765713 (M.R.P.)

Received: 13 February 2020; Accepted: 20 March 2020; Published: 27 March 2020

**Abstract:** The growing interest towards textile-based drug delivery systems is due to their potential innovative medical and well-being applications. In recent years, the technique of encapsulation or inclusion of the medicine/active principle into a polymer functional matrix has been employed in order to obtain textile materials with controlled drug release. In this study, a sol–gel-based coating was developed and used as an entrapping polymeric cross-linked network for a *N*-Palmitoyl-ethanolamine (PEA) derivative, 2-methyl-pentadecanoic acid (4-nitro-phenyl)-amide or *N*-Palmitoyl-(4-nitro-phenyl)-amine (PNPA), whose anti-inflammatory and antioxidant properties have already been shown. A wide series of chemical-physical methods have been used to characterize the silica-based functional sol and to ascertain the efficient and temporary deposit of PNPA on the sol–gel coated cotton fabrics. The medicine release system achieved was shown to ensure biocompatibility, PNPA reservoir and its subsequent releasing under the action of cutaneous stimuli, thus providing useful insights in the design of medical textiles.

**Keywords:** sol–gel coating; medical textiles; antioxidant; anti-inflammatory; PEA derivative; drug release

## 1. Introduction

The research field dealing with the development of controlled drug delivery systems has been of relevant scientific interest since the 1970s and has grown and diversified rapidly in recent years, in particular thanks to the benefits it brings to healthcare; furthermore, it covers a large market segment [1]. In general, the effectiveness of drug therapy is the main objective of controlled release systems [1,2], with a corresponding (i) reduction of the number of drug administrations; (ii) improvement in therapeutic activity [3,4]; (iii) consequent reduction of the intensity of side effects; and (iv) elimination of specialized drug administration [5]. This pharmaceutical technology, especially in recent years, has seen application in other fields ranging from cosmetics [6] to agriculture [7], including textiles [8,9], as an interesting and innovative application. Indeed, textile fabrics, thanks to their biocompatibility, breathing structure and absorptive capacity, are of great interest as a medium (for ex vivo applications) for controlled release of drugs, active principles or aroma substances of particular comfortability [10]. Several examples are nowadays in common use, such as the well-known transdermal patches or textile costumes, generally characterized by different layers in which the release of a specific drug substance, deposited on the textile surface, is activated by stimuli such as temperature, humidity, enzyme, perspiration types or friction [11].

In general, in controlled release systems, biocompatibility and controllability are important features; furthermore, in terms of biocompatibility, the carcinogenicity, toxicity, teratogenicity and mutagenicity are important elements to be controlled [10]. The use of textile fabrics for the realization of controlled release systems presents several advantages but on the other hand, some disadvantages with respect to the oral administration of substances. Indeed, this administration represents an attractive and easier therapy for the patient due to its efficacy since the drug avoids both the digestive apparatus and the hepatic metabolism that reduce the concentration of the drug. Furthermore, it requires lower dosages due to the higher diffusion through tissues, which correspond to lower social costs of therapies [11]. On the other hand, the disadvantages are related to the diffusion rate of the drug as function of its molecular structure and body surface administration [11]. Different methods have been employed in order to develop better textile-based delivery systems (i.e., bandages, patches), with good controllability, biocompatibility and active species entrapment/release by use of host-guest molecules (cyclodextrins [12], aza-crown ethers, fullerenes) or doping functional molecules (ion-exchange; drug-loaded hollow; nanoparticles; bioactive) [10]. Several C–C polymer heteroatom-containing (i.e., N, P, Si) backbones for controlled release applications have been tested and considered in order to improve the drug therapy effectiveness [5]. All the developed release polymeric systems have been shown to act through mechanisms of temporal controlled release, such as drug-delayed dissolution, diffusion-controlled, and drug solution flow control after interaction with environmental water or by reacting to specific skin stimuli.

The sol–gel method has been shown to be a useful method in the preparation of functional nanostructured coatings for textiles, thus combining the entrapment/encapsulation of bioactive compounds, biomolecules and their controlled release [13].

In our previous studies, we have already shown that nano-hybrid sol–gel-based coatings feature abrasion resistance, tensile strength and elongation properties of the treated fabrics [13,14]. These peculiar characteristics may be combined with a proper doping molecule, such as a dye [15–21], an antimicrobial [22], a hydrophobic [23–25] or a flame-resistant molecule [26,27], with the aim of improving the textile surface properties and making a functional nano-hybrid coating.

With the aim of developing a functional sol–gel-based coating suitable for medical application, we thought worthwhile to make a silica sol containing the 3-glycidoxypropyltriethoxysilane (GPTES, hereafter "G"), as silica cross-linker precursor, and a PEA derivative, the *N*-Palmitoyl-(4-nitro-phenyl)-amine (hereafter PNPA), whose anti-inflammatory and antioxidant properties have already been tested and compared with other analogue molecules [28]. The sol was successfully applied on cotton surfaces and, after drying and curing, a stable and uniform PNPA-containing silica-based coating was obtained, as confirmed by morphological studies (SEM and AFM microscopy). As already shown in previous studies for halochromic dyestuff [20], the PEA derivative results firmly encapsulated into the 3D hybrid silica layer in absence of external stimuli (i.e., variable pH conditions), thanks to non-covalent and weak interactions (i.e., hydrogen bonds and van der Waals interactions) acting between the non-polar active molecule and the alkoxysilane hosting network.

Finally, the functionally coated cotton samples were employed for in vitro diffusion studies with the aim of testing their ability to release the synthesized PEA derivative in a controlled manner compared to a standard solution of the molecule. This effectively prepared functional hybrid system, based on the non-covalently immobilized PNPA, showed good results so that it can be considered a suitable scaffold for fabrics in drug release applications, thus providing useful insights in the design and the development of medical textiles.

## 2. Materials and Methods

*N*-Palmitoyl-(4-nitro-phenyl)-amine (PNPA) was synthetized according to a synthetic strategy described in previous researches [28]. 3-glycidyloxypropyltriethoxysilane (GPTES) and methanol were purchased from Wacker and Aldrich, respectively, and used without further purification. Two cotton scoured and bleached 100% plain-weave textile fabrics (coded $CO_L$ and $CO_H$) kindly supplied by Albini S.p.A. (Albino, Italy) and Mascioni S.p.A. (Cuvio, Italy), respectively, were used for this research. The fabrics showed different mass per unit area ($CO_L$ = 119 g/m$^2$ and $CO_H$ = 331 g/m$^2$, respectively). Cotton fabrics were washed before treatment at pH 7 and at 40 °C for 20 min in a non-ionic detergent, rinsed several times with de-ionized water and then dried. The cleaned samples were conditioned at 20 (± 1) °C and under standard atmospheric pressure at 65 (± 2)% relative humidity for at least 24 h prior to all experiments.

PNPA (25 mg) was dissolved in 40 mL of methanol through ultrasonication and left under continuous stirring. Then 2 mL of a 1 M aqueous sol–gel solution of GPTES were added drop by drop to the clear methanol solution of antioxidant molecule, thus resulting in a final GPTES sol concentration of 0.05 M (1:0.034 molar ratio with respect to PNPA). The obtained solution (G-PNPA sol) was ultrasonicated and left at room temperature under stirring for at least 90 min. The same reaction was also carried out in absence of the antioxidant molecule in order to obtain a reference GPTES sol sample. Both solutions were applied separately onto cotton textile (10 cm × 10 cm) through a two-roll laboratory padder (Werner Mathis, Zurich, Switzerland) at a nip pressure of 2 bar, then dried (80 °C for 5 min) and cured (100 °C for 1 min) in an electric laboratory oven.

PNPA, G-PNPA and G sols were fully investigated through FT-IR spectroscopy and Nuclear magnetic resonance (NMR). Untreated and treated textiles were characterized by FT-IR spectroscopy, Scanning Electron Microscopy (SEM) coupled to energy dispersive X-ray (EDS) and Atomic Force Microscopy (AFM).

The PNPA controlled release from the two prepared textiles, $CO_L$_G-PNPA and $CO_H$_G-PNPA, was investigated by performing in vitro diffusion studies using Franz diffusion cells and according to the experimental protocol reported in a previous work [29]. For this purpose, Strat-M® membranes (25 mm discs, Cat. No. SKBM02560, Merck Millipore, Darmstadt, Germany) were positioned between the donor and the receptor compartments of each Franz cell and the experiments were carried out at 37 ± 0.5 °C. The two tested items, $CO_L$_G-PNPA and $CO_H$_G-PNPA, were placed on the Strat-M® membrane with the GPTES layer facing towards the acceptor chamber. Then, the Franz cell compartments were fixed together and filled with 0.5 and 5.5 mL of phosphate buffer at pH 7.4 ($10^{-3}$ M), respectively. The content of the receptor chamber was withdrawn at different times, such as 1, 2, 4, 6 and 24 h, for UV-Vis analysis and replaced with phosphate buffer. The same experimental conditions were applied to a control sample consisting of a standard PNPA solution. The in vitro diffusion studies were carried out in triplicate and the obtained results were expressed as diffused amount (%).

FT-IR spectra were performed by a Thermo Avatar 370 equipped with an attenuated total reflection (ATR) accessory and using a diamond crystal as internal reflectance element. Spectra were acquired with 32 scans and in the range from 4000 to 550 cm$^{-1}$ with a resolution of 4 cm$^{-1}$.

One- and two-dimensional NMR experiment were recorded in methanol-$d_4$ at 298.2 (±0.1) K on Bruker ARX-300, equipped with a 5 mm gradient probe and operating at 300.1 MHz for 1H nucleus. All chemical shifts are shown in parts per million (δ/ppm), downfield to tetramethylsilane ($Me_4Si$) as an internal standard (δ = 0.0 ppm), or referenced to the residual protiated solvent signal such as in methanol-$d_4$ ($^1$H NMR: 3.30 ppm). $^1$H NMR signals were assigned by means of two-dimensional homonuclear NMR gradient experiments (gCOSY, gNOESY), acquired using standard Bruker pulse sequences.

SEM morphology and SEM-EDS of the investigated samples were obtained using a FEI Quanta FEG 450 microscope. An operating voltage of 5 kV in low vacuum was used for SEM images. EDS analysis was conducted with an operating voltage of 20 kV, always in low vacuum. Samples were fixed on aluminum sample holders by means of a graphitic adhesive.

AFM characterization was performed using a stand-alone SMENA head by NTMDT, equipped with a Bruker silicon probe model NCHV working in semi-contact mode. The samples were fixed onto metallic stubs using a small piece of double-sided scotch tape and studied at RT.

## 3. Results

### 3.1. Sol–Gel Synthesis and Coating Application

To obtain a controlled drug release fabric, 3-glycidoxypropyltriethoxysilane (GPTES) was employed in a sol–gel synthesis in the presence of an antioxidant/anti-inflammatory molecule, the N-Palmitoyl-(4-nitro-phenyl)-amine (PNPA) [28]. The simultaneous presence of both an epoxy group and a triethoxysilane functionalities makes this specific silica precursor able to cross-link to other GPTES molecules, to entrap the antioxidant doping molecule and still to coat the textile surface, respectively. In this regard, a GPTES-based sol was synthesized in acidic aqueous medium by addition of slight amount of HCl, as catalyst, and then added to a methanol solution containing the PNPA molecule. As reported in previous studies [15–21], the sol–gel synthesis leads to the formation of a hybrid polymeric 3D network through the opening of the epoxy ring of GPTES and interaction of the triethoxysilane end to form an extended polyethylene oxide network (PEO) [20], in whose holes the PNPA is physically and stably entrapped (Scheme 1).

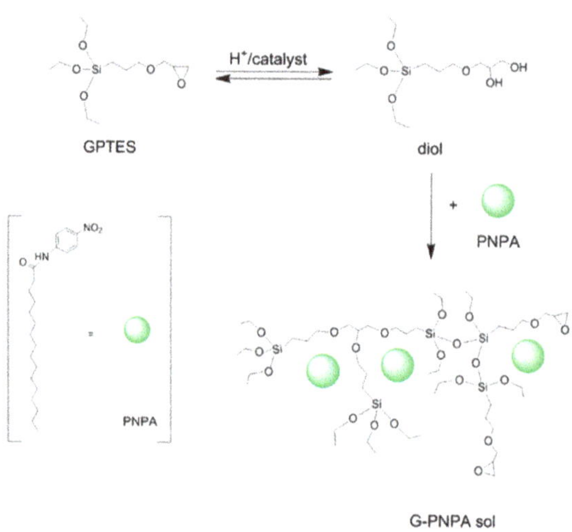


**Scheme 1.** Reaction pathways toward the formation of the G-PNPA sol, as obtained in methanol solution at room temperature in slight acid conditions.

The so obtained G-PNPA functional sol was applied by padding on cotton fabrics and cured thermally with the aim to prepare a nano-hybrid coating for controlled release application (Scheme 2).

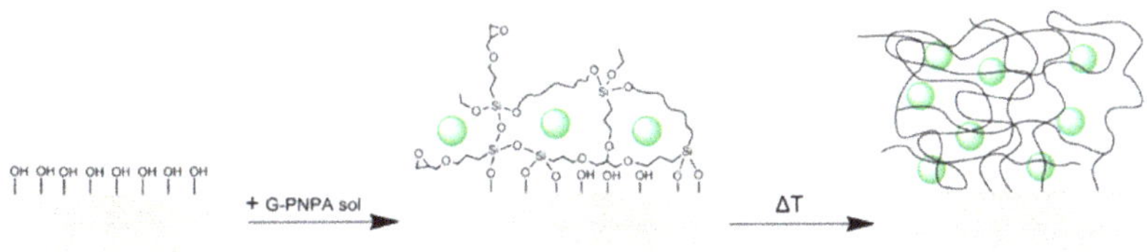


**Scheme 2.** Schematic representation of the application of the G-PNPA sol on cotton surface and the formation of the coating xerogel.

### 3.2. *Sol NMR Characterization*

A reaction mixture relative to the G-PNPA sol as obtained in methanol for coating application, has been achieved in situ (1:0.1 = [GPTES]:[PNPA] molar ratio) in methanol-$d_4$ and characterized by means of $^1$H one- and two-dimensional NMR spectroscopy. Figure 1 shows the $^1$H NMR spectra as recorded at time zero and after 24 h (in methanol-$d_4$ at 298 K, 300 MHz). The comparison of the aliphatic regions of the $^1$H NMR spectra of the reaction mixtures recorded at different times clearly reveal the presence of the protonic pattern expected for the diol/PEO silylated (red squares in Figure 1) derivatives and the starting GPTES (black squares in Figure 1, [20]) as shown in Scheme 1. In particular, the aliphatic regions of the $^1$H NMR spectra in Figure 1 clearly show: (i) the presence of the expected protonic pattern for the GPTES open ring derivative, bringing a hydroxyl and an ether group bonded to two vicinal carbon $C_e$ and $C_f$ atoms (δ = 0.68, $CH_{2a}$; 1.71, $CH_{2b}$; 3.48, $CH_{2c}$ + $CH_{2d}$ + $CH_{2f}$; 3.88 $CH_{2e}$, red squares in lower spectrum) [30] (ii) the proton peaks relative to the starting GPTES in a decreased concentration (δ = 0.67, $CH_{2a}$; 1.68, $CH_{2b}$; 2.61+2.78, $CH_{2f}$; 3.16, $CH_{2e}$; 3.32 + 3.76, $CH_{2d}$; 3.49, $CH_{2c}$, black squares in upper spectrum); (iii) the presence of the upper-field methylene and the methyl proton resonances relative to free ethanol moieties, compared to those relative to the ethylic groups of GPTES (δ = 3.63, $CH_2$, 1.19, $CH_3$ vs 3.84, $CH_2$, 1.22, $CH_3$; cut signals in both spectra).

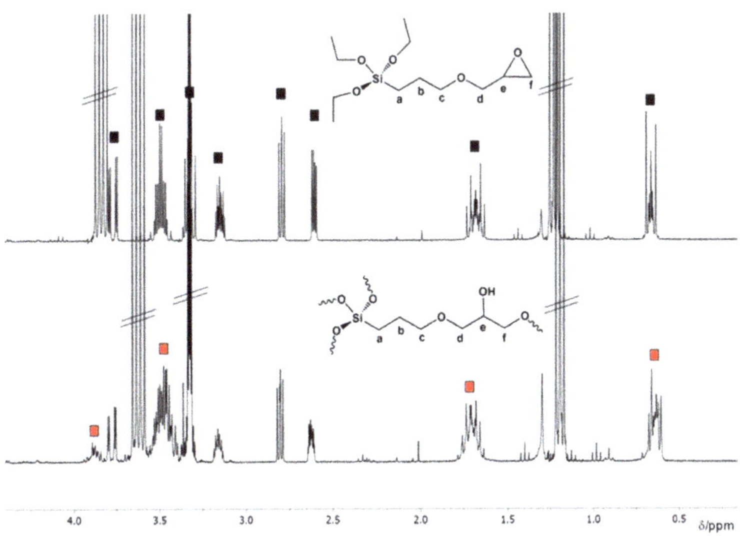


**Figure 1.** $^1$H NMR spectra relative to solutions of the G-PNPA sol, as obtained in methanol-$d_4$ at 298 K, 300 MHz, at time zero (upper spectrum) and after 24 h reaction time in the presence of slight amount of HCl (lower spectrum).

Unfortunately, the long penta-decanoic protonic chain is buried under other signals, also due to the strong molar ratio exceed of GPTES. The aromatic region shows the expected and unchanged pattern of broad signals for para-substituted phenyl ring of the PNPA molecule.

The $^1$H NMR experiments clearly show that the GPTES epoxy ring opening reaction by a suitable nucleophilic group of the PNPA molecule is not occurring, as well as the formation of a stable ether covalent bond. As previously shown [20], the stable encapsulation of the PNPA molecule into the PEO silylated GPTES derivative is most likely due to the formation of weak bonds (i.e., van der Waals or electrostatic) between the polymerized GPTES (i.e., ether oxygen and hydroxyl group) and the N-Palmitoyl-(4-nitro-phenyl)-amine (i.e., nitrogen, oxygen, long alkyl chain, phenyl group).

### 3.3. ATR FT-IR

FT-IR spectra of untreated and treated cotton fabrics were analyzed after the normalization at 1362 $cm^{-1}$, absorption band relative to the CH bending of cellulose. In Figure 2, FT-IR of $CO_L$ untreated and treated samples with GPTES sol and G-PNPA sol ($CO_L$_UT, $CO_L$_GPTES and $CO_L$_G-PNPA, respectively) are reported and compared with those relative to $CO_H$ untreated and treated fabrics with the same solutions ($CO_H$_UT, $CO_H$_GPTES and $CO_H$_G-PNPA, respectively). In particular, in all spectra, it is possible to distinguish clearly the characteristic absorption bands relative to cellulose moieties, such as: the broad bands around to 3331 $cm^{-1}$ and 2894 $cm^{-1}$ (stretching mode of OH and CH, respectively), and the absorption bands between 1097 $cm^{-1}$ and 895 $cm^{-1}$ relative to the asymmetric in-plane ring stretch and to C–O stretch and asymmetric out-of-phase ring stretch ($C_1$–O–$C_4$) [31]. In particular, in the inset in Figure 2A,B the presence of a silica coating is evident through the increase of the bands in the range 1145 $cm^{-1}$–895 $cm^{-1}$ for the $CO_L$, assigned to the asymmetric stretching of the -Si–O–Si, and in the range between 995 $cm^{-1}$–760 $cm^{-1}$ for the $CO_H$, due to the Si–O–Si absorption bending (852 $cm^{-1}$) and stretching (790 $cm^{-1}$).

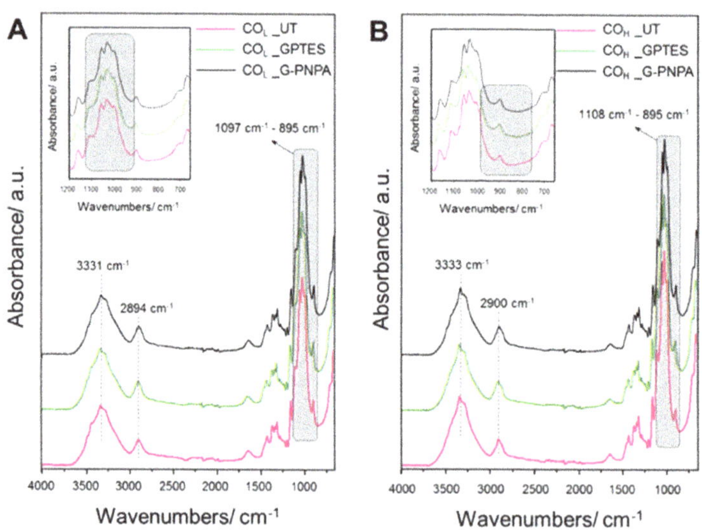


**Figure 2.** FT-IR spectra of $CO_L$ and $CO_H$ untreated and treated with GPTES sol and G-PNPA sol (**A** and **B**, respectively). (**A**) FT-IR spectra of $CO_L$ untreated and treated with GPTES sol and G-PNPA sol; (**B**) FT-IR spectra of $CO_H$ untreated and treated with GPTES sol and G-PNPA sol.

With the aim of investigating the chemical structure and interactions between silica precursor and antioxidant molecule without the interference of the intense absorption bands of cotton fabrics, the FT-IR spectra of each sol was carried out on the pure xerogel. The latter was obtained by depositing a few drops of each solution onto optical glass slides further subjected to a thermal treatment (100 °C) to remove the solvent [26]. Indeed, by analyzing the FT-IR spectra of PNPA, it was possible to identify the characteristic peaks in order to establish its interaction with the GPTES sol. Spectra of the GPTES sol (Figure 3, green line) clearly shows the formation of the inorganic network due to the absorption bands relative to: asymmetric (1093–1010 cm$^{-1}$) and symmetric stretching of Si–O–Si (756 cm$^{-1}$) and its bending mode (850 cm$^{-1}$) [17,18,20]. Furthermore, the absorption bands at 3072–3000 cm$^{-1}$, 1255 cm$^{-1}$ and 906–850 cm$^{-1}$ were assigned to asymmetric and symmetric C–H stretch, ring breathing, as well as the asymmetric and symmetric ring deformation, respectively [17,18], indicating that some unopened epoxy ring remain in GPTES sol.

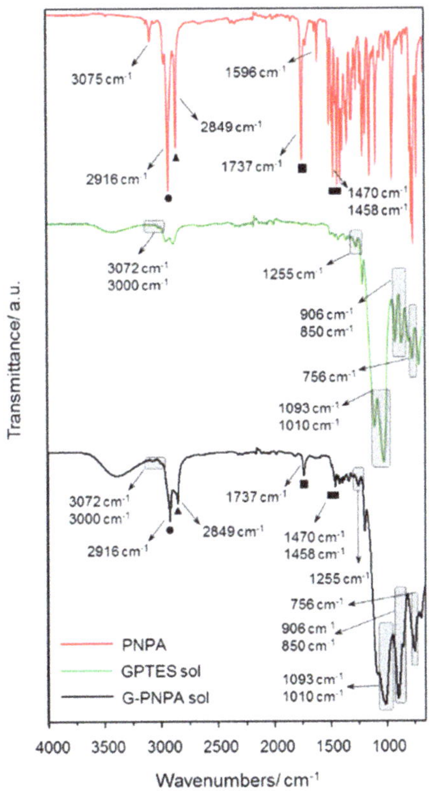


**Figure 3.** FT-IR spectra of PNPA, GPTES sol and G-PNPA sol.

These peaks relative to the GPTES sol are also evidenced in the sol containing PNPA (Figure 3, black line), in which the main characteristic absorption bands of the antioxidant molecules are present. As shown in Figure 2 (red line), PNPA is featured by the asymmetric and symmetric $CH_2$ stretching mode of alkyl chain (2916 and 2849 cm$^{-1}$, respectively) with relative bending (1470–1458 cm$^{-1}$), the C=O and C–N stretching of the secondary amide (1737 and 3075 cm$^{-1}$, respectively) and the C–N stretching mode (1596 cm$^{-1}$) [32].

## *3.4. Morphological Characterizations*

All untreated and treated cotton samples were investigated by SEM and AFM microscopy, in order to characterize the morphology of the coated textile fabric and to underline structural differences before and after the coating application.

SEM analysis shows that the two analyzed tissues have two different weaves: $CO_H$_UT samples have a larger weave than $CO_L$_UT ones, and the fibers appear less ordered compared with $CO_L$_UT ones and more threadbare (compare Figure 4a,b). Both samples do not show modifications after treatment and the morphology appears unchanged also at higher magnification (Figure 5). This demonstrates that, after treatment, no changes at micrometer scale occur.

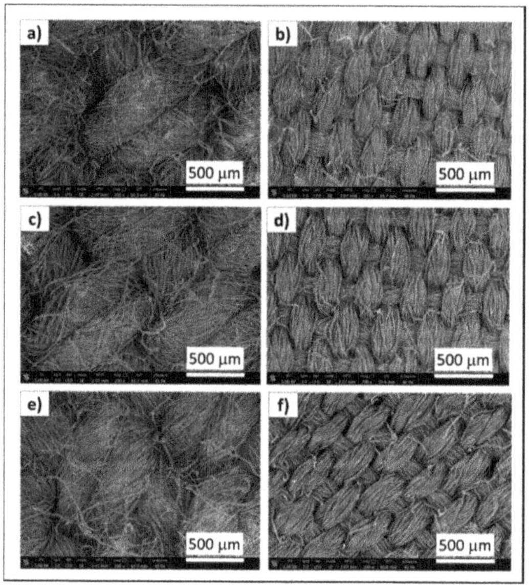


**Figure 4.** SEM analysis of investigated samples: $CO_H$_UT (**a**), $CO_L$_UT (**b**), $CO_H$_GPTES (**c**), $CO_L$_GPTES (**d**), $CO_H$_G-PNPA (**e**), $CO_L$_G-PNPA (**f**).

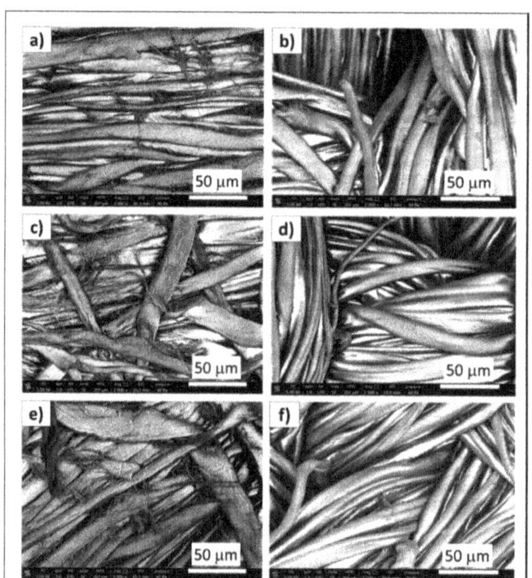


**Figure 5.** SEM analysis of investigated samples at higher magnification: $CO_H$_UT (**a**), $CO_L$_UT (**b**), $CO_H$_GPTES (**c**), $CO_L$_GPTES (**d**), $CO_H$_G-PNPA (**e**), $CO_L$_G-PNPA (**f**).

EDS analysis shows Si peak only in samples treated with GPTES ($CO_H$_GPTES, $CO_L$_GPTES) and subsequently with PNPA ($CO_H$_G-PNPA, $CO_L$_G-PNPA) (Figure 6, only two treated samples are reported for briefness, because the other ones exhibited a similar spectrum). The other peaks refer to carbon, present in the analyzed tissues but also in graphite adhesive used for SEM analysis, oxygen, present also in vapor-low vacuum atmosphere, and aluminum, derived from the SEM stub.

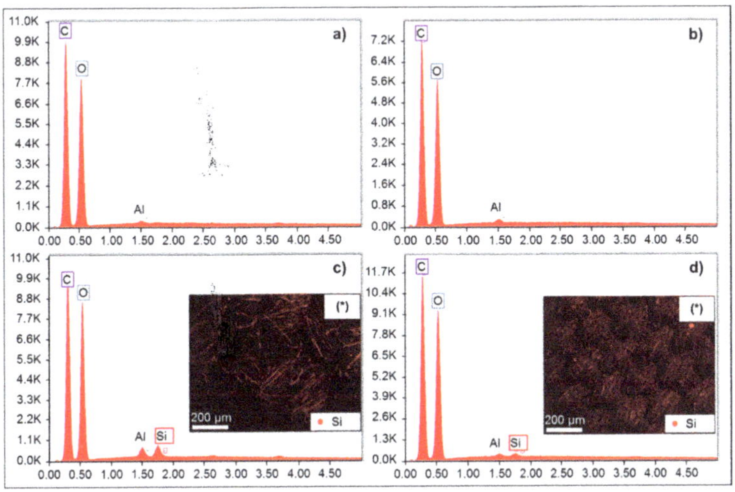


**Figure 6.** EDS analysis of investigated samples: $CO_H$_UT (**a**), $CO_L$_UT (**b**), $CO_H$_G-PNPA (**c**), $CO_L$_G-PNPA (**d**), and mapping of $CO_H$_G-PNPA (inset **c***) and $CO_L$_G-PNPA (inset **d***).

The mapping presented in Figure 6 (inset c* and d*) shows that distribution of Si element (present in GPTES) is very uniform and no phase separation occurs.

### *3.5. AFM Characterizations*

Figure 7 shows that all the studied sampled are characterized by the typical filamentary structure of the tissue fibers on the nanoscale.

Thanks to the high resolution achieved by the AFM technique, it is possible to notice that both the two sample pairs $CO_L$_GPTES (Figure 7c), $CO_H$_GPTES (Figure 7d) and $CO_L$_G-PNPA (Figure 7e), $CO_H$_G-PNPA (Figure 7f) are rougher than the untreated tissues $CO_L$_UT (Figure 7a), $CO_H$_UT (Figure 7b), indicating that the GPTES and PNPA intimately and homogeneously wrap the fibers. This result is in good agreement with the EDS Si mapping. In the lower right corner of Figure 7f, it is possible to observe the presence of a lack in the coating; the study of the line profiles in that area allows us to evaluate the thickness of the coating that ranges between 2.5 and 4 nm.

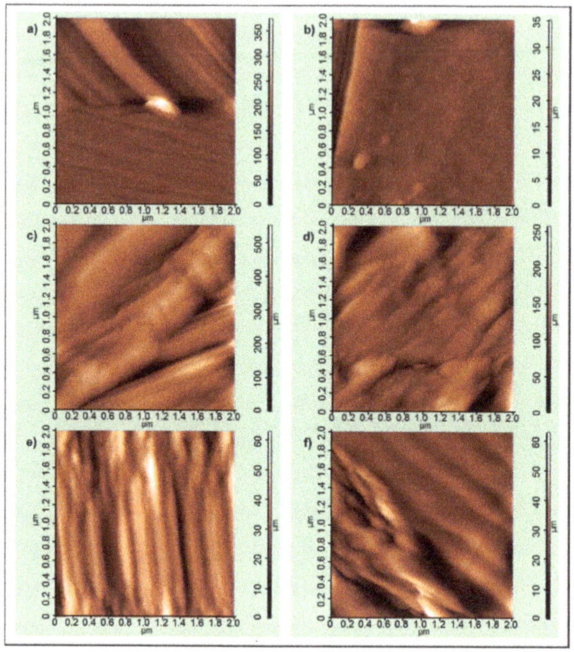

**Figure 7.** AFM micrograph of $CO_L$_UT (**a**), $CO_H$_UT (**b**), $CO_L$_GPTES (**c**), $CO_H$_GPTES (**d**), $CO_L$_G-PNPA (**e**), and $CO_H$_G-PNPA (**f**).

### *3.6. PNPA in Vitro Diffusion Studies*

In vitro diffusion studies on the developed cotton-based textiles were carried out with the aim of evaluating their ability to release in a controlled manner the synthesized PEA derivative compared to a standard solution of the molecule (Scheme 3).

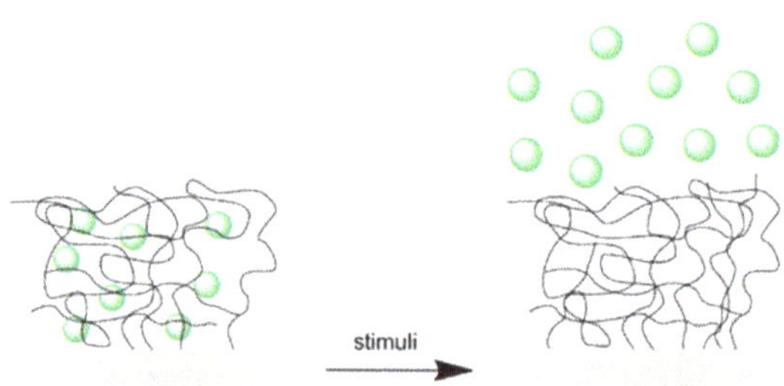


**Scheme 3.** Schematic representation on the PNPA controlled release from the functional sol–gel coated cotton by stimuli effect.

In the performed experiments, Strat-M® membranes were used as a synthetic alternative, which is predictive of the diffusion process occurring through human skin. The studies involved two different textiles, such as $CO_L$_G-PNPA and $CO_H$_G-PNPA, prepared employing the obtained G-PNPA functional sol. The obtained results are expressed as cumulative diffused amount (%) and the diffusion profiles for the tested items are reported in Figure 8.

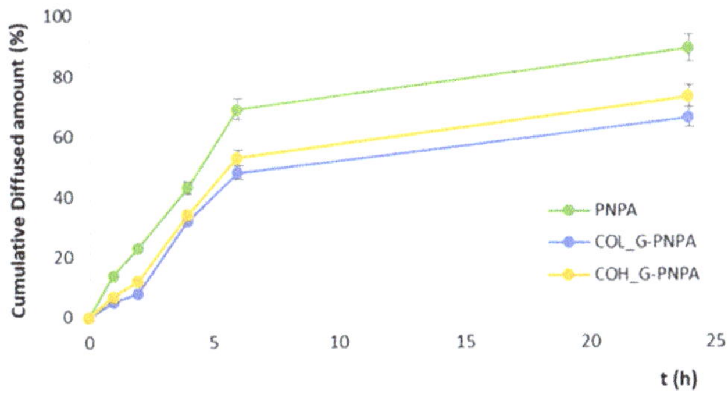


**Figure 8.** In vitro diffusion profiles.

The two G-PNPA-coated cotton textiles, $CO_L$_G-PNPA and $CO_H$_G-PNPA, show similar results in the performed in vitro diffusion studies. In the case of the $CO_L$_G-PNPA sample, indeed, the amount of released PNPA is equal to 5% within the first hour reaching the 48% and 66% in 6 and 24 h, respectively, while the $CO_H$_G-PNPA sample exhibits a value of 7% after the first hour achieving the 53% and 73% at the time points of 6 and 24 h. On the other hand, the 14% of PNPA is released after the first hour from the control solution reaching the 69% and 89% at 6 and 24 h. The experimental data confirm the ability of both the developed textiles to release in a controlled way the synthesized PEA derivative. These results could be related to the rate-limiting steps in drug release reported in the literature [33,34]: the drug diffusion within the polymer matrix and the rate of polymer swelling. Depending on the presence of sweat, simulated with the buffer solution in contact with the PNPA-treated fabric, the weak interactions between the treated fabric and the antioxidant molecule slowly disappear until the drug is completely released. The presence of the sol–gel matrix increases the bonding interactions between the network and the drug, slowing down its release.

## 4. Conclusions

The N-Palmitoyl-(4-nitro-phenyl)-amine, PNPA, a PEA derivative that already showed good anti-inflammatory and antioxidant properties, was firmly entrapped in a sol–gel-based matrix obtained by polymerization reaction of the epoxy-alkoxysilane GPTES, as cross-linker compound. NMR studies run on the sol evidenced that the GPTES epoxy-ring opening and the subsequent polymerization give rise to a polyethylene oxide 3D network grafted on textile with the PNPA immobilized into it. After the textile drying and curing the xerogel was morphologically studied on the treated cotton samples by SEM and AFM microscopy, revealing that no changes in the fiber morphology occurred at the micrometer scale, and that GPTES and PNPA intimately and homogeneously wrap the fibers. Furthermore, SEM mapping revealed a uniform distribution of the silica-based coating on the cotton fibers. In vitro diffusion studies were realized on the developed functionalized cotton-based textiles in order to check their ability to release the PEA derivative in a controlled manner in comparison to a standard molecule solution. As a matter of fact, this functional textile has been shown to be a suitable system for PNPA release, thanks to the chemical binding weakening of PNPA with the sol–gel polymer matrix by mean of medium effect, thus opening the way to the design of similar functional hybrid coatings for biomedical application. Although parameters of sol particles (e.g., hydrodynamic radius and electrokinetic potential) as well as the porosity of systems were not investigated, the obtained results confirmed the potential of the nanoengineered procedure as a versatile method for preparing stable and tunable drug-releasing materials. The combination of the functionality and transparency provided by the hybrid coating with its easy processability could represent an innovative route to fabricate biomaterials for healthcare. Future work will focus on the comparison of silica sols characteristics with release properties of this delivery system.

**Author Contributions:** Conceptualization was done by M.S.S., G.R. and M.R.P. Investigation was performed by F.P., O.I.P., C.S., M.G.B., V.T., G.R., C.T., C.M., E.P., M.S.S., D.I., M.R.P. and D.D. Project administration was directed by M.S.S., G.R. and M.R.P. Resources were provided by F.P., O.I.P., C.S., M.G.B., V.T., G.R., C.T., C.M., E.P., M.S.S., D.I., M.R.P. and D.D. Supervision was taken care of by M.S.S., G.R. and M.R.P. Validation was carried out by F.P. and O.I.P. Visualization and original draft writing was done by M.S.S., G.R. and M.R.P. All authors have read and agreed to the published version of the manuscript.

**Funding:** This research received no external funding.

**Acknowledgments:** MURST: CNR and MIUR are gratefully acknowledged for financial support.

**Conflicts of Interest:** The authors declare no conflict of interest.

## References

1. Langer, R. Drug delivery and targeting. *Nature* **1998**, *392*, 5–10.
2. Brouwers, J.R.B.J. Advanced and controlled drug delivery systems in clinical disease management. *Pharm. World Sci.* **1996**, *18*, 153–162. [CrossRef]
3. Sinicropi, M.S.; Iacopetta, D.; Rosano, C.; Randino, R.; Caruso, A.; Saturnino, C.; Muià, N.; Ceramella, J.; Puoci, F.; Rodriquez, M.; et al. N-thioalkylcarbazoles derivatives as new anti-proliferative agents: Synthesis, characterisation and molecular mechanism evaluation. *J. Enzyme Inhib. Med. Chem.* **2018**, *33*, 434–444. [CrossRef]
4. Saturnino, C.; Caruso, A.; Longo, P.; Capasso, A.; Pingitore, A.; Caroleo, M.C.; Cione, E.; Perri, M.; Nicolò, F.; Nardo, V.M.; et al. Crystallographic study and biological evaluation of 1,4-dimethyl-N-alkylcarbazoles. *Curr. Top. Med. Chem.* **2015**, *15*, 973–979. [CrossRef]
5. Uhrich, K.E.; Cannizzaro, S.M.; Langer, R.S.; Shakesheff, K.M. Polymeric systems for controlled drug release. *Chem. Rev.* **1999**, *99*, 3181–3198. [CrossRef]
6. Brannon-Peppas, L. Controlled Release in the Food and Cosmetics Industries. In *Polymeric Delivery Systems—Chapter 3*; El-Nokaly, M.A., Piatt, D.M., Charpentier, B.A., Eds.; American Chemical Society: Washington, DC, USA, 1993; pp. 42–52.
7. Levy, R.; Nichols, M.A.; Miller, T.W., Jr. Encapsulated Systems for Controlled Release and Pest Management. In *Polymeric Delivery Systems—Chapter 13*; El-Nokaly, M.A., Piatt, D.M., Charpentier, B.A., Eds.; American Chemical Society: Washington, DC, USA, 1993; pp. 202–212.
8. Radu, C.D.; Parteni, O.; Popa, M.; Muresan, I.E.; Ochiuz, L.; Bulgariu, L.; Munteanu, C.; Istrate, B.; Ulea, E. Comparative study of a drug release from a textile to skin. *J. Pharm. Drug Deliv. Res.* **2015**, *4*. [CrossRef]
9. Hashemikia, S.; Hemmatinejad, N.; Ahmadi, E.; Montazer, M. A novel cotton fabric with anti-bacterial and drug delivery properties using SBA-15-NH 2/polysiloxane hybrid containing tetracycline. *Mater. Sci. Eng. C Mater.* **2016**, *59*, 429–437. [CrossRef]
10. Ten Breteler, M.R.; Nierstrasz, V.A.; Warmoeskerken, M.M.C.G. Textile slow-release systems with medical applications. *Autex Res. J.* **2002**, *2*, 175–189.
11. Radu, C.D.; Popa, M.; Parteni, O.; Salariu, M.; Lupușoru, E.C.; Ghiciuc, C.; Foia, L.; Chiriac, A.; Lupusoru, R.; Oproiu, L.; et al. Achievements and limits on the controlled release of a drug from a textile fabric to dermis. *Open Conf. Proc. J.* **2014**, *5*, 1–8. [CrossRef]
12. Radu, C.D.; Parteni, O.; Ochiuz, L. Applications of cyclodextrins in medical textiles—Review. *J. Control Release* **2016**, *224*, 146–157. [CrossRef]
13. Mahltig, B.; Haufe, H.; Böttcher, H. Functionalisation of textiles by inorganic sol-gel coatings. *J. Mater. Chem.* **2005**, *15*, 4385. [CrossRef]
14. Plutino, M.R.; Colleoni, C.; Donelli, I.; Freddi, G.; Guido, E.; Maschi, O.; Mezzi, A.; Rosace, G. Sol-gel 3-glycidoxypropyltriethoxysilane finishing on different fabrics: The role of precursor concentration and catalyst on the textile performances and cytotoxic activity. *J. Colloid Interf. Sci.* **2017**, *506*, 504–517. [CrossRef] [PubMed]
15. Caldara, M.; Colleoni, C.; Guido, E.; Re, V.; Rosace, G. Development of a textile-optoelectronic pH meter based on hybrid xerogel doped with methyl red. *Sens. Actuators B Chem.* **2012**, *171–172*, 1013–1021. [CrossRef]
16. Van der Schueren, L.; De Clerck, K.; Brancatelli, G.; Rosace, G.; Van Damme, E.; De Vos, W. Novel cellulose and polyamide halochromic textile sensors based on the encapsulation of Methyl Red into a sol-gel matrix. *Sens. Actuators B Chem.* **2012**, *162*, 27–34. [CrossRef]

17. Guido, E.; Colleoni, C.; De Clerck, K.; Plutino, M.R.; Rosace, G. Influence of catalyst in the synthesis of a cellulose-based sensor: Kinetic study of 3-glycidoxypropyltrimethoxysilane epoxy ring opening by Lewis acid. *Sens. Actuators B Chem.* **2014**, *203*, 213–222. [CrossRef]
18. Caldara, M.; Colleoni, C.; Guido, E.; Re, V.; Rosace, G. Optical monitoring of sweat pH by a textile fabric wearable sensor based on covalently bonded litmus-3-glycidoxypropyltrimethoxysilane coating. *Sens. Actuators B Chem.* **2016**, *222*, 213–220. [CrossRef]
19. Plutino, M.R.; Guido, E.; Colleoni, C.; Rosace, G. Effect of GPTMS functionalization on the improvement of the pH-sensitive methyl red photostability. *Sens. Actuators B Chem.* **2017**, *238*, 281–291. [CrossRef]
20. Rosace, G.; Guido, E.; Colleoni, C.; Brucale, M.; Piperopoulos, E.; Milone, C.; Plutino, M.R. Halochromic resorufin-GPTMS hybrid sol-gel: Chemical-physical properties and use as pH sensor fabric coating. *Sens. Actuators B Chem.* **2017**, *241*, 85–95. [CrossRef]
21. Trovato, V.; Colleoni, C.; Castellano, A.; Plutino, M.R. The key role of 3-glycidoxypropyltrimethoxysilane sol-gel precursor in the development of wearable sensors for health monitoring. *J. Sol-Gel Sci. Technol.* **2018**, *87*, 27–40. [CrossRef]
22. Poli, R.; Colleoni, C.; Calvimontes, A.; Polášková, H.; Dutschk, V.; Rosace, R. Innovative sol-gel route in neutral hydroalcoholic condition to obtain antibacterial cotton finishing by zinc precursor. *J. Sol-Gel Sci. Technol.* **2015**, *74*, 151–160. [CrossRef]
23. Avnir, D.; Coradin, T.; Lev, O.; Livage, J. Recent bio-applications of sol-gel materials. *J. Mater. Chem.* **2006**, *16*, 1013–1030. [CrossRef]
24. Cardiano, P. Hydrophobic properties of new epoxy-silica hybrids. *J. Appl. Polym. Sci.* **2008**, *108*, 3380–3387. [CrossRef]
25. Cardiano, P.; Lo Schiavo, S.; Piraino, P. Hydrorepellent properties of organic–inorganic hybrid materials. *J. Non-Cryst. Solids* **2010**, *356*, 917–926. [CrossRef]
26. Brancatelli, G.; Colleoni, C.; Massafra, M.R.; Rosace, G. Effect of hybrid phosphorus-doped silica thin films produced by sol-gel method on the thermal behavior of cotton fabrics. *Polym. Degrad. Stabil.* **2011**, *96*, 483–490. [CrossRef]
27. Grancaric, A.M.; Colleoni, C.; Guido, E.; Botteri, L.; Rosace, G. Thermal behaviour and flame retardancy of monoethanolamine-doped sol-gel coatings of cotton fabric. *Prog. Org. Coat.* **2017**, *103*, 174–181. [CrossRef]
28. Saturnino, C.; Popolo, A.; Ramunno, A.; Adesso, S.; Pecoraro, M.; Plutino, M.R.; Rizzato, S.; Albinati, A.; Marzocco, S.; Sala, M.; et al. Anti-inflammatory, antioxidant and crystallographic studies of N-Palmitoyl-ethanol Amine (PEA) derivatives. *Molecules* **2017**, *22*, 616. [CrossRef]
29. Parisi, O.; Scrivano, L.; Amone, F.; Malivindi, R.; Ruffo, M.; Vattimo, A.F.; Pezzi, V.; Puoci, F. Interconnected PolymerS TeChnology (IPSTiC): An effective approach for the modulation of 5α-reductase activity in hair loss conditions. *J. Funct. Biomater.* **2018**, *9*, 44. [CrossRef]
30. Castellano, A.; Colleoni, C.; Iacono, G.; Mezzi, A.; Plutino, M.R.; Malucelli, G.; Rosace, G. Synthesis and characterization of a phosphorous/nitrogen based sol-gel coating as a novel halogen- and formaldehyde-free flame retardant finishing for cotton fabric. *Polym. Degrad. Stabil.* **2019**, *162*, 148–159. [CrossRef]
31. Chung, C.; Lee, M.; Choe, E.K. Characterization of cotton fabric scouring by FT-IR ATR spectroscopy. *Carbohyd. Polym.* **2004**, *58*, 417–420. [CrossRef]
32. Peng, Y.D.; Wang, F.; Gao, L.; Dong, W.K. Structurally characterized dinuclear Zinc(II) bis(salamo)-type tetraoxime complex possessing square pyramidal and trigonal bipyramidal geometries. *J. Chin. Chem. Soc. Taipei* **2018**, *65*, 893–899. [CrossRef]
33. Ma, Z.H.; Yu, D.G.; Branford-White, C.J.; Nie, H.L.; Fan, Z.X.; Zhu, L.M. Microencapsulation of tamoxifen: Application to cotton fabric. *Colloid Surf. B* **2009**, *69*, 85–90. [CrossRef] [PubMed]
34. Messaritaki, A.; Black, S.J.; van der Walle, C.F.; Rigby, S.P. NMR and confocal microscopy studies of the mechanisms of burst drug release from PLGA microspheres. *J. Control Release* **2005**, *108*, 271–281. [CrossRef] [PubMed]

applied sciences

MDPI

*Article*

# Nutraceuticals Obtained by SFE-$CO_2$ from Cladodes of Two *Opuntia ficus-indica* (L.) Mill Wild in Calabria

**Domenico Iacopetta [1], Noemi Baldino [2,*], Anna Caruso [1], Valentina Perri [2], Francesca Romana Lupi [2], Bruno de Cindio [2], Domenico Gabriele [2] and Maria Stefania Sinicropi [1]**

[1] Department of Pharmacy, Health and Nutritional Sciences, University of Calabria, 87036 Arcavacata di Rende (CS), Italy; domenico.iacopetta@unical.it (D.I.); anna.caruso@unical.it (A.C.); s.sinicropi@unical.it (M.S.S.)

[2] Department of Informatics, Modeling, Electronics and System Engineering, University of Calabria Via P. Bucci – Cubo 39C, I-87036 Arcavacata di Rende (CS), Italy; valentina.perri@unical.it (V.P.); francesca.lupi@unical.it (F.R.L.); bruno.decindio@unical.it (B.d.C.); domenico.gabriele@unical.it (D.G.)

* Correspondence: noemi.baldino@unical.it; Tel.: +39-0984-496-640

**Abstract:** The aim of the present study was to evaluate the possibility to extract, by supercritical fluids, nutraceuticals as polyphenolic compounds, able in the prevention and in the treatment of a series of chronic-degenerative diseases, from plant matrices like the cactus pear. Supercritical fluid technology is an innovative method to extract nutraceuticals from natural matrices. This method offers numerous advantages that include the use of moderate temperatures, solvents with good transport properties (high diffusivity and low viscosity), and cheap and nontoxic fluids. Fresh cladodes from two different wild ecotypes of *Opuntia ficus-indica* (L.) Mill. were extracted both with methanol and with SFE-$CO_2$ using different samples preparations, to maximize the % yields and the selectivity of extraction of polyphenols. The biggest contents of phenolics, evaluated by *Folin-Ciocalteu* assay, has been observed with the sample dehydrated of *O. ficus-indica* cultivar that shows, as well, the best yield % (m/m) of extraction with both methanol and SFE-$CO_2$. Better results were obtained with the samples of *O. ficus-indica* cult. (*OFI cult.*), in spite of the *O. ficus-indica* s.l. (*OFI s.l.*); the two different ecotypes of *OFI* showed dissimilar phytochemicals profile. We noticed that the reduction of both quantity and quality of polyphenols was drastic with the increase of pressure at 250 bar; this shows that high pressures result in a loss of bioactive principles, like polyphenols. By changing the variables of extraction processes with SFE-$CO_2$ and by varying the preventive treatments of the natural matrices, it was possible to increase the selectivity and the purity of the products. Thus, the optimization of this useful and green technique allowed us to increase the value of the *Opuntia* cladodes, a by-product very diffused in Calabria, which is an extraordinary source of nutraceuticals. These extracts could be used directly as functional foods or as starting material in the pharmaceutical, nutraceutical or cosmetic companies; they are safe and without any solvents traces and it is possible to obtain it in a few hours respect to the conventional extraction that requires longer extraction time.

**Keywords:** cladodes; *Opuntia*; nutraceuticals; antioxidants; polyphenols; SFE-$CO_2$; rutin; iso-quercitrin; nicotiflorin; narcissin



**Citation:** Iacopetta, D.; Baldino, N.; Caruso, A.; Perri, V.; Lupi, F.R.; de Cindio, B.; Gabriele, D.; Sinicropi, M.S. Nutraceuticals Obtained by SFE-$CO_2$ from Cladodes of Two *Opuntia ficus-indica* (L.) Mill Wild in Calabria. *Appl. Sci.* **2021**, *11*, 477. https://doi.org/10.3390/app11020477

Received: 10 December 2020
Accepted: 2 January 2021
Published: 6 January 2021



## 1. Introduction

Nowadays, the consumption of natural extracts from plants and seeds present in nature and the discovery of the many benefits connected to their intake, have favoured and promoted research in the field of extraction and use.

For this purpose, in recent years, extract of plant parts (seeds, cladodes, stem, fruits, etc.) of the family *Opuntiaceae* have been investigated, as for instance: *Opuntia joconostle* seeds for cholesterol-lowering properties [1]; *Opuntia humifusa* cladodes, for cytotoxic activity against the human breast cancer cell lines MCF-7 and human colon SW-480 cells, and stem and fruits able to inhibit the growth of U87MG glioblastoma cells [2]; *Opuntia stricta* for antioxidant, anti-inflammatory and cytotoxic activities [3,4]. Instead, our attention was focused on the cladodes of the cactus pear (*Opuntia ficus-indica* (L.)) Mill., a tropical or subtropical invasive plant, widespread and well adapted to Mediterranean area due to the climate.

*Opuntia* was an important source of the agricultural economy and diet of the ancient Mexicans of the Aztec empire [5] and it was also traditionally used to differentiate the properties of the peasants and to curb fires; the cladodes have high water content, 95% in mass [6].

In some countries, like Mexico, Italy and South Africa, the most commercial variety *Opuntia ficus-indica* (L.) Mill. is cultivated on the considerable surface for industrial purposes; in Mexico, its culture stretches on a surface of 300,000 ha [7]. The immature cladodes of Opuntia ficus indica are a part of the edible plant and also used in Mexican cuisine for many traditional dishes and in folk medicine for certain diseases, such as obesity, gastrointestinal disorders [8], ulcers and healing of wounds [8].

*Opuntia ficus indica* fruits are used as a laxative in Turkey, to reduce kidney stones, rheumatism pains, and as a sedative [9]. It was also been reported that the extracts of fruits and stems exhibit hypoglycemic, anti-ulcer [10], anti-allergic, analgesic and antioxidant activity [11]. From the fruits and the stem extracts was isolated the $\beta$-sitosterol, an active anti-inflammatory principle [12], whereas the fruit and flower infusions significantly increase diuresis [13].

Several researches show that the prickly pear is a plant very rich in vitamins, minerals, amino acids and sugars [14]. In particular, its fruit, with a pleasant flavor, has a high content of potassium, phosphorus, magnesium, sodium, calcium, vitamins C and E, dietary fibers, pigments betalains (betacyanins and betaxanthins), polyphenols and their glycosides, glutamine, proline and taurine; the cladodes, consumed as vegetables and used also as animal feed too [15], are rich of mucilages, pectins, sterols, vitamins and polyphenols.

The blades of the prickly pear are however very perishable and can be stored for about 5 days at room temperature or up to 10 days in refrigerated environments [16]. They contain high quantities of pectins and fibers that are capable of increasing fecal mass and intestinal motility, which therefore improve plasma levels of cholesterol and glucose. [17].

The young cladodes of OFI, also known as nopalitos, contain functional polyphenols (Figure 1), like Ruthin, Iso-quercitrin, Narcissin and Nicotiflorin [18] and a series of polysaccharides with high molecular weight and important functional properties, as rheological [19], medicinal and nutritional [20].

Rutin is found in many fruits and vegetables and has been used in over 130 therapeutic medicinal preparations that have been registered as drugs worldwide [21]. In literature, it has been found that Rutin improves the memory of mice in Alzheimer's as it reduces the levels of A$\beta$-oligomer and attenuates oxidative stress and neuroinflammation. [22].

Isoquercitrin, as glycosylated flavonoid, is rapidly absorbed and transformed, in the gastro-intestinal tract, in glucuronidated Quercetin [23,24] that was found to be the major form in plasma after oral administration of Isoquercitrin in rats [25]. Nicotiflorin, as well, shows protective effects on memory dysfunction in multi-infarct dementia model rats [26]. Narcissin is known for its biological proprieties, such as hepatoprotective, antioxidant, anti $\alpha$-glucosidase, and cytotoxicity against human myelogenous erythroleukemia cells [27].

In general, flavonoids can be found in many varieties of fruit, vegetables and cereals [28–31]. They play a very important role in a series of pathologies, both in their control and in their prevention, such as in Alzheimer's, cancer, etc., perhaps because they act eliminating free radicals. [32–36].

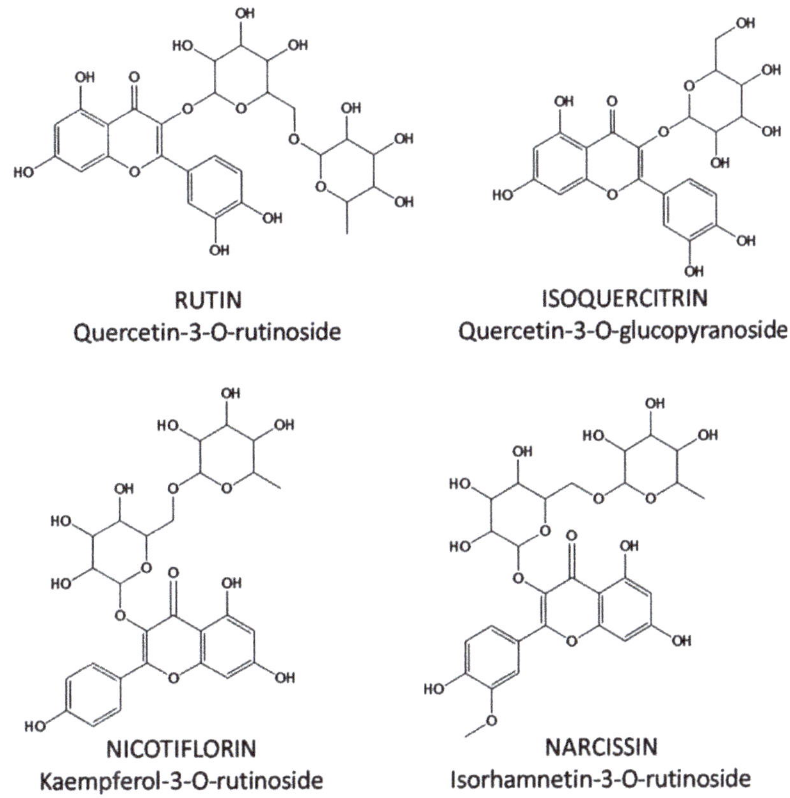


**Figure 1.** Polyphenols present in cladodes of *Opuntia ficus-indica*.

Conventionally the used techniques to obtain polyphenols extracts, such as Soxhlet, maceration [37], organic solvent extraction [38–40], autoclave treatment and microwave [37] or ultrasound-assisted extraction [39–44], usually require several hours or days and spending a large volume of solvents, frequently toxics, and with problematic garbage disposal. The correct separation between solute and solvent is important, as a possible degradation of the thermolabile components would compromise the benefits of the extract. Furthermore, any residual solvent could lower the quality and quantitative yield of the extraction.

Instead, for the extraction of polyphenols from cladodes, in this work it was used a green extraction technique: the supercritical extraction (SFE) with $CO_2$ [37], generally used to obtain oils from natural products [45–48].

This method has been used for the purpose of investigating the possibility of phytochemicals extraction without any solvent and to optimize the selectivity between phenolics and other substances in the cladodes of two ecotypes of *Opuntia ficus-indica* wild in Calabria, *O. ficus-indica cult.* (*OFI cult.*) and *O. ficus-indica s.l.* (*OFI s.l.*).

Supercritical fluid technology is the most innovative method for preparing bioactive products from plants used as supplements for functional foods [49] and it results as promising technology both in food farming and pharmaceutical industry. Supercritical fluids have high solvation capabilities, which are similar to those of liquids, as well as diffusion properties similar to those of gases. For these reasons, the extraction using supercritical fluids is particularly suitable for the extraction of biocompounds from plant matrices [50].

Although there are several usable supercritical fluids, supercritical $CO_2$ is the most commonly used. This application success is due to its low critical constants (Tc = 31.1 °C, Pc = 7.38 MPa), to the fact that it is non-toxic and non-flammable and because it is available at high purity at low cost. Moreover, other advantages are a high diffusion coefficient and low viscosity; being gas at atmospheric condition, the $CO_2$ immediately seeps out when brought to the environment, so the products obtained are free from the "extraction solvent" and thermal degradation compounds.

The extracts obtained by SFE-$CO_2$ are also considered as *Generally Recognized As Safe* (GRAS) for the *American Food and Drug Administration*, being possible to add them to all food without undesirable effects for health. Some data have also been reported on the application of SFE-$CO_2$ from vegetable by-products [51].

Today, new chemistry knowledge at the molecular level regarding the functional and structural properties of active principles may allow a better selection of the products and extracts that can satisfy the request of the market, also according to the specific needs of both food and pharmaceutical industries [52]. Thus, with this innovative and advantageous technique, we extracted the Rutin, Isoquercitrin, Nicotiflorin and Narcissin from hard to manipulate plant matrices, namely the cladodes, with the aim to use them as safe supplements in the food and pharmaceuticals companies.

## 2. Materials and Methods

### 2.1. Plant Material

Two cladodes ecotypes of *Opuntia ficus-indica* (L.) Mill. (Figure 2) were analysed in this study: one supposed to be a cultivar with the hybrid origin, which often escapes, from cultivation and behave like an invasive species, it is almost spineless and it was accepted with the name *Opuntia ficus-indica cult.* (*OFI cult.*); and the second one with white and hard spines, long about 3–4 cm, sharing some characters with *Opuntia amyclaea* Ten., *O. maxima* Miller and *O. ficus-barbarica*. Hereafter we will use for the spiny ecotype the name *Opuntia ficus-indica* (L.) Mill. *s.l.* (*OFI s.l.*), because the studied local area lacks of taxonomic and nomenclatural study of the genus *Opuntia*.

**Figure 2.** Two ecotypes of *Opuntia ficus-indica* (L.) Mill cladodes: (**A**) *Opuntia ficus indica cult.* and (**B**) *Opuntia ficus-indica* (L.) Mill s.l.

The cladodes from both *O ficus-indica* were collected between June–August 2012 in Calabria (South of Italy). All the cladodes, covered by spines and multicellular hairs or trichomes, were manually cleaned, cut into small pieces and then homogenized (Figure 3) with a TYPE HR 2064 PHILIPS-600 W and frozen until analysis.

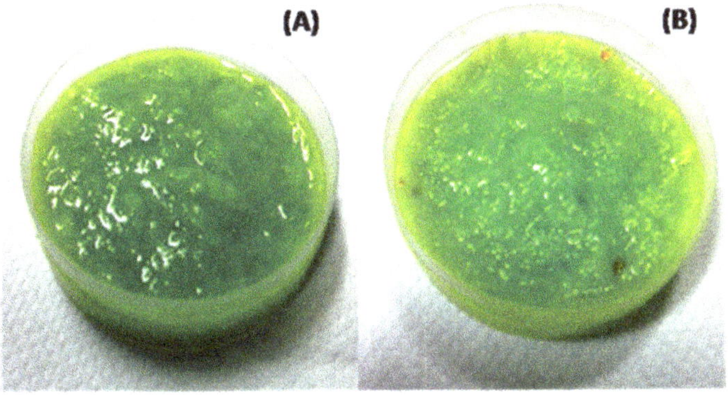


**Figure 3.** Homogenized from fresh samples of *OFI cult.* (**A**) and *OFI s.l.* (**B**).

*2.2. Chemicals*

Solvents as methanol (analytical grade and for HPLC), trifluoroacetic acid, water (analytical grade and for HPLC) and reagents such as *Folin-Ciocalteu* and Chlorogenic acid were purchased from Sigma-Aldrich (St. Louis, MO, USA).

Ottawa Sand and Spe-ed$^{TM}$ PSE Matrix-Hydroscopic Samples Dispersing Agent (Diatomee Sand), used as SFE-$CO_2$ accessories, were purchased from Labservice Analytica S.R.L. (Italy).

Standards, for HPLC analysis and quantification, including Rutin, Isoquercitrin, Nicotiflorin and Narcissin were purchased from Extrasynthese (Lyon, France).

*2.3. Preliminary Analysis*

The pH value of the homogenized system (pH 700, Eutech Instruments, Germany), the activity level (Novasina AW Sprint–TH 500, Switzerland), the Brix degrees (ATAGO, Hand Refractometer N Type Series, Japan) and humidity (Mettler Toledo Moisture Analyzer HB43-S, Switzerland) were measured. The analyses were repeated in triplicate and the results are reported as mean value and standard deviation.

*2.4. Exhaustive Extraction and Total Phenolics Content*

The fresh cladodes of *Opuntia ficus-indica* both ecotypes were extracted with methanol (48 h × 3 times) at 4 °C. The same procedures were followed for a sample of *OFI cult.* and *OFI s.l.* Moreover, the exhaustive analysis was performed on samples dried in the oven for 360 min at 32 ± 1 °C; removing the 30% of the weight.

The extraction solutions were filtered in synthetic cloth, concentrated and dried under *vacuum* at 35 ± 1 °C for the thermolability of the polyphenols compounds.

The total phenolic content of the cladodes extracts was quantified using *Folin-Ciocalteu* reagent and chlorogenic acid used as standards. The absorbance was measured at 726 nm (Perkin Elmer Lambda 40 UV/VIS spectrophotometer) and the total content was expressed (mean ± S.D. of three determinations) as mg of chlorogenic acid equivalents on 100 g of fresh raw material (Singleton Rossi, 1965).

*2.5. Extraction of Polyphenols with Supercritical Fluids*

The supercritical extractions by $CO_2$ were performed on a *Spe-ed* SFE 4 extractor (Applied Separation, Allentown, PA, USA) and following the necessary steps: loading in the 50 mL stainless steel vessel, pressurization with $CO_2$ and waiting for a transitory state of temperature, which is kept constant by an oven module. After the reaching of target temperature and pressurization, in order to guarantee the intimate contact between the $CO_2$ and the matrix, the system was kept closed for 20 min (static phase), and only after this time the micrometric valve was opened until a constant flow of 1 was reached. 5 L/min (dynamic phase), measured by a ball float rotameter. A graphical scheme of *Spe-ed* SFE 4 extractor is given in Figure 4.

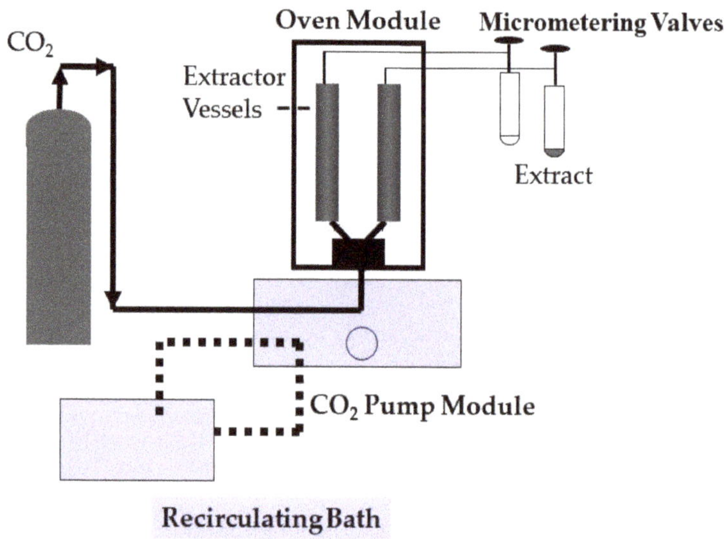


**Figure 4.** Scheme of Spe-ed SFE 4 extractor unit.

The extract left the vessel through a valve, which was in advance thermostated to avoid $CO_2$ solidification, due to the expansion. Experiments were carried out at 40 °C, 110 bar and 250 bar by varying the preventive sample preparation with PSE accessories, as Ottawa Sand and an Spe-ed$^{TM}$ PSE Matrix (Hydroscopic Samples Dispersing Agent). All the extracts were stored at −18 °C before HPLC analysis.

This extraction method has always attracted because it is safe and eco friendly. In addition, the supercritical fluid, in the specific case $CO_2$, has properties similar to those of gas in the supercritical state so that it can extract in a very short time compared to the classic extraction types, returning an extract already separated from the solvent after depressurization. Supercritical fluids have a high diffusion coefficient and a low viscosity which favours their intimate contact with the matrix. On the other hand, however, this method has limits, especially in relation to the use of the supercritical solvent because $CO_2$ has low polarity. Therefore this type of extraction technique is generally used for the extraction of oils, fats or in any case polar substances. Only the use of other solvents, such as ethanol or water, increases the solvent power of $CO_2$ in the supercritical state.

The different methods to prepare the samples and pressure are listed in Table 1.

**Table 1.** Different preparation of samples before SFE separation.

| Sample | Ref. | Preventive Preparation | P (bar) |
|---|---|---|---|
| *OFI cult.* | CULT_20D | 20% Diatomee Sand | 110 |
| *OFI cult.* | CULT_20P | 20% Diatomee Sand | 250 |
| *OFI cult.* | CULT_90E_20O | Dried until 90% + 20% Ottawa Sand | 110 |
| *OFI cult.* | CULT_30E_20O | Dried until 30% + 20% Ottawa Sand | 110 |
| *OFI cult.* | CULT_C_20D | Centrifuged + 20% Diatomee | 110 |
| *OFI s.l.* | SL_30E_20D | Dried until 30% + 20% Diatomee Sand | 110 |
| *OFI s.l.* | SL_90E_20O | Dried until 90% + 20% Ottawa Sand | 110 |
| *OFI s.l.* | SL_90E_20D | Dried until 90% + 20% Diatomee Sand | 110 |

All samples were cleaned, homogenized and frozen after collection, to be subsequently analysed.

Some samples analysed were centrifuged after defrosting at 3500 rpm for 10 min and only the precipitated part was used to extract the polyphenols while other samples were partially dried at 32 °C in the oven to reduce the amount of water.

The extraction time was 1 h. The final extracts were collected in glass tubes covered with an aluminium foil and frozen until analysis. Each extraction was carried out in duplicate.

### 2.6. HPLC Analysis

The analysis of all the extracts, both from solvent extraction and belonged from SFE, was carried out by means of HPLC using a Smartline HPLC system (Knauer, Germany). Chromatographic separation was carried out using a 2.0 mm ID × 150 mL, with pre-column, C-18 TSKgel ODS-100 V, 21810 (TOSOH BIOSCIENCE), both thermostatically at 40 °C. The operative conditions: mobile phase, flow rate and gradient of elution utilized are reported in Table 2. For the mobile phase, methanol and water with trifluoroacetic acid (TFA) were used.

**Table 2.** The mobile phase, flow rate and gradient of elution utilized.

| Time (min) | Methanol [%] | Water + 0.1% TFA [%] | Flow (mL/min) |
|---|---|---|---|
| 0 | 0 | 100 | 0.2 |
| 2 | 20 | 80 | 0.2 |
| 55 | 100 | 0 | 0.2 |
| 65 | 0 | 100 | 0.2 |

Absorbance spectra were recorded every 2 s, between 200 and 450 nm, with a bandwidth of 4 nm, and chromatograms were acquired at 254 and 280 nm. HPLC analysis was performed in duplicate.

A wavelength of 280 nm was used for quantification [18], while the calibration line was obtained from the integration of the absorption peaks obtained from a series of dilutions of Rutin, Isoquercitrin, Nicotiflorine and Narcissin.

## 3. Results and Discussion

### 3.1. Preliminary Analysis

Cladodes by *Opuntia* genus showed a weakly acid pH and this allows easier conservation of the homogenized system. The results are shown in Table 3.

**Table 3.** Level of acidity, water activity, Brix degrees and humidity of fresh samples of *FI cult.* and *OFI s.l.*

| Plant | pH (-) | $a_W$ (-) | °Brix (°B) | Humidity (*w*/*w*) |
|---|---|---|---|---|
| *OFI cult.* | 4.45 ± 0.03 | 0.935 ± 0.03 | 6.5 ± 0.1 | 96.7 ± 1.5 |
| *OFI s.l.* | 4.50 ± 0.02 | 0.933 ± 0.02 | 7.8 ± 0.2 | 96.8 ± 1.3 |

The higher amount of saccharides was found in the sample of *Opuntia ficus-indica* s.l., than that one of *Opuntia ficus indica* cult., and this is possible due to different factors such as the variety and age of the plant, soil, and climate. The pH was measured, because it can influence the viscosity of the mucilage and it could hamper the extraction, and it is similar for the two types of plant. The water activity and humidity values are quite similar, guarantying the same times of exposition to airflow and heat during the dehydration.

### 3.2. Exhaustive Extraction, Total Phenolics Content and HPLC Analysis

The high content of total phenolics was observed with the *Folin–Ciocalteu* assay in the samples dried in the oven, Cult_AD and Sl_AD; these show as well the better quantitative yield of extraction, calculated using the following equation [45]:

$$Yield\ \% = \frac{[W_{d.e.}\ (g)]}{[W_m\ (g)]} * 100 \quad (1)$$

where $W_{d.e.}$ is the weight of the dry extract obtained and $W_m$ the mass of the plant macerated. The data for the yield of extraction at different sample preparation and the total phenolics content are shown in Table 4.

**Table 4.** Data for the yield of extraction at different sample preparation and total phenolic content expressed as mg of phenols on 100 g of fresh raw material.

| Plant and Time of Collection | Ref. | Sample Preparation | Yield of Methanolic Extraction (%w/w) | Total Phenolics Content (mg/100 g Raw Material) |
|---|---|---|---|---|
| *OFI cult.* (June) | Cult_J | Fresh macerated | 2.79 ± 0.29 | 170.01 ± 2.98 |
| *OFI cult.* (August) | Cult_AD | Dried macerated | 5.01 ± 0.63 | 320.13 ± 6.54 |
| *OFI cult.* (August) | Cult_AF | Fresh frozen macerated | 2.14 ± 0.22 | 170.01 ± 2.65 |
| *OFI s.l.* (August) | Sl_AD | Dried macerated | 4.39 ± 0.48 | 414.90 ± 9.58 |
| *OFI s.l.* (August) | Sl_AF | Fresh frozen macerated | 2.19 ± 0.25 | 221.29 ± 5.93 |

In Table 4 it is possible to point out that samples dried give a higher yield of extraction and total phenolics content with respect to the fresh macerated ones, showing that the total polyphenol contents vary depending on the type of treatment and extraction.

Comparing the same fresh sample collected in two different months, it is possible to observe that the extraction yield was slightly higher in the sample Cult_J with respect to the Fresh frozen macerated of August, but the total phenolic content is almost unchanged between samples of Cult_J and Cult_AF. This indicates that the different maturation period does not influence the content of secondary metabolites, such as phenols.

Moreover, the total phenolic content is higher in the dried samples Cult_AD and Sl_AD, probably because drying process is designed to dehydrate the matrix in order to stop the common enzymatic processes; the aqueous environment of the cytoplasm of plant cells could damage the active compounds [52].

The highest content of total phenols was found in the sample of Sl_AD, perhaps because being a wild ecotype is less affected by climate change, more adaptable than a cultivated plant.

In Figure 5 are reported the data, obtained by the HPLC analysis, of extractions with solvent.

Rutin is one of the polyphenols presents in greater amount in the analysed species. Moreover, it is possible to observe that by dring the sample it is possible to obtain extracts with a greater amount of some polyphenols (Cult_AD and Sl_AD).

As reported in literature rutin is already described to be present in cladode extracts of different Opuntia species [8,20].

Isoquercitrin is another bioactive compound present in our extract in similar quantity in the two species analysed while the quantity of Nicotiflorin and Narcissin are very low in the extracts. The different exposure to the sun or climate change can cause the secondary metabolism of the plant. As a consequence, the antioxidant profile can be influenced.

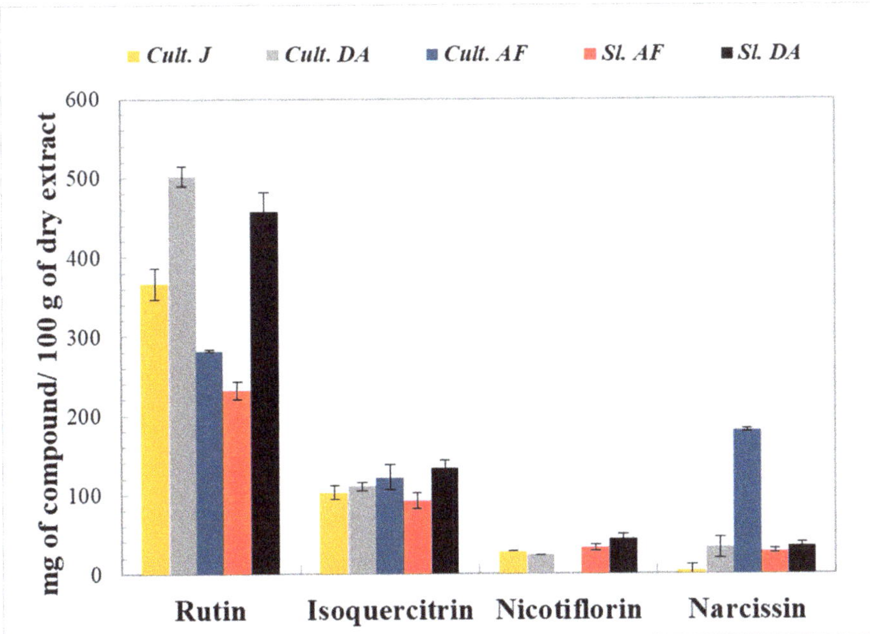


**Figure 5.** Quantification in mg of standards in 100 g of dry extract.

*3.3. Extraction with Supercritical Fluids and HPLC Analysis*

The yields extractions are calculated by using the following equation according to Yieddes et al. [45]:

$$Yield\ \% = \frac{[W_{extract}\ (g)]}{[W_{load}\ (g)]} * 100$$

where $W_{extract}$ is the weight of the extract obtained, which is the difference between the mass of glass trap with extract and the mass of empty glass trap and $W_{load}$ is the mass of sample load in the column before the extraction. The results are reported in Table 5; instead, the results of HPLC quantification on mg of the compound for 100 g of material loaded (mean ± S.D. of two determinations) for *OFI cult.* are shown in Figure 6, instead of *OFI s.l.* in Figure 7.

**Table 5.** Yield % of extraction (mean ± S.D. of two determinations) of different samples by SFE-$CO_2$.

| Plant | Ref. | Yield of Extraction (%) |
|---|---|---|
| *OFI cult.* | CULT_20D | 0.48 ± 0.07 |
| *OFI cult.* | CULT_20P | 1.00 ± 0.13 |
| *OFI cult.* | CULT_90E_20O | 0.88 ± 0.11 |
| *OFI cult.* | CULT_30E_20O | 2.14 ± 0.35 |
| *OFI cult.* | CULT_C_20D | 1.81 ±0.15 |
| *OFI s.l.* | SL_30E_20D | 0.10 ± 0.03 |
| *OFI s.l.* | SL_90E_20O | 0.25 ± 0.05 |
| *OFI s.l.* | SL_90E_20D | 0.02 ± 0.01 |

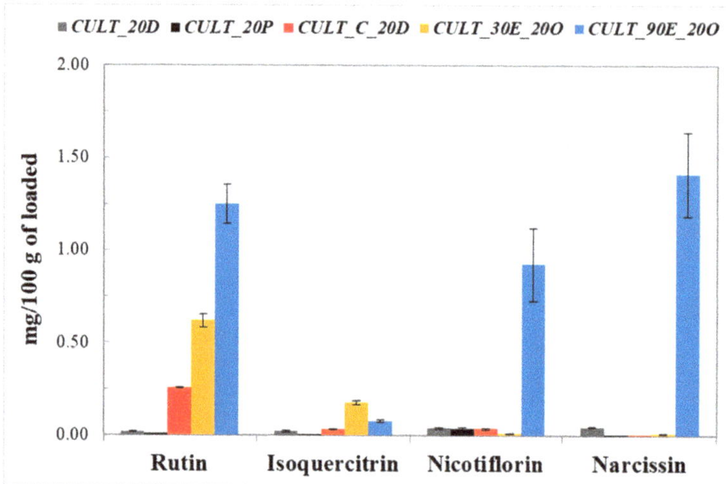


**Figure 6.** Quantification in mg of four different polyphenols for 100 g of loaded material for *OFI cult*.

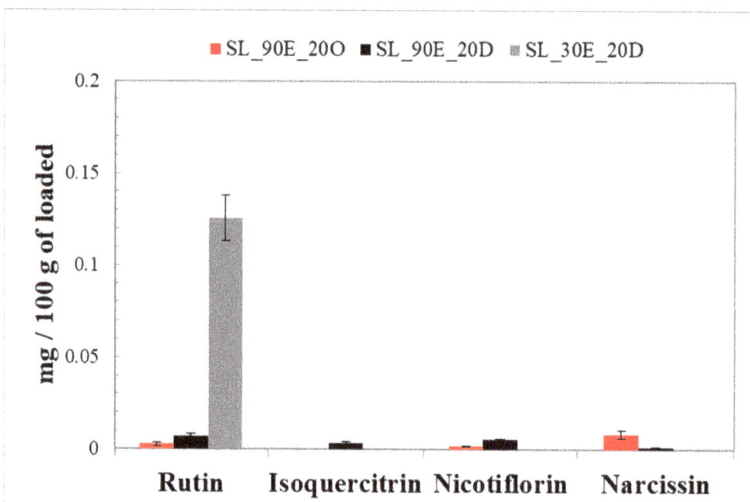


**Figure 7.** Quantification in mg of four different polyphenols for 100 g of loaded material for *OFI s.l.*

The yields of extraction with SFE-$CO_2$ are lower than that one with solvent as well as the amount of polyphenols, but the resulting extracts do not need to be separated from solvent, they are purer and cleaner.

It is possible to observe that the better results were obtained with the samples of *OFI cult.* despite the *OFI s.l.* In particular, the best yield of extraction was 2.14% $\pm$ 0.35, obtained with the sample CULT_30E_20O of *OFI cult* dehydrated at 30% and added with 20% (of the total weight loaded for the extraction) of Ottawa Sand at 110 bar. Given the results obtained from the extractions, it is possible to assume that Ottawa sands, hydrophobic natural silica particles, guarantee a better dispersion than diatomaceous earth, which also acts as a drying agent, being hydrophilic. The Ottawa sands, therefore, by favoring a homogeneous dispersion, under the same initial conditions (type of sample and initial drying conditions), guarantee a tighter contact between the matrix and the supercritical fluid. At the increase of the dehydration to 90% for the sample CULT_90E_20O, a decrease in the yields' % follows, but an improvement is obtained concerning the selective extraction of polyphenols. In this sample, it is possible to observe the highest quantities of Rutin, Narcissin and Nicotiflorin, in accordance with the extractions by maceration, where the greatest quantity of polyphenols was obtained for the samples first dried and then macerated in methanol (Table 4).

The treatment of *OFI cult.* with 20% of Diatomee Sand (sample CULT_20D) implies a reduction of both quantity and selectivity of extraction of polyphenols, perhaps because the Diatomee Sand is hygroscopic. This sand absorbs water from the surrounding environment reducing the function of co-solvent of water naturally present in the matrix.

Being polyphenols polar compounds, the pressure was then raised to 250 bar for the sample CULT_20P, maintaining the same sand, trying to increase the polarity of $CO_2$, but the effect of the sand was stronger. It was obtained a reduction of both yield % and selectivity of polyphenols; this shows that high pressures result in a loss of bioactive principles sensitive as polyphenols.

When the sample is centrifuged (sample CULT_C_20D) to remove water, it was obtained a good increase of the yield % of extraction and a little improvement on the selectivity of polyphenols.

The yields % of extractions of the *OFI s.l.* were very low, ranging between 0.02 and 0.25%; but the selectivity and purity of the extracts derived therefrom are very high, as can be seen in Figure 8 for the sample SL_30E_20D.

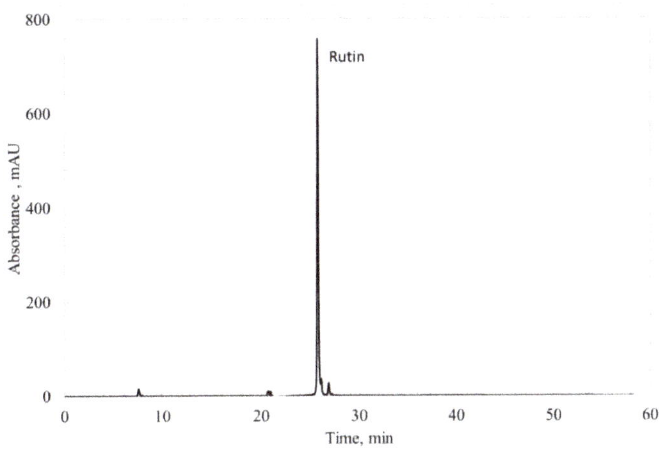


**Figure 8.** Chromatogram resulting from the analysis of the extract by SFE-$CO_2$ of *OFI s.l.* dried to 30% and added 20% of Diatomee Sand.

## 4. Conclusions

There is a global trend toward the use of natural flavonoids present in fruits, vegetables, oilseeds and herbs as antioxidants or functional foods.

In the present work, two different ecotypes of *O. ficus-indica,* showing dissimilar phytochemicals profile, were studied. The total phenolic content in the extracts with solvents is almost unchanged between the fresh samples collected in June and in August. This indicates that the different maturation period, if the plant still did not produce the fruit, does not interfere with the content of secondary metabolites, such as phenols, that is almost the same. The polyphenol present in greater amount in all species is Rutin; the dried samples (Cult_AD and Sl_AD) contain a greater amount than the fresh.

Being polyphenols polar compounds, the yields % quantitative of extraction with SFE-$CO_2$ are not high even if the extractions are very selective and the extracts nearly pure.

Concerning the SFE-$CO_2$ better results were obtained with the samples of *OFI cult.* in spite of the *OFI s.l.*, probably because the ecotype *OFI s.l.* was much mucilaginous, impeding to complete some extractions.

The better SFE-$CO_2$ extraction results were obtained with samples preventively dried, as the sample CULT_90E_20O. Thus, the reduction of water, responsible for the degradation of active compounds, in presence of the Ottawa sand, produces highly selective extracts.

The treatment with Diatomee Sand implies a reduction of both quantity and selectivity of extraction of polyphenols, perhaps because the Diatomee Sand is hygroscopic and it adsorbs water from the surrounding environment, reducing the function of co-solvent of the water naturally present in the matrix. By changing the % hydration, it is possible to modulate the selectivity of extraction.

The polyphenols are polar compounds, therefore the pressure was increased to 250 bar to increase the polarity of $CO_2$. Nevertheless, a reduction of both the % yield quantitative and selectivity of extraction was obtained, demonstrating that the high pressures result in a loss of bioactive principles, like polyphenols. Our results evidenced as, for the first time, by improving the parameters of SFE-$CO_2$, which is an innovative, safe and cheap technique, it is possible to extract qualitatively and quantitatively the polyphenolic fraction from plant matrices hard to treat, namely the *Opuntia* cladodes. These results increase their value as good nutraceuticals sources, instead to be considered only a by-product from prickly pear cultivation in Calabria.

Finally, the cladodes extracts could be used as food supplements or starting materials in the pharmaceuticals industries.

**Author Contributions:** Conceptualization, B.d.C.; methodology, N.B.; validation, D.G.; formal analysis, F.R.L., V.P. and A.C.; investigation, N.B. and V.P.; writing—original draft preparation, D.I. and N.B.; writing—review and editing, D.I. and M.S.S.; supervision, M.S.S. All authors have read and agreed to the published version of the manuscript.

**Funding:** This research received no external funding.

**Institutional Review Board Statement:** Not applicable.

**Informed Consent Statement:** Not applicable.

**Data Availability Statement:** Data is contained within the article.

**Acknowledgments:** The authors are grateful to the Dimitar Uzunov for the collection of the plants material and the identification of species. We also thank the Laboratory of Rheology and Food Engineering of University of Calabria, the Department of Informatics, Modeling, Electronics and Systems Engineering and the Department of Pharmacy, Health and Nutritional Sciences-University of Calabria.

**Conflicts of Interest:** The authors declare no conflict of interest.

## References

1. Osorio-Esquivel, O.; Ortiz-Moreno, A.; Garduno-Siciliano, L.; Alvarez, V.B.; Hernandez-Navarro, M.D. Antihyperlipidemic effect of methanolic extract from *Opuntia joconostle* seeds in mice fed a hypercholesterolemic diet. *Plant Foods Hum. Nutr.* **2012**, *67*, 365–370. [CrossRef] [PubMed]
2. Kim, J.; Jho, K.H.; Choi, Y.H.; Nam, S.Y. Chemopreventive effect of cactus (*Opuntia humifusa*) extracts: Radical scavenging activity, pro-apoptosis, and anti-inflammatory effect in human colon (SW480) and breast cancer (MCF7) cells. *Food Funct.* **2013**, *4*, 681–688. [CrossRef] [PubMed]
3. Izuegbuna, O.; Otunola, G.; Bradley, G. Chemical composition, antioxidant, anti-inflammatory, and cytotoxic activities of *Opuntia stricta* cladodes. *PLoS ONE* **2019**, *14*, e0209682. [CrossRef] [PubMed]
4. Surup, F.; Minh Thi Tran, T.; Pfutze, S.; Budde, J.; Moussa-Ayoub, T.E.; Rohn, S.; Jerz, G. Opuntisines, 14-membered cyclopeptide alkaloids from fruits of *Opuntia stricta* var. *dillenii* isolated by high-performance countercurrent chromatography. *Food Chem.* **2021**, *334*, 127552. [CrossRef] [PubMed]
5. Sibhatu, K.T.; Qaim, M. Rural food security, subsistence agriculture, and seasonality. *PLoS ONE* **2017**, *12*, e0186406. [CrossRef]
6. Del Socorro Santos Diaz, M.; Barba de la Rosa, A.-P.; Helies-Toussaint, C.; Gueraud, F.; Negre-Salvayre, A. *Opuntia* spp.: Characterization and Benefits in Chronic Diseases. *Oxid. Med. Cell. Longev.* **2017**, *2017*, 8634249. [CrossRef]

7. Mounir, B.; Younes, E.G.; Asmaa, M.; Abdeljalil, Z.; Abdellah, A. Physico-chemical changes in cladodes of *Opuntia ficus-indica* as a function of the growth stage and harvesting areas. *J. Plant Physiol.* **2020**, *251*, 153196. [CrossRef]
8. Santos-Zea, L.; Gutierrez-Uribe, J.A.; Serna-Saldivar, S.O. Comparative analyses of total phenols, antioxidant activity, and flavonol glycoside profile of cladode flours from different varieties of *Opuntia* spp. *J. Agric. Food Chem.* **2011**, *59*, 7054–7061. [CrossRef]
9. Aboura, I.; Nani, A.; Belarbi, M.; Murtaza, B.; Fluckiger, A.; Dumont, A.; Benammar, C.; Tounsi, M.S.; Ghiringhelli, F.; Rialland, M.; et al. Protective effects of polyphenol-rich infusions from carob (*Ceratonia siliqua*) leaves and cladodes of *Opuntia ficus-indica* against inflammation associated with diet-induced obesity and DSS-induced colitis in Swiss mice. *Biomed. Pharm.* **2017**, *96*, 1022–1035. [CrossRef]
10. Akkol, E.K.; Ilhan, M.; Karpuz, B.; Genc, Y.; Sobarzo-Sanchez, E. Sedative and Anxiolytic Activities of *Opuntia ficus indica* (L.) Mill.: An Experimental Assessment in Mice. *Molecules* **2020**, *25*, 1844. [CrossRef]
11. Lee, M.H.; Kim, J.Y.; Yoon, J.H.; Lim, H.J.; Kim, T.H.; Jin, C.; Kwak, W.-J.; Han, C.-K.; Ryu, J.H. Inhibition of nitric oxide synthase expression in activated microglia and peroxynitrite scavenging activity by *Opuntia ficus indica* var. *saboten*. *Phytother. Res.* **2006**, *20*, 742–747. [CrossRef] [PubMed]
12. Kaur, M. Pharmacological actions of *Opuntia ficus indica*: A Review. *J. Appl. Pharm. Sci.* **2012**, *2*. [CrossRef]
13. Dok-Go, H.; Lee, K.H.; Kim, H.J.; Lee, E.H.; Lee, J.; Song, Y.S.; Lee, Y.-H.; Jin, C.; Lee, Y.S.; Cho, J. Neuroprotective effects of antioxidative flavonoids, quercetin, (+)-dihydroquercetin and quercetin 3-methyl ether, isolated from *Opuntia ficus-indica* var. *saboten*. *Brain Res.* **2003**, *965*, 130–136. [CrossRef]
14. Galati, E.M.; Tripodo, M.M.; Trovato, A.; Miceli, N.; Monforte, M.T. Biological effect of *Opuntia ficus indica* (L.) Mill. (Cactaceae) waste matter. Note I: Diuretic activity. *J. Ethnopharmacol.* **2002**, *79*, 17–21. [CrossRef]
15. El-Mostafa, K.; El Kharrassi, Y.; Badreddine, A.; Andreoletti, P.; Vamecq, J.; El Kebbaj, M.H.; Latruffe, N.; Lizard, G.; Nasser, B.; Cherkaoui-Malki, M. Nopal Cactus (*Opuntia ficus-indica*) as a Source of Bioactive Compounds for Nutrition, Health and Disease. *Molecules* **2014**, *19*, 14879–14901. [CrossRef] [PubMed]
16. Galati, E.; Lanuzza, F.; Occhiuto, F.; Monforte, M.; Tripodo, M.; D'Angelo, V. Antioxidant phytochemicals of *Opuntia ficus-indica* (L.) Mill. cladodes with potential anti-spasmodic activity. *Pharmacogn. Mag.* **2017**, *13*, 424. [CrossRef]
17. Piga, A.; D'Aquino, S.; Agabbio, M.; Emonti, G.; Farris, G.A. Influence of Storage Temperature on Shelf-life of Minimally Processed Cactus Pear Fruits. *Food Sci. Technol.* **2000**, *33*, 15–20. [CrossRef]
18. Galati, E.M.; Tripodo, M.M.; Trovato, A.; d'Aquino, A.; Monforte, M.T. Biological Activity of *Opuntia ficus indica* Cladodes II: Effect on Experimental Hypercholesterolemia in Rats. *Pharm. Biol.* **2008**, *41*, 175–179. [CrossRef]
19. Guevara-Figueroa, T.; Jiménez-Islas, H.; Reyes-Escogido, M.L.; Mortensen, A.G.; Laursen, B.B.; Lin, L.-W.; De León-Rodríguez, A.; Fomsgaard, I.S.; Barba de la Rosa, A.P. Proximate composition, phenolic acids, and flavonoids characterization of commercial and wild nopal (*Opuntia* spp.). *J. Food Compos. Anal.* **2010**, *23*, 525–532. [CrossRef]
20. Robert, P.; Vergara, C.; Silva-Weiss, A.; Osorio, F.A.; Santander, R.; Sáenz, C.; Giménez, B. Influence of gelation on the retention of purple cactus pear extract in microencapsulated double emulsions. *PLoS ONE* **2020**, *15*, e0227866. [CrossRef]
21. Medina-Torres, L.; Brito-Da La Fuente, E.; Torrestiana-Sanchez, B.; Katthain, R. Rheological properties of the mucilage gum (*Opuntia ficus indica*). *Food Hydrocoll.* **2000**, *14*, 417–424. [CrossRef]
22. Chua, L.S. A review on plant-based rutin extraction methods and its pharmacological activities. *J. Ethnopharmacol.* **2013**, *150*, 805–817. [CrossRef] [PubMed]
23. Xu, P.-X.; Wang, S.-W.; Yu, X.-L.; Su, Y.-J.; Wang, T.; Zhou, W.-W.; Zhang, H.; Wang, Y.-J.; Liu, R.-T. Rutin improves spatial memory in Alzheimer's disease transgenic mice by reducing Aβ oligomer level and attenuating oxidative stress and neuroinflammation. *Behav. Brain Res.* **2014**, *264*, 173–180. [CrossRef]
24. Iacopetta, D.; Grande, F.; Caruso, A.; Mordocco, R.A.; Plutino, M.R.; Scrivano, L.; Ceramella, J.; Muià, N.; Saturnino, C.; Puoci, F.; et al. New insights for the use of quercetin analogs in cancer treatment. *Future Med. Chem.* **2017**, *9*, 2011–2028. [CrossRef] [PubMed]
25. Grande, F.; Parisi, O.I.; Mordocco, R.A.; Rocca, C.; Puoci, F.; Scrivano, L.; Quintieri, A.M.; Cantafio, P.; Ferla, S.; Brancale, A.; et al. Quercetin derivatives as novel antihypertensive agents: Synthesis and physiological characterization. *Eur. J. Pharm. Sci.* **2016**, *82*, 161–170. [CrossRef] [PubMed]
26. Chang, Q.; Zuo, Z.; Chow, M.S.S.; Ho, W.K.K. Difference in absorption of the two structurally similar flavonoid glycosides, hyperoside and isoquercitrin, in rats. *Eur. J. Pharm. Biopharm.* **2005**, *59*, 549–555. [CrossRef]
27. Huang, J.-L.; Fu, S.-T.; Jiang, Y.-Y.; Cao, Y.-B.; Guo, M.-L.; Wang, Y.; Xu, Z. Protective effects of Nicotiflorin on reducing memory dysfunction, energy metabolism failure and oxidative stress in multi-infarct dementia model rats. *Pharmacol. Biochem. Behav.* **2007**, *86*, 741–748. [CrossRef]
28. Kim, S.; Yoo, G.; Oh, Y.; Yang, H.; Kim, T.; Sung, S. Efficient preparation of narcissin from *Opuntia ficus-indica* fruits by combination of response surface methodology and high-speed countercurrent chromatography. *Pharmacogn. Mag.* **2018**, *14*, 338. [CrossRef]
29. Cappello, A.R.; Dolce, V.; Iacopetta, D.; Martello, M.; Fiorillo, M.; Curcio, R.; Muto, L.; Dhanyalayam, D. Bergamot (*Citrus bergamia* Risso) Flavonoids and Their Potential Benefits in Human Hyperlipidemia and Atherosclerosis: An Overview. *Mini Rev. Med. Chem.* **2016**, *16*, 619–629. [CrossRef]
30. Ceramella, J.; Loizzo, M.R.; Iacopetta, D.; Bonesi, M.; Sicari, V.; Pellicanò, T.M.; Saturnino, C.; Malzert-Fréon, A.; Tundis, R.; Sinicropi, M.S. *Anchusa azurea* Mill. (Boraginaceae) aerial parts methanol extract interfering with cytoskeleton organization induces programmed cancer cells death. *Food Funct.* **2019**, *10*, 4280–4290. [CrossRef]

31. Fazio, A.; Iacopetta, D.; La Torre, C.; Ceramella, J.; Muià, N.; Catalano, A.; Carocci, A.; Sinicropi, M.S. Finding solutions for agricultural wastes: Antioxidant and antitumor properties of pomegranate *Akko* peel extracts and β-glucan recovery. *Food Funct.* **2018**, *9*, 6618–6631. [CrossRef] [PubMed]
32. Tundis, R.; Iacopetta, D.; Sinicropi, M.S.; Bonesi, M.; Leporini, M.; Passalacqua, N.G.; Ceramella, J.; Menichini, F.; Loizzo, M.R. Assessment of antioxidant, antitumor and pro-apoptotic effects of *Salvia fruticosa* Mill. subsp. *thomasii* (Lacaita) Brullo, Guglielmo, Pavone & Terrasi (Lamiaceae). *Food Chem. Toxicol.* **2017**, *106*, 155–164. [PubMed]
33. Pradeep, S.R.; Guha, M. Effect of processing methods on the nutraceutical and antioxidant properties of little millet (*Panicum sumatrense*) extracts. *Food Chem.* **2011**, *126*, 1643–1647. [CrossRef] [PubMed]
34. Li, R.-S.; Wang, X.-B.; Hu, X.-J.; Kong, L.-Y. Design, synthesis and evaluation of flavonoid derivatives as potential multifunctional acetylcholinesterase inhibitors against Alzheimer's disease. *Bioorganic Med. Chem. Lett.* **2013**, *23*, 2636–2641. [CrossRef]
35. Ravishankar, D.; Rajora, A.K.; Greco, F.; Osborn, H.M.I. Flavonoids as prospective compounds for anti-cancer therapy. *Int. J. Biochem. Cell Biol.* **2013**, *45*, 2821–2831. [CrossRef]
36. Rizza, P.; Pellegrino, M.; Caruso, A.; Iacopetta, D.; Sinicropi, M.S.; Rault, S.; Lancelot, J.C.; El-Kashef, H.; Lesnard, A.; Rochais, C.; et al. 3-(Dipropylamino)-5-hydroxybenzofuro[2,3-f] quinazolin-1(2H)-one (DPA-HBFQ-1) plays an inhibitory role on breast cancer cell growth and progression. *Eur. J. Med. Chem.* **2016**, *107*, 275–287. [CrossRef]
37. Caruso, A.; Barbarossa, A.; Tassone, A.; Ceramella, J.; Carocci, A.; Catalano, A.; Basile, G.; Fazio, A.; Iacopetta, D.; Franchini, C.; et al. Pomegranate: Nutraceutical with Promising Benefits on Human Health. *Appl. Sci.* **2020**, *10*, 6915. [CrossRef]
38. Khaled, S.; Dahmoune, F.; Madani, K.; Urieta, J.S.; Mainar, A.M. Supercritical fractionation of antioxidants from algerian *Opuntia ficus-indica* (L.) Mill. seeds. *J. Food Process. Preserv.* **2019**, *44*. [CrossRef]
39. Chbani, M.; Matthäus, B.; Charrouf, Z.; El Monfalouti, H.; Kartah, B.; Gharby, S.; Willenberg, I. Characterization of Phenolic Compounds Extracted from Cold Pressed Cactus (*Opuntia ficus-indica* L.) Seed Oil and the Effect of Roasting on Their Composition. *Foods* **2020**, *9*, 1098. [CrossRef]
40. Han, E.H.; Lim, M.K.; Lee, S.; Lee, S.H.; Yun, S.M.; Yu, H.-J.; Ryu, S.-H.; Lim, Y.-H. Efficacy of Ethanolic Extract of *Opuntia ficus-indica* var. *saboten* Stems for Improving Cognitive Function in Elderly Subjects 55–85 Years of Age: A Randomized, Double-Blind, Placebo-Controlled Study. *J. Med. Food* **2020**, *23*, 1146–1154. [CrossRef]
41. Missaoui, M.; D'Antuono, I.; D'Imperio, M.; Linsalata, V.; Boukhchina, S.; Logrieco, A.F.; Cardinali, A. Characterization of Micronutrients, Bioaccessibility and Antioxidant Activity of Prickly Pear Cladodes as Functional Ingredient. *Molecules* **2020**, *25*, 2176. [CrossRef] [PubMed]
42. Melgar, B.; Dias, M.I.; Barros, L.; Ferreira, I.C.F.R.; Rodriguez-Lopez, A.D.; Garcia-Castello, E.M. Ultrasound and Microwave Assisted Extraction of *Opuntia* Fruit Peels Biocompounds: Optimization and Comparison Using RSM-CCD. *Molecules* **2019**, *24*, 3618. [CrossRef] [PubMed]
43. Espino-Manzano, S.O.; León-López, A.; Aguirre-Álvarez, G.; González-Lemus, U.; Prince, L.; Campos-Montiel, R.G. Application of Nanoemulsions (W/O) of Extract of *Opuntia oligacantha* C.F. Först and Orange Oil in Gelatine Films. *Molecules* **2020**, *25*, 3487. [CrossRef] [PubMed]
44. Yeddes, N.; Chérif, J.K.; Jrad, A.; Barth, D.; Trabelsi-Ayadi, M. Supercritical SC-$CO_2$ and Soxhlet *n*-Hexane Extract of Tunisian *Opuntia ficus indica* Seeds and Fatty Acids Analysis. *J. Lipids* **2012**, *2012*, 1–6. [CrossRef]
45. Khémiri, I.; Bitri, L. Effectiveness of *Opuntia ficus indica* L. *inermis* Seed Oil in the Protection and the Healing of Experimentally Induced Gastric Mucosa Ulcer. *Oxidative Med. Cell. Longev.* **2019**, *2019*, 1–17.
46. Fernández-Ponce, M.T.; Casas, L.; Mantell, C.; Rodríguez, M.; Martínez de la Ossa, E. Extraction of antioxidant compounds from different varieties of *Mangifera indica* leaves using green technologies. *J. Supercrit. Fluids* **2012**, *72*, 168–175. [CrossRef]
47. Francomano, F.; Caruso, A.; Barbarossa, A.; Fazio, A.; La Torre, C.; Ceramella, J.; Mallamaci, R.; Saturnino, C.; Iacopetta, D.; Sinicropi, M.S. β-Caryophyllene: A Sesquiterpene with Countless Biological Properties. *Appl. Sci.* **2019**, *9*, 5420. [CrossRef]
48. Liza, M.S.; Abdul Rahman, R.; Mandana, B.; Jinap, S.; Rahmat, A.; Zaidul, I.S.M.; Hamid, A. Supercritical carbon dioxide extraction of bioactive flavonoid from *Strobilanthes crispus* (Pecah Kaca). *Food Bioprod. Process.* **2010**, *88*, 319–326. [CrossRef]
49. Liu, W.; Fu, Y.-J.; Zu, Y.-G.; Tong, M.-H.; Wu, N.; Liu, X.-L.; Zhang, S. Supercritical carbon dioxide extraction of seed oil from *Opuntia dillenii* Haw. and its antioxidant activity. *Food Chem.* **2009**, *114*, 334–339. [CrossRef]
50. Romo-Hualde, A.; Yetano-Cunchillos, A.I.; González-Ferrero, C.; Sáiz-Abajo, M.J.; González-Navarro, C.J. Supercritical fluid extraction and microencapsulation of bioactive compounds from red pepper (*Capsicum annum* L.) by-products. *Food Chem.* **2012**, *133*, 1045–1049. [CrossRef]
51. Msaddak, L.; Abdelhedi, O.; Kridene, A.; Rateb, M.; Belbahri, L.; Ammar, E.; Nasri, M.; Zouari, N. *Opuntia ficus-indica* cladodes as a functional ingredient: Bioactive compounds profile and their effect on antioxidant quality of bread. *Lipids Health Dis.* **2017**, *16*. [CrossRef] [PubMed]
52. Capasso, F.; Grandolini, G.; Izzo, A.A. *Fitoterapia: Impiego Razionale delle Droghe Vegetali*; Springer-Verlag Italia Srl: Milan, Italy, 2006; Volume XXIX, 1031p.

*applied sciences*

MDPI

*Review*

# Pomegranate: Nutraceutical with Promising Benefits on Human Health

**Anna Caruso [1,†], Alexia Barbarossa [2,†], Antonio Tassone [1], Jessica Ceramella [1], Alessia Carocci [2,*], Alessia Catalano [2], Giovanna Basile [1], Alessia Fazio [1], Domenico Iacopetta [1], Carlo Franchini [2] and Maria Stefania Sinicropi [1]**

1 Department of Pharmacy, Health and Nutritional Sciences, University of Calabria, 87036 Arcavacata di Rende, Italy; anna.caruso@unical.it (A.C.); antonio.tassone1986@libero.it (A.T.); jessicaceramella@gmail.com (J.C.); biologanutrizionistagb@gmail.com (G.B.); alessia.fazio@unical.it (A.F.); domenico.iacopetta@unical.it (D.I.); s.sinicropi@unical.it (M.S.S.)

2 Department of Pharmacy-Drug Sciences, University of Bari Aldo Moro, 70126 Bari, Italy; alexia.barbarossa@uniba.it (A.B.); alessia.catalano@uniba.it (A.C.); carlo.franchini@uniba.it (C.F.)

* Correspondence: alessia.carocci@uniba.it

† These authors equally contributed to this work.

Received: 9 September 2020; Accepted: 29 September 2020; Published: 2 October 2020

**Abstract:** Pomegranate is an old plant made up by flowers, roots, fruits and leaves, native to Central Asia and principally cultivated in the Mediterranean and California (although now widespread almost all over the globe). The current use of this precious plant regards not only the exteriority of the fruit (employed also for ornamental purpose) but especially the nutritional and, still potential, health benefits that come out from the various parts composing this one (carpellary membranes, arils, seeds and bark). Indeed, the phytochemical composition of the fruit abounds in compounds (flavonoids, ellagitannins, proanthocyanidins, mineral salts, vitamins, lipids, organic acids) presenting a significant biological and nutraceutical value. For these reasons, pomegranate interest is increased over the years as the object of study for many research groups, particularly in the pharmaceutical sector. Specifically, in-depth studies of its biological and functional properties and the research of new formulations could be applied to a wide spectrum of diseases including neoplastic, cardiovascular, viral, inflammatory, metabolic, microbial, intestinal, reproductive and skin diseases. In this review, considering the increasing scientific and commercial interest of nutraceuticals, we reported an update of the investigations concerning the health-promoting properties of pomegranate and its bioactive compounds against principal human pathologies.

**Keywords:** pomegranate; *Punica granatum* L.; pomegranate skin extract; pomegranate fruit extract; nutraceutical properties; biological properties

## 1. Introduction

In the last few years, interest in pomegranate (*Punica granatum* L. Punicaceae) has risen, because of the nutritional and medical benefits, the external appearance of the fruit, and also for cosmetic and pharmaceutical purposes [1].

Among the varieties known to date, the five most popular in the world are: *Wonderful*, of American origin; *Hicanzar*, of Turkish origin; *Acco*, of Israeli origin; *Bagua*, of Indian origin; and *Mollar de Elche* and *Valenciana*, of Spanish origin [2]. Moreover, there is a dwarf variety of pomegranate, called *Punica granatum Nana*, characterized by a smaller size and inedible fruits, which is usually used as a small decorative pot plant [3]. Thanks to its different potential health properties, several components of the pomegranate plant, such as fruits, bark, flowers, roots, and leaves, have been employed for medicinal uses for a wide range of pathologies and health disorders [4].

The chemical composition of the fruit can vary and depends, above all, on the cultivation area, climate, ripeness, cultivation practices, and storage conditions [4]. The pomegranate is a source of numerous chemical compounds of high biological and nutraceutical value (e.g., phenolic acids, tannins, vitamins, antioxidants, and lipids), that are present in skin, carpellary membranes, arils, and seeds. The most important product derived from it is the juice derived from the arils or the whole fruit [5].

About 50% of the total weight of the pomegranate is made up of the skin and skin membranes that represent a significant source of flavonoids, ellagitannins, proanthocyanidins, mineral salts as potassium, phosphorus, sodium, magnesium, and iron. On the other hand, the grains are made up of water (85%), sugars (10%), particularly fructose and glucose, vitamins (including C, A, and group B vitamins), antioxidant substances, organic acids, as ascorbic, citric and malic acid, lipids. The seeds contain fatty acids, whose content ranges from 12 to 20% of the total weight (dry weight). Amongst them, a higher presence of alpha-linolenic acid (omega 3), linoleic acid (omega 6), and oleic acid (omega 9) has been detected, together with stearic acid, which may lower cholesterol levels, and palmitic acid [6]. Seeds are rich in protein, crude fiber, vitamins, minerals, pectins, sugars, polyphenols, isoflavones (especially genistein), coumestrol, sex steroids as estrone [5].

Today, the nutraceutical properties of the pomegranate arouse considerable interest in the scientific community and literature data reported several studies in which functional activities of the pomegranate and its derivatives, such as juice, seed oil, peel, etc., are highlighted [7]. This review gives an insight to the update of the present knowledge of the potential health benefits of pomegranate.

## 2. Pomegranate Health-Promoting Properties

### 2.1. Antioxidant Activity

Reactive oxygen species (ROS), formed in normal cellular metabolic processes or generated from exposure to ionizing or xenobiotic radiation, are held concausal factors in a large amount of chronic diseases. The toxicity of ROS is attributable to the ability of damaging essential biological substrates, such as DNA, RNA, proteins, and membrane lipids. ROS comprise superoxide radicals, lipoperoxide oxides, hydrogen peroxide, and hydroxyl free radicals [8]. An antioxidant is generally defined as a natural (fruit and vegetables) or artificial substance that can neutralize or protect a biological system from free radicals, such as oxygen, nitrogen, and lipid radicals [9,10]. These antioxidant properties make fruit and vegetables elements with good health properties, avoiding or decreasing the risk of suffering from determined degenerative diseases [11–18].

Anthocyanins, phenols [19], and vitamins as A (14), C (15), and E (16) [20] confer the high antioxidant power to pomegranate as reported by several authors in both in vitro and in vivo models [21].

Some authors [22] state that the antioxidant capacity of phenolic compounds is a consequence of their ability to capture free radicals and their chelating ability of metal cations.

Gil et al. [23] reported that the antioxidant effect possessed by pomegranate juice is 3 times higher than that of red wine or green tea and 2, 6, and 8 times superior than that found in red berries, grapefruit, and orange juice, respectively.

In 2013, a parallelism of total phenolic content and antioxidant properties between several extraction solvents of pomegranate seed (PS) and pomegranate defatted seed (PDS) was carried out. Data revealed this trend, in decreasing order, for the used solvents regarding the radical scavenging activity methanol > water > acetone > butanol > ethyl acetate > hexane ($EC_{50}$ antiradical potential amounting to 0.14 for PS and 0.19 µg/g for PDS). Similarly, the reducing activity test, decreed that methanol extract of PS an PDS possessed the greater reducing strength [24].

Derakhshan et al. in 2018 investigated the antioxidant activity and the total phenolic content of pomegranate peels, juice, and seeds from three regions of Natanz, Shahreza, and Doorak using as solvent ethanol. The best antioxidant activity was obtained by Doorak's seed and peel, as well as for the higher total phenolic content [25].

More recently, the analysis of five pomegranate juices genotypes (*Mollar*, *Kingdom*, *Dente di Cavallo*, and two old populations *Francofonte* and *Santa Tecla*) assessed that the total phenolic content ranged between 741.9 ± 55.8 and 424.2 ± 47.5 mg GAE/100 mL and the *Francoforte* genotype exhibited the higher amount. Furthermore, twenty-three phenolic compounds were detected. In particular, cyanidin-3,5-*O*-diglucoside and pelargonidin-3,5-*O*-diglucoside were the kind of anthocyanins present in all genotypes; the *Santa Tecla* population possessed the richest amount of these anthocyanins with values of 97.64 and 40.29 $mgL^{-1}$, respectively, whereas in the *Francoforte* population, ferulic acid hexoside was the most abundant compound (391.18 $mgL^{-1}$). The antioxidant activity values ranged between 221.5 and 36.73 µmol Trolox equivalents/100mL of juice and the higher one was observed for the *Santa Tecla* pomegranate population [26].

Preliminary studies performed by Bernabucci et al. explored the possible beneficial properties of pomegranate peel extracts (PPE) on bovine mammary epithelial cells BME-UV1. Their outcomes evidenced the ability of peel extract to decrease ROS production and malondialdehyde induced by the presence of hydrogen peroxide or lipopolysaccharide. Furthermore, PPE lowered pro-inflammatory cytokines expressions, exhibiting an anti-inflammatory effect on BME-UV1 treated with lipopolysaccharide [27].

A new and interesting study conducted by Hanani et al. addressed this issue from another perspective. Indeed, pomegranate peel powder was integrated into fish gelatin film-forming solution (FFS) with the aim to create an active packaging film. They found that the addition of 1% of pomegranate peel powder significantly increased the DPPH radical-scavenging activity to 59.74%, whereas the addition of 5% led to a percentage of 71.82%, and the fish gelatin film without pomegranate peel powder used as control achieved only the 53% [28].

Similarly, Bertolo et al. aimed to incorporate PPE at different doses to chitosan/gelatin gels. As a consequence, the antioxidant activity of the generated mixture was significantly improved. Moreover, concerning the rheological properties, PPE extended linear viscoelastic range and enabled the samples to easily flow under the applied shear rate boosting, in this way, the functional characteristics of chitosan/gelatin-based materials were improved [29]. In agreement with all these studies are the findings reported by Jalal et al., whose intention was to compare antioxidant activity of pomegranate peel and seed powder extracts through both DPPH and ABTS methods. The two assays evidenced the higher antioxidant capacity in peel powder samples with respect to seed powder samples. Furthermore, the methanolic extracts achieved the best results compared to water extract [30].

Surek et al. analyzed and compared the antioxidant activities of co-products from industrial pasteurized pomegranate nectar (PN), processing the peel (PP), press cake (PC), and precipitate after clarification (PAC) in comparison with raw material (arils) and final products (CON and PN) using DPPH, CUPRAC (Cupric ion reducing antioxidant capacity), FRAP (Ferric Antioxidant Power), and ABTS (2,2′-azino-bis(3-ethylbenzothiazoline-6-sulfonic acid)) methods. PP has shown to possess the greater antioxidant capacity in all the assays. This effect could be due to the presence of a rich amount of phenols, flavonoids, and tannins, compared with the other tested extracts. Consequently, it is plausible to think that discarding the peel would cause a decrease of antioxidant potential [31].

In 2019, deeper analyses were conducted on the principal antioxidants enzymes (peroxisomal catalase and superoxide dismutase (SOD) isozymes) and the NADPH-regenerating system of pomegranate and their possibility to vary for different cultivar and genotypes. In the reported study, there were analyzed seeds and juices from two pomegranate varieties (*Valenciana* and *Mollar*) grown in two diverse Spanish locations. The evaluations of the isoenzymatic superoxide dismutase (SOD) activity pattern showed one Mn-SOD and five CuZn-SODs (I–V) whose richness depended on the variety while immunoblot assays exhibited at least one additional Fe-SOD with a subunit size of about 23 kDa in both varieties. Moreover, a strong metabolism of ROS could be due to the presence of the $H_2O_2$-scavenging peroxisomal catalase in seeds and juice [32].

The research works mentioned in this section have proven that the major factor responsible for pomegranate antioxidant activity is the presence of polyphenols and anthocyanins. In particular, it has been demonstrated that peel extract is the richest of these elements, which makes it suitable as a natural source of anti-free radicals.

### 2.2. Anticancer Properties

Cancer is the principal cause of death in both developed and developing countries. The mortality rate of this pathology is superior among low and middle-income populations. The higher mortality rates in less well-off countries are mainly due to the lack of adequate health systems [33–36]. Reactive oxygen species' (ROS) overproduction is held liable to be one of the key factors for the development of several diseases, including cancer. Tumor biology has revealed that most neoplasms have a much higher amount of reactive oxygen species than healthy ones, such as superoxide anion, $H_2O_2$, and hydroxyl radicals [37]. These reactive oxygen-containing chemicals react with nucleic acids, proteins, and lipids, contributing significantly to tumor cells proliferation, DNA alterations, apoptosis, metastasis, and angiogenesis [38–41].

As a consequence of the troubles produced by the existing chemotherapeutic agents, nowadays there is an increasing concern in the search for herbal formulation with cancer preventive effect. Indeed, studies are focusing particularly on fruits rich in polyphenols due to their anticancer potential [42].

In order to investigate the effectiveness of pomegranate and its derivatives as anti-proliferative, anti-invasive, and pro-apoptotic agents, several studies have been conducted on different cell lines, such as breast cancer lines (MCF-7 and MDA-MB-231), uterine cancer lines (HeLa and Ishikawa), colorectal adenocarcinoma lines (RKO), and animal models [43–45]. Some research groups have shown that the simultaneous use of skin, seed, and pomegranate juice extracts has a synergistic action in inhibiting cell proliferation in several in vitro models, such as: Human breast MCF-7, uterine HeLa, human prostate DU 145, and PC-3 cancer cells. [46,47]. This result has been confirmed by Hong et al. [48], who demonstrated that pomegranate extracts and juice components have a more potent action than the individual isolated polyphenols, suggesting that it is a synergistic and additive effect of many phytochemical compounds, among which proanthocyanidins, anthocyanins, flavonoids, and ellagitannins. The proanthocyanidins are strong antioxidant compounds that, after acid hydrolysis, can release catechins as (+)-(2R,3S)-catechin (1). They act synergistically with ascorbic acid, which gives therapeutic potential against neoplastic and cardiovascular diseases, since it is able to suppress the action of free radicals, as well as to protect the body from the development and metastasis of cancer (which in part depends on the damage caused to the DNA). The anthocyanins are derivatives of phenyl-4H-benzopyran-4-one and are present in the vacuoles of the epidermal cells of many plants [49]. These are characterized by a high chemical reactivity and very low toxicity. They have, in fact, an anti-free radical action and they modulate the arachidonic acid cascade by inhibiting cAMP-phosphodiesterase [49]. The main anthocyanins present in the pomegranate juice are: cyanidin (2), delphinidin (3), and pelargonidine-3-O-glucoside (4). It was shown that anthocyanins decrease the proliferation of colon cancer cells, HT-29, in a dose-dependent manner [5].

The flavonoids, present in the bark and skin responsible for the red color, have a very high antioxidant activity and are useful for blood circulation [50]. The main flavonoids are quercetin (5), which, in addition to an antioxidant action [51], also showed antiviral and cardioprotective effects [52,53], kaempferol (6), with anticancer properties, and the rutin (7), a molecule with antithrombotic properties [54]. Among the ellagitannins, ellagic acid (8), a highly thermostable molecule, can be extracted from pomegranate skin and possesses important biological activities including antitumor, antiviral, and antimicrobial [55]. In addition, it was shown that ellagic acid (8) induces cell lysis, apoptosis, and thus decreases cell viability due to DNA breakage and alterations in the cell cycle. Gonzalez–Sarrias et al. [56] have demonstrated that ellagic acid and its secondary products can contribute to the prevention of colon cancer by regulating the expression of multiple genes implicated in crucial processes linked to the development of cancer.

Another tannin, the punicalagin (9), is present almost exclusively in the skin. It has several pharmacological properties, among which are anti-inflammatory, anti-proliferative, pro-apoptotic, and anti-genotoxic [57]. It also induces apoptosis in colon cancer cells (cell lines: HT-29, HCT116) and prostate cancer cells at a concentration of 100 mg/mL [5]. A study with 46 patients with prostate cancer under experimentation showed, for 16 of them, a considerable decrease in PSA (prostate-specific antigen) during treatment with pomegranate juice [58].

Koyama et al. [59] report that treatment of LAPC4 prostate cancer cells with 10 μg/mL pomegranate extract, obtained from seedless arils and skin, and standardized to a 37% ellagitannin content in punicalagin, blocks cell proliferation and triggers apoptosis. Furthermore, pomegranate skin extract (PoPx) with a high concentration of ellagitannins provokes apoptosis in human breast cancer cells (MCF-7), estrogen receptor (ER)-positive (ER+).

In previous investigations, PoPx and genistein have shown significant inhibitory effects on the proliferation of breast cancer cells MCF-7. In addition, PoPx can impede cell proliferation and the expression of angiogenesis markers, phosphorylation of p38, and C-Jun protein kinases activated by mitogens, and the activation of pro-survival signaling pathways. PoPx inhibits the nuclear factor NF-kB gene expression, connected with proliferation, invasion, and motility in aggressive breast cancer phenotypes [60]. Aiyer et al. reported that treatment at a dose of 300 mg/mL of pomegranate fruit extract (PoMx) in combination with 1 μM tamoxifen is able to reverse the resistance [61]. Moreover, recently Peng et al. identified PoMx as an interesting agent to cope up with oral cancer metastasis. Indeed, it was able to block wound healing migration, transwell migration, and matrix gel invasion. Analyzing the molecular mechanism, they also demonstrated that PoMx was able to downregulate matrix metalloproteinase MMP-2 and MMP-9 activities and expressions as well as epithelial-mesenchymal transition (EMT) signaling in HSC-3 cells [62].

PoPx prevent melanocyte proliferation and melanin synthesis by suppressing tyrosinase activity ($IC_{50}$ = 182.2 mg/mL). The amount of inhibition is comparable to arbutin, a glycoside with isoquinoline structure capable of being hydrolyzed in glucose and hydroquinone. An extensive amount of research has confirmed the ability of PoPx and ellagitannins to block the generation of free radicals in Ultraviolet (UV)A- and B-irradiated human skin, thus defending it from DNA fragmentation, from skin burns and depigmentation [63,64].

In 2010, Dai et al. reported that pomegranate extract (PE) is able to inhibit in a time- and concentration-dependent manner (already at 10 μg/mL) the viability and proliferation of a mouse mammary cancer cell line (WA-4, resulting from mouse MMTV-Wnt-1 mammary tumors) characterized by the presence of several cells possessing stem cells features. In particular, the arrest of cellular growth in the G0/G1 phase and an increased level of caspase 3 enzyme was observed, suggesting the mechanism of cell death by apoptosis.

Further investigations assessed the outcomes of individual phytochemicals derived from PE. Indeed, ellagic acid, ursolic acid (10), and luteolin (11) could contribute to the inhibitory potential of PE, with an $IC_{50}$ value of 10 μM. Instead, caffeic acid seemed to be inefficient [65].

Subsequently, in 2011, the same research group verified the arrest of cell growth due to PE but in the G2 phase by using as an in vitro model the human pancreatic cells PANC-1 and AsPC-1. The inhibition of cellular growth was already observed with an $IC_{50}$ amounting to 50 μg/mL in both the lines, but individual phytochemicals were slightly responsible for it. However, data showed increased proportion levels of cells lacking of CD44 and CD24 (connected with high tumor-initiating ability), demonstrating that PE can modify cell phenotypes, reducing the tumorigenicity [66].

Motaal and Shaker assessed the antioxidant and anticancer properties of different PEs, highlighting that the peel extract possesses the best antioxidant activity with an $IC_{50}$ value of 0.50 ± 0.9 mg/mL, and promising anticancer action toward MCF-7 breast cancer cells and HCT-116 colon cancer with $IC_{50}$ values of 7.7 ± 0.01 and 9.3 ± 0.06 μg/mL, respectively [67].

In 2015, El-Awady and co-workers investigated the possible antitumor activity of two different extracts of *Punica Granatum* grown in Saudi Arabia. Their findings evidenced a good percentage of cellular growth inhibition at the maximum concentration tested (100 μg/mL) from both seeds and husks extracts, tested on hepatocellular carcinoma HepG2 (values amounting to 95.8 and 98.3%, respectively) and colon cancer CACO cells (values amounting to 99 and 97%, respectively). The whole cytotoxic profile exhibited a dose-dependent trend. The percentages of cellular growth inhibition permitted to determinate the corresponding $IC_{50}$ values. They were 45 and 40 μg/mL on both HepG2 and CACO cells for pomegranate seeds and husks extracts, respectively [68].

Another study, performed by Modaeinama et al. in 2015, documented that the peel methanolic extract possesses a potent anti-proliferative action toward several tumor cell lines as: MCF-7 (breast adenocarcinoma), A549 (lung non-small cell cancer), SKOV3 (ovarian cancer), and PC-3 (prostate adenocarcinoma). The cytotoxicity has been investigated at different concentrations; the best antitumor effect was detected on MCF-7, with a percentage of survival inhibition amounting to 83.7% at doses of 5 μg/mL. Instead, for the other cell lines, this percentage was 77.87, 76.54, and 63.41% for PC3, A549, and SKOV3, respectively, at the same dosage used for MCF-7 [69].

Research based on the possible effect of a PE on chronic myeloid leukemia surveys that it is able to suppress cell growth by activating apoptosis and inducing cell cycle arrest. The findings of the research group led by Asmaa, outlined that the peel extract of pomegranate could inhibit K562 cell growth in a dose-dependent manner, with total cell death at the maximum concentration tested and an $IC_{50}$ value of 100 ± 0.05 μg/mL. Further investigations have assessed the ability of the extract to induce the cell cycle arrest in the G2/M phase and apoptosis. Indeed, an upregulation of proapoptotic proteins caspases 9, 7, 3, and cytochrome c, and a downregulation of antiapoptotic protein Bcl-2 were detected [70].

In depth studies of genetics appraised the possible influence of PE on immunoregulatory-related miRNAs level of adipose-derived MSCs (Ad-MSCs) obtained from adipose tissue. Data suggested that the principal mechanism of action could be due to the capability to decrease the expression of PI3K\AKT1\NF-kB genes involved in inflammatory pathways via miRNAs expression. It plays a role in the immune modulation in Ad-MSCs. In other words, PE could possess a key role in regulating the immunomodulatory function of stem cells [71].

A new approach in the use of phytochemicals for the cancer fighting lies in the knowledge of preparing metal-based nanoparticles from the extract of different plant parts, using green synthesis procedures. Indeed, silver nanoparticles possess the capability to up- and downregulate several cellular mechanisms, and this has propelled to the forefront in investigations to use them as a possible delivery system for other chemicals or as real control systems for the cellular growth [72–74].

In 2018, Sarkar and Kotteeswaran carried out the green synthesis of silver nanoparticles from the aqueous leaf extract of pomegranate, in order to test their capability as anticancerous toward human cervical cancer cells (HeLa). Their findings have proven that this kind of delivery system is able to inhibit cell growth in a dose-dependent manner with an $IC_{50}$ value of 100 μg/mL. Moreover, they assessed the ability of these pomegranate nanoparticles to induce necrosis and apoptosis by detecting increased levels of lactate dehydrogenase and DNA fragmentation [75].

Recently, Yusefi et al. employed four different weight % of *Punica granatum* fruit peel extract as green stabilizers to produce iron oxide nanoparticles (IONPs) and investigated their potentiality as anticancer agents. Among all the cell lines tested, IONPs have been shown to have a powerful inhibitory effect against nasopharyngeal carcinoma (NPC) cell line, HONE1. In particular, the most active IONPs, containing 2 and 4 wt% of peel extract, exhibited $IC_{50}$ values amounting to 197.46 and 85.06 mg/mL respectively [76].

All the data reported contribute to a vision of the state of the art in this field, remarking the potential antiproliferative effect of different parts of this fruit on a large panel of tumoral cells. However, in vivo studies, nowadays lacking, could be important to upgrade the researches.

### 2.3. Anti-Inflammatory Properties

Inflammation represents the first physiological response against injuries caused by physical agents, poisons, and others. [5,77–79]. The defense system, the so-called primary inflammation, neutralizes infectious microorganisms, removes irritation, and maintains ordinary physiological functions [80]. The phlogosis is triggered by various chemical and biological agents, including pro-inflammatory enzymes and cytokines, such as eicosanoids or degradation products of inflamed tissue.

According to recent reports, pomegranate exhibited potential as an anti-inflammatory medicine in both in vitro and in vivo studies. Some extracts of pomegranate, especially the cold-pressed seed extract, decrease the action of cyclooxygenase and lipoxygenase enzymes in vitro. Cyclooxygenases are useful enzymes for the degradation of arachidonic acid into prostaglandins, key mediators of inflammation. Lipoxygenase, on the other hand, mediates the transformation of arachidonic acid into leukotrienes and it is reduced by extracts of pomegranate seeds [81,82].

Bousetta et al. [83] report that punicic acid (12), a fatty acid contained in pomegranate seed oil, has an anti-inflammatory activity due to the inhibition of neutrophil activation, and therefore limits lipid peroxidation.

Lee et al. [80] analyzed some hydrolyzable tannins, including punicalagin (9) and punicalin (13), isolated from pomegranate by fractionation. Following in vitro studies, each of these compounds exerted a dose-dependent inhibitory effect on nitric oxide synthesis with an important anti-inflammatory effect [84].

De Nigris et al. [85] have shown that the dietary inclusion of pomegranate fruit extract caused an important reduction in expression of the markers of vascular inflammation, thrombospondin, and cytokine transforming growth factor (TGF-β1) in obese Zucker rats, a model of metabolic syndrome.

Larossa et al. [86] indicated that pomegranate composition could prevent colon inflammation before and during the disease process. They assessed the consequences of pomegranate intake and its main microbiota-derived metabolite urolithin-A (UROA) on colon inflammation. Results suggested that URO-A could represent the most powerful anti-inflammatory compound derived from pomegranate ingestion in healthy subjects, whereas in colon inflammation, the effects could be caused by the nonmetabolized ellagitannin-related fraction. Additionally, pomegranate extract supplementation has been shown to lead to decreased levels of prostaglandin E2 in the colon mucosa due to reduced cyclooxygenase-2 (COX-2) overexpression and diminished levels of prostaglandin synthetase E (PTGES) due to high ellagic acid (8) content [87].

Park et al. analyzed the effects of pomegranate peel extract (PPE) on THP-1 monocytic cells exposed to PM10, air-borne particulate matter with a diameter of <10 μM (PM10) that are known to induce cytotoxicity and ROS production and also to increase the expression and secretion of inflammatory cytokines, such as TNF-α, IL-1β. This study demonstrated that PPE at 10–100 $\mu g m L^{-1}$ is able to lower the production of ROS and the expression of TNF-α, IL-1β, MCP-1, and ICAM-1 thus preventing inflammatory events due to particulate matter [88].

A recent study by Xu et al. explore the inflammation effects of pomegranate flower (PFE) ethanol extract in lipopolysaccharide (LPS) -induced RAW264.7 cells. LPS is a component of the Gram-negative bacteria cell wall that has been often used in inflammatory response because it can activate macrophages. Their findings suggest that PFE is able to suppress the production of NO, PGE-2, and pro-inflammatory cytokines (TNF-α, IL-6, IL-1β), as well as the protein expression of iNOS and COX-2 in LPS-stimulated RAW264.7 macrophages. Furthermore, PFE treatment markedly inhibited LPS-induced NF-kB activation through blocking nuclear translocation of NF-kB and IkBα degradation and PFE treatment also inhibited the phosphorylation of mitogen-activated protein kinases (MAPKs) [89].

Kim et al. demonstrated the advantageous protective outcomes of pomegranate beverages, in comparison with those of mango in a preclinical model of colitis. The results obtained suggested that extracts rich in gallo- and ellagitannins work on several molecular targets in the protection against ulcerative colitis. Mango polyphenols inhibited the IGF-1R- AKT/mTOR via, and pomegranate polyphenols downregulated the mTOR downstream pathway, decreasing the ERK1/2 expression [90].

The systemic effects of PE on the formalin-induced nociceptive behavior and against gastric injury generated by non-steroidal anti-inflammatory drugs and ethanol in mice have been investigated. PE provide antinociceptive and anti-inflammatory activities without damaging the stomach and even displaying gastroprotection, possibly by modulation at central and peripheral levels [91].

The topical anti-inflammatory effect of a standardized pomegranate rind extract (SPRE), toward a mouse model of contact dermatitis has been evaluated, in relation to its marker compound ellagic acid, by Mo et al. The results highlighted the powerful anti-inflammatory activity of topical application of SPRE, ellagic acid being responsible for the extract activity as its major antioxidant constituent. The study suggests the topical formulation of SPRE as a promising therapy for contact dermatitis and as an alternative therapy for cutaneous disorders [92].

All these researches highlight the ability of the different components present throughout the pomegranate fruit to act as anti-inflammatory agent. However, further studies are required to completely exploit pomegranate's preventive and therapeutic potential in vivo.

### 2.4. Antidiabetic Activity

Diabetes is the principal widespread metabolic disease in the world and its occurrence has a rising trend. According to the World Health Organization, it is the third most common disorder after cardiovascular and oncological diseases [93].

Diet allows for the control of diabetes mellitus, and pomegranate fruit and its derivatives are part of it [56].

Among antidiabetic molecules, polyphenols, contained in pomegranate, are able to lower glycemic values, including via the reduction of glucose absorption through the intestine or peripheral tissues, although the most likely mechanism is the decrease of the enzyme glucosidase [94].

Li et al. [95] have hypothesized that pomegranate flower extract improves post-prandial hyperglycemia in type 2 diabetes and obesity, at least partially, as a result of the suppression of intestinal $\alpha$-glucosidase activity.

Huang et al. [96] instead reported a plausible mechanism of the antidiabetic activity of the PFE involving the activation of PPAR-$\gamma$. In addition, caffeic acid (17) or 3,4-dihydroxycinnamic acid another component of PE, increases the uptake of glucose by adipocytes in rats and myoblasts in mice [5].

Finally, McFarlin et al. [97] explored the impact of pomegranate seed oil on fat storage in mice noticing an enhancement of insulin sensitivity.

Parmar and Kar went into the potential effect of pomegranate peel extract on tissue lipid peroxidation (LPO), the concentration of thyroid hormones, insulin, and glucose in male rats starting from in vitro evaluations, which proved an inhibition of $H_2O_2$-induced LPO in red blood cells of rats by 0.25, 0.50, 1.0, and 2.0 μg/mL in a concentration-dependent manner. The maximum was achieved at 2.0 μg/mL. In vivo, *P. granatum* reduced LPO in hepatic, cardiac, and renal tissues and serum glucose concentration. These data suggested a potential regulatory role of this peel extract on thyroid function and glucose metabolism [98].

Another study assessed the ability of pomegranate fruit extract to decrease the serum resistin levels (adipocytokine, seen as a possible connection between obesity and type 2 diabetes) in ovariectomized mice, an animal pattern presenting exalted resistin levels in serum and upregulated resistin mRNA expression in white adipose tissue. Furthermore, PoMx was able to reduce the secretion and intracellular levels of resistin in differentiated murine 3T3-L1 adipocytes, without modifying resistin mRNA expression. Other findings suggested that PFE stimulated the degradation of resistin. All these activities could be mainly attributed to ellagic acid [99].

Several studies agree that oxidative stress triggered by diabetes mellitus gives rise to brain damage. Indeed, Cambay et al. indicated that the co-administration of pomegranate flower extract and antidiabetics resulted into improvements in learning and memory performances of diabetic rats. Moreover, in test subjects in which LPO was increased and glutathione (GSH) content was diminished in hippocampal tissue, the supplementation of pomegranate flower extract PGF was able to restore these levels. Besides, daily PGF intake decreased glial-fibrillar acidic protein contents induced by diabetes in the hippocampus [100].

On the contrary, in 2016, a research team questioned these findings, corroborating that daily consumption of pomegranate seed oil in patients suffering from diabetes mellitus 2 did not substantially impact the levels of parameters such as fasting blood sugar (FBS), insulin, HbA1c, alanine transferase, and homeostasis model assessment-insulin resistance [101].

PFE has been, over the years, part of the Chinese diet as an ally against type 2 diabetes mellitus (T2DM). For this reason, Tang et al. tried to analyze the kind of activity implied in the antidiabetic effect. The rat model (male Sprague–Dawley (SD) rats) diet was implemented with PFE polyphenols extract at doses of 50 and 100 mg/kg for a period of 4 weeks. Several assays as oral glucose tolerance test (OGTT), insulin tolerance test (ITT) and homeostasis model assessment of insulin resistance (HOMA-IR) were performed, revealing an improvement of insulin sensitivity. Deeper molecular investigations established that insulin-signaling activity was enhanced with an elevation in insulin-stimulated phosphorylation of insulin receptor substrate (IRS-1), Akt, and GSK-3b. Additionally, they assessed a decrease, after the treatment, of endoplasmic reticulum (ER) stress signals including phosphorylation of inositol-requiring kinase1 (IRE1) and activation of X box-binding protein (XBP-1) splicing. In addition, regarding the blood lipid profile, liver glycogen content and antioxidant status were improved by PFE in the rats [102].

Similar results were found for pomegranate fruits aqueous extract as a conspicuous decrease in fasting blood glucose (FBG) by 28.1 and 67.9% in short-term and long-term treatment models (alloxan-diabetic male Wistar rats) and a raise in the mRNAs expression levels of IRS-1, Akt, Glut-2, and Glut-4, suggesting an improvement of glucose uptake and storage [103].

Recently, the potential combination of pomegranate juice and *Lactobacillus casei* NRRL-B-1922 (used as fermenting agent) to create a functional juice able to combat T2DM has been documented. The results highlighted that pomegranate could support the growth of *L. casei* even without nutrient supplementation. Further analysis demonstrated that the glucose and fructose contents were gradually lowered with consumption rates of 0.51 and 0.37 g/L/h, respectively, in the bioreactor. Moreover, the possible antidiabetic mechanism of action could be attributable to the dipeptidyl peptidase-4 (DPP-4) inhibition (with a rate of 80%). Indeed, DPP4 is a well-known target of many antidiabetic agents [104].

A recent clinical trial demonstrated that the assumption of pomegranate juice in patients with T2D can change and lower erythropoietin levels just after 3 h of ingestion whereas no modifications were observed in healthy patients [105]. However, further studies to assess the long-term antidiabetic effects of the pomegranate consumption are in progress.

### 2.5. Antimicrobial Activities

Antimicrobial agents are mainly applied on microorganisms that cause food poisoning (infected or toxin-producing agents) and on microorganisms that alter food by producing final metabolic products (catabolites) or enzymes with bad odor, unpleasant taste, problems of persistence, different coloring, and/or health risk [106].

The antimicrobial effect of pomegranate and its products has been demonstrated in a large amount of research [107], in particular, polyphenols, flavonoids, and condensed and hydrolysable tannins derived from the fruit have been studied as promising agents to treat or prevent a broad range of infections [108].

The antimicrobial mechanisms of phenolic compounds engage the reaction of phenols with sulfhydryl groups of membrane proteins of the microbial cell, inducing a bactericidal effect due to membrane protein precipitation and inhibition of enzymes such as glycosyltransferase [109].

Food-borne disorders and urinary tract infections are treated typical in the Indian subcontinent through PoPx, while ellagitannins, punicalagin (9), ellagic acid (8), and gallic acid (18) in the pomegranate skin, as natural antimicrobial agents, have been extensively used against *Staphylococcus aureus* and *Escherichia coli* hemorrhages for their capability to precipitate membrane proteins and block enzymes such as glycosyltransferase, causing lysis. In vivo and in situ application of an 80% pomegranate skin (PoPx) methanol extract showed a growth reduction against *Listeria monocytogenes*, *Staphylococcus aureus*, *Escherichia coli*, and *Yersinia enterocolitica*. However, it has been reported that higher doses of PoPx (24.7 mg/mL) represent the lowest bactericidal concentration of *Listeria monocytogenes* [110].

Another study evidenced the powerful effect of *Punica granatum* L. peels' ethanolic extract (Tunisian Nana variety) against two Salmonella strains, *Salmonella Enteritidis* and *Salmonella Kentucky*, microorganisms resistant to the majority of antibiotics. Minimal inhibitory concentration (MIC) values ranged from 10.75 to 12.5 mg/mL regarding both strains. In particular, the inhibitory effect achieved by peel extract (using hydroethanolic and hydromethanethanolic mixtures) on *Salmonella Kentucky* achieved inhibition diameters of 22.2 and 22 mm, respectively [111].

Višnjevec et al. explored and evaluated the antimicrobial properties of ethanol and water extracts of pomegranate exocarp and mesocarp from Istria. The best antibacterial activity was recorded, here too, for ethanolic extract. Indeed, the microorganisms more susceptible to this extract were *C. albicans*, *C. parapsilosis*, *R. mucilaginosa*, *E. dermatitidis*, and *S. aureus*, with MIC ranging from 0.156 to 1.25 mg/mL. Instead, the best antimicrobial activities of the exocarp and mesocarp water extracts were found toward *S. aureus*, followed by *E. coli* without any antifungal activity [112].

Another use of pomegranate peel extract as antimicrobial agent involved its participation in the synthesis of silver nanoparticles. The latter, due to the use of transmission electron microscopy (TEM) and scanning electron microscopy (SEM), demonstrated to be evenly distributed in the solution, with a spherical shape and size ranging from 20 to 40 nm and with an average particle size of 26.95 nm. Subsequently, tests evidenced the capability of nanoparticles to substantially inhibit Gram-negative and Gram-positive bacteria as *E. coli*, *P. aeruginosa*, *P. vulgaris*, *S. typhi*, *S. aureus*, *S. epidermidis*, and *K. pneumonia* already at 25 and 50 μg/mL [113].

As can be evinced from this section, pomegranate and, in particular, the peel extract, exhibited antimicrobial effects against a wide spectrum of bacteria. Unfortunately, in order to become effectively beneficial for the pharmaceutical field, much more has to be done regarding the deepening of the mechanism of action.

### 2.6. Prevention of Cardiovascular Diseases

The main risk factors for the occurrence of coronary diseases is dyslipidemia, characterized by an excessive increase in low-density lipoproteins (LDL) and/or low levels of high-density lipoproteins (HDL) [114,115].

It is known that LDL oxidation contributes to atherosclerosis and the development of cardiovascular diseases [53,116].

Inhibition of LDL oxidation is considered a promising way to avoid the storage of foaming cells and, ultimately, cholesterol deposits in the arteries.

Due to its formidable antioxidant capacity, the extract obtained from the pomegranate skin has the power to suppress LDL oxidation and thus delay the advancement of atherosclerosis with an important decrease in foaming cell levels at the artery level.

The polyphenols present in pomegranate, punicalagin (9), gallic acid (18) and, to a lesser extent, ellagic acid (8), increase the expression and secretion of the liver enzyme paraoxonase 1 in a dose-dependent manner, thus reducing the risk of developing atherosclerosis [117].

Riaz and Khan in 2016 gave an insight into the anticoagulant, antiplatelet, and antianemic effects of *Punica granatum* juice by studying the hematological profile of the rabbits used as an in vivo model. Their findings evidenced a substantial increase in erythrocyte count, mean corpuscular hemoglobin concentration after 30 days of treatment, whereas red cell distribution width was significantly reduced. Instead, leucocyte count, hematocrit (Hct), mean corpuscular volume (MCV), mean corpuscular hemoglobin (MCH), and platelet count (PLT) were not modified at any dose. Furthermore, there was an important increase in bleeding time both after 30 and 60 days, and a significant rise in thrombin time (TT) and activated partial thromboplastin time (αPTT) after 60 days, whereas prothrombin time did not change. Comparing the pomegranate ability in lowering cholesterol and the prolonged αPTT time, they speculated that the cholesterol reduction induced by *P. granatum* may lead to a decrease in the concentration of coagulation factors, influencing, in this way, αPTT. In conclusion, data illustrated that *P. granatum* significantly inhibited aggregation of platelets induced by adenosine diphosphate ADP and epinephrine (Epi) in the blood samples of animals treated already at 2 mL/kg for 30 and 60 days [118].

The hypothesis that pomegranate ameliorates the cardiac function has been also supported by Razani et al. Indeed, they administered 220 mL of pomegranate juice to one hundred hospitalized patients suffering from unstable angina or myocardial infarction for 5 days in order, a the end, to evaluate cardiac markers and parameters. The results indicated a significant reduction in the intensity, occurrence, and duration of angina pectoris in patients with unstable angina. Moreover, the measurement of serum levels of troponin and malondialdehyde revealed a decrease of these factors. Nevertheless, other parameters like interleukin-6,tumor necrosis factor alpha, blood pressure, and heart rate were not affected [119].

More recently, the possible interaction between ethanolic peel extract of *Punica granatum* and doxorubicin (well known for its cardiotoxicity) was evaluated. The assessment was performed by measuring heart weight/body weight ratio, biochemical parameters, and histopathological changes. After the treatment with 100 mg/kg body weight, an increased heart weight/body weight ratio, in comparison with the doxorubicin-treated group, was observed as well as blood and tissue glutathione, superoxide dismutase, and catalase levels. On the contrary, creatine kinase, lactate dehydrogenase enzyme, and malondialdehyde levels appeared significantly reduced. Concerning the histopathological morphology, the only assumption was that doxorubicin modified myocardium through vacuolar alterations in the heart muscle and necrosis of heart muscle with remote cells increasing in the size in between the necrotic and fragmented muscle fibers. Instead, treatment with *P. granatum* peel extract displayed a protective effect with less histological changes such as irregular and spread vacuolation limited to subendocardial layers [120].

A recent clinical trial evidenced that the consumption of pomegranate juice improves the lipid profile and oxidative and inflammatory biomarkers of hemodialysis patients. The results outlined that triglycerides were decreased in the pomegranate juice condition and augmented in the control groups. On the other hand, high-density lipoprotein cholesterol was increased after PJ assumption, whereas total and low-density lipoprotein cholesterol were not affected. Systolic and diastolic blood pressure dramatically fell in the PJ condition, while total antioxidant capacity increased [121].

These data were also supported by Sohrab et al., who found in their study a decrement of systolic blood pressure and diastolic blood pressure; whereas, total cholesterol, low-density lipoprotein cholesterol decreased significantly compared to pre-trial values within the intervention group [122].

All these findings demonstrate the importance of pomegranate juice dietary supplementation in modulating cholesterol levels and preventing cardiovascular diseases.

### 2.7. Antiviral Activities

The hydrolysable tannins of pomegranate, including punicalin (13), punicalagin (9), gallic acid (18), and ellagic acid (8), have antiviral properties capable of modulating respiratory infections and influenza.

The antiviral properties of the polyphenolic extract of pomegranate are due to the inhibition of influenza virus RNA replication [117,123].

Similarly, phenols in the skin inactivate viruses through direct structural damage and indirect intercellular inhibition of viral replication. The virucidal effects of pomegranate phenolic compounds imply their interaction with the antigenic glycoprotein, hemagglutinin, present on the surface of some viruses (es. influenza viruses), which produce a loss of red blood cell agglutination [124].

In particular, Haidari et al. argued that punicalagin (9) is the effective component of pomegranate polyphenol extract able to block replication of the virus RNA, inhibit agglutination of chicken red blood cells (RBCs) by the virus of influenza H3N2, and have viricidal activity. They also investigated the potential synergistic effect of pomegranate polyphenol extract and oseltamivir (a well-known antiviral agent), outlining that oseltamivir in association with pomegranate polyphenol extract amplificated its anti-influenza power [125].

Sundararajan et al. in 2010 confirmed and implemented these data, strengthening that the direct anti-influenza activity of pomegranate polyphenols (PPs) is principally an outcome of PP-induced virion structural damage. Indeed, these components of pomegranates quickly block the influenza virus through a direct effect on the viral particle. They also suggested that this action might be separate from effects on hemagglutinin (HA) function. Moreover, they demonstrated the efficacy of pomegranate PPs against H1N1 and H3N2 influenza viruses and against the reassortant H5N1 virus rg-VN/04 [126].

Furthermore, another study attributed to punicalagin the power to reduce the viral cytopathic effect on rhabdomyosarcoma cells, calculating an $IC_{50}$ value of 15 μg/mL. This effect was also assessed in vivo, noticing a decrease of mortality in mice treated with a lethal dose of enterovirus 71 [126].

Reddy et al. focused on the potential effect against hepatitis C virus (HCV) of the ellagitannins extracted from pomegranate (*Punica granatum*) fruit peel. In particular, pure compounds punicalagin, punicalin, and ellagic acid exhibited in vitro the capability of blocking the HCV NS3/4A protease activity in a concentration-dependent manner with $IC_{50}$ values of less than 0.1 mM (for punicalagin and punicalin), whereas $IC_{50}$ for ellagic acid was achieved at 1.0 mM. These data were confirmed through in silico studies and observing a consistent reduction of HCV replication in cell culture systems. Data ex vivo pointed out the optimal bioavailability and the toxicity absence of these compounds [127].

In addition, a good activity of pomegranate extract toward adenovirus has been found. The $IC_{50}$ and $CC_{50}$ (50% cytotoxicity concentration) were estimated on HeLa cells with values of 165 ± 10.1 and 18.6 ± 6.7μg/mL, respectively, whereas the selectivity index (SI, calculated as the ratio of $CC_{50}$ and $IC_{50}$) on adenovirus amounted to 8.9 [128].

In their analysis, Houston et al. investigated the co-administration effects of pomegranate rind extract in conjunction with zinc (II) salts in order to challenge herpes simplex virus HSV-1 and its aciclovir-resistant. Data showed a potentiation factor by up to 5.5-fold. Regarding aciclovir-resistance, pomegranate rind extract exhibited an $EC_{50}$ value amounting to 0.02 μg $mL^{-1}$, whereas acyclovir displayed no activity [129].

Given the several reported studies that have proven the antiviral effects of pomegranate, Arunkumara and Rajarajanb surmised the possible inhibitory effect of *Punica granatum* against Herpes simplex virus-2 (HSV-2). In effect, their findings assessed that ethanolic peel extract suppresses HSV-2 at a concentration of 62.5 μg/mL. In order to understand the real responsibility of this action, the extract was subjected to bioactive compounds separation by bioassay-guided fractionation. Then, the single components were tested as antivirals. The key component for the activity seemed to be punicalagin, showing a total inhibition rate at 31.25 μg/mL. Moreover, bioactive compound analysis using the ADMET tool, established that human intestinal absorption (HIA) properties of acyclovir (ACY), gallic acid (GA), and ellagic acid (EA) had moderate adsorption values of 63.77, 53.69, and 61.39%, respectively, and punicalagin presented very strong plasma protein binding. In addition, docking studies highlighted that the most active component could interact with HSV-2 amino acids through several hydrogen bonds [130].

The global pandemic due to the spread of the severe acute respiratory syndrome coronavirus 2 (SARS-CoV-2) has dramatically shaped queries on the necessity to find as soon as possible a cure for this disease. With this in mind, many efforts, also with the aid of artificial intelligence, have been made to discover molecules able to interact with the viral proteins and cause the Sars-CoV-2 inhibition. Several compounds of natural origin were found to be able to do this. Among them, docking studies revealed that the ellagic acid, one of the already mentioned main components of pomegranate, has the potentiality to interact with important proteins involved in Sars-CoV-2 as RNA-dependent RNA polymerase (RdRp), angiotensin-converting enzyme 2 (ACE2), spike glycoprotein (SGp), and main protease (3CLpro) [131,132].

Furthermore, another study conducted on the viral main protease (3-chymotrypsin-like cysteine enzyme), held responsible for the COVID-19 control of duplication and life cycle management, assessed the potential of hydrolysable tannins (present also in pomegranate) as its natural inhibitors. Indeed, punicalin seems to establish H-bonds with the crucial catalytic residues of pocket spatial position [133].

From this research, it is deduced that pomegranate (especially peel extract) possess a prophylactic potential against viral epidemics and pandemics, specifically influenza. This may open up new avenues for research in the nutritional and medical science fields if the studies are made complete with in vivo experiments.

### 2.8. Other Properties

#### 2.8.1. Obesity

Obesity is a chronic disorder of multiple factors characterized by an extensive accumulation of fat or general hypertrophy of adipose tissue in the body [134]. Hence, obesity can be so described when the natural energy reserve of humans or other mammals stored in the form of body fat increases to such an extent that it is associated with a series of complications, health problems, diseases, and even an increase in the mortality rate [135].

Studies have been carried out showing that the consumption of pomegranate leaf extracts, in particular punicalagin (9), diminishes the sense of hunger and body weight [136], and that the extract inhibits the incidence of obesity and hyperlipidemia. These activities seem to be due partially to inhibition of pancreatic lipase and partly to a decreased calorie intake [137].

Khabeer et al. went through the anti-obesity potential of SHAMstat3pg, a fatty acid composite extracted from pomegranate seed oil, made up of punicic acid, oleic acid, and linoleic acid. Their findings established that the treatment of 10 μg/mL of SHAMstat3pg (24 h) inhibited in a dose-dependent manner adipogenesis of human adipose derived mesenchymal stem cells (HADMSC), improved inflammation, attenuated ATP production, and glucose uptake. In addition, the extract favorably modulated the mRNA expression of the studied obesity-associated gene transcripts [138].

Several positive effects on fat reduction have been underlined, maybe for the presence of anthocyanins, tannins, and numerous antioxidant compounds. However, given that obesity is classified as one of the main risk factors for the development many disorders (metabolic and not), human studies in this field are still limited and need more attention to understand the preventive role of pomegranate in order to make it a potent nutraceutical ally.

#### 2.8.2. Intestinal Regulation

Pomegranate derivative intake leads to a considerable accumulation of ellagitannins in the large intestine, where they react with the intestinal microflora [139]. Bialonska et al. [140] reported that the beneficial effects on microflora are mainly due to PoPx. A well-known cause of intestinal damage is represented by high-intensity exercise. In order to overcome this risk, Chaves et al. studied the consequences of the assumption of fermented milk supplemented with whey protein (approximately 80% protein), probiotic (*Bifidobacterium animalis* subsp. lactis BB12), and pomegranate juice (*Punica granatum* L.) on the physical performance, intestinal motility and villi structure, inflammatory markers, and intestinal microbiota of in vivo model Wistar rats under high-intensity acute exercise. The group only subjected to exercise went through changes in the intestinal villi interspace, in the proportion of *Lactobacillus* species and an increase in *Clostridium* species, as well as a decrease in intestinal motility. The treated group, instead, ameliorated intestinal motility and preserved the intestinal villi interspace and the natural microbiota proportions, but the physical performance was not enhanced [141].

In 2019, George et al. demonstrated the capability of pomegranate peel extract to constrain the pathogenicity of *Citrobacter rodentium (Cr)* infections. The results obtained underlined a discrepancy between mice treated with pomegranate peel extract and the control group regarding the composition of the microbiome. Indeed, in treated animals, a reduced proportion of Firmicutes/Bacteroidetes was found by increasing Bacteroidetes and decreasing Firmicutes amounts, which led to a large reduction in *Lactobacillus*. These alterations have led to the belief that the intake of pomegranate could play a protective function against various intestinal infections [142].

Similar findings have been evidenced by Smith et al. which assessed that the treatment pomegranate peel extract was able to reduce Cr infections, weight loss, and mortality with respect to whom assumed just water as control groups. Instead, the treatment did not alter Cr colonization of the colon, whereas it reduced the colonization of the spleen. Moreover, it lowered the degree of Cr-induced colon damage, associated with mortality [143].

Nevertheless, the advances made to understand the capability of pomegranate to influence the microbiota remain limited still and need to be deepened in vivo.

#### 2.8.3. Effects on the Male Reproductive System

According to Türk et al. [144], the consumption of pomegranate juice produces an increase in the concentration of sperm in the epididymis, higher sperm mobility and density, and a reduction of poor quality sperm compared to the reference or control group. In a more recent study, this same group of researchers suggested that ellagic acid (8) has a protective effect on both testicles and sperm. This effect may be connected to the powerful action of ellagic acid against oxidative stress [145]. Concerning erectile dysfunction or impotence, the continued inability to achieve or maintain an erection sufficient to maintain satisfactory sexual intercourse, Forest et al. [146] stated that after four weeks of treatment with pomegranate juice, patients showed better erectile activity than others who had received a product with a placebo effect.

However, this field the knowledge is still limited, and other studies are necessary to assess the veracity of this hypothesis.

#### 2.8.4. Antidiarrheal Effects

Qnais et al. [147] assessed the antidiarrheal effects of the water extract of pomegranate fruit skin on rats. The outcomes demonstrated that the extract dose-dependently inhibits spontaneous ileum movements and attenuates acetylcholine-induced contractions.

In other studies, the antidiarrheal effects of fruit skin on rats were evaluated by administering an oral dose of 400 mg/kg. The findings showed that pomegranate extract decreases the number of defecations and stool weight [148].

These previous findings were also supported by Zhao et al. in 2018 by studying the aqueous extract of pomegranate peels and, in particular, its bioactivity-guided fractions and bioactive components. The fraction considered responsible of the antidiarrheal activity was the ethyl acetate one, mainly composed by punicalagin, corilagin, and ellagic acid. Data also revealed that the administration of the ethyl acetate fraction at 100, 200, and 400 mg/kg was able to decrease gastrointestinal transit in charcoal meal tests in mice as well as inhibit castor oil-induced enteropooling compared to control animals. From histopathological evaluations, it emerged that small intestine lesions of mice treated with the ethyl acetate fraction were relieved in comparison to those in mice treated with castor oil [149].

Although the studies in this sector are insufficient, they suggest the potentiality of pomegranate in regulating intestinal motility.

#### 2.8.5. Effects on Oral Health

Currently, science recognizes that chronic periodontal inflammatory disorder is strictly linked to the worsening of cardiovascular disease [150].

DiSilvestro et al. [151] have demonstrated that an oral rinse based on extracts of pomegranate would significantly reduce (about 84%) the number of microorganisms from the dental plaque.

Sastravaha et al. [152] have highlighted the efficacy of a toothpaste containing pomegranate extracts as an additional treatment to complement routine periodontal therapies and have demonstrated that pomegranate flavonoids have an in vitro antibacterial property against the microorganisms responsible for gingivitis.

Nonetheless, there is still ample scope for research in this field.

#### 2.8.6. Effects on Pregnant and Breast-Feeding Women

Recent studies demonstrated the importance of a nutritional supplementation with pomegranate.

In particular, Al-Wazni et al. assessed the potential beneficial effects of three types of pomegranate extracts to fight urinary tract infections. Indeed, their findings demonstrated that the pomegranate peel extracts were able to inhibit *S. aureus* and *E. coli* bacteria, pathogens common to the urinary tract, in comparison to ciprofloxacin, used as reference molecule, and the negative control was represented by distilled water [153].

In 2017, Manouchehrian et al., through a clinical trial conducted on 80 healthy pregnant women, demonstrated the potential effect of adding pomegranate paste to pregnant women's diets on the incidence of neonatal jaundice. Indeed, the results revealed decreased values of bilirubin in the group fed with pomegranate paste with respect to control group [154].

A study carried out on pomegranate juice maternal supplementation, famous to be enriched with antioxidants and bioactive polyphenols, hypothesized the potential improvement of uterine and umbilical artery function with the consequent enhancement of fetal growth in the $eNOS^{-/-}$ mouse model of fetal growth restriction (FGR) without increasing fetal weight [155].

Similarly, Henning et al. envisaged the presence of one of the main pomegranate constituents, ellagic acid and its metabolites, in breast milk after the consumption of 8 ounces per day of pomegranate juice. As expected, they found detectable levels of ellagic acid and metabolites in breast milk after 14 days of nutritional supplementation. This could mean that the phenolic compounds delivering through breast milk could enhance infant health and development [156].

Needless to say, pomegranate represents an inexhaustible source of benefits, but for a safe use as therapeutic agent on pregnant and breast-feeding women other studies must be carried out.

#### 2.8.7. Effects on Nervous System

Prior research considered neuroinflammation as the principal factor responsible for neurodegenerative diseases, including Alzheimer's and Parkinson diseases, as a consequence of neuronal degeneration after over-activating microglia in the brain [157].

Recent studies have promoted the neuroprotective activity of pomegranate due to the presence of several well-known anti-inflammatory components previously mentioned in the text (see Section 2.2).

In this context are well inserted the investigations carried out by DaSilva et al. in 2017. Their findings substantiated that urolithins, ellagitannin-gut microbial-derived metabolites present in PE, were able to lower important neuroinflammatory mediators such as NO, IL-6, PGE-2, and TNF-α from lipopolysaccharide (LPS)-stimulated BV-2 microglia. Furthermore, they established the protection of SH-SY5Y and BV-2 (in vitro neuronal models) cell viability by hampering apoptosis and caspase 3/7 and 9 released from $H_2O_2$-induced oxidative stress [158].

Subsequently, Velagapudi et al. delved into urolithin mechanism of action by assessing not only the decrease of the previously mentioned mediators (IL-6, TNF-α, NO) but underlining the relation between inhibition of neuroinflammation and Sirt-1 activation in BV2 microglia. Indeed, SIRT-1 levels, after treatment with 5 and 10 μM of urolithin, were much higher than control. This could suggest that this compound should activate SIRT-1 in order to perform its neuroprotective activity effect [159].

More recently, Césedas et al. confirmed these evidences toward another in vitro model, neuro-2a cells, by inducing oxidative stress. Data outlined that urolithin was able to improve mitochondrial activity (MTT assay), redox state (ROS formation, lipid peroxidation), and the activity of antioxidant enzymes (CAT: catalase, SOD: superoxide dismutase, GR: glutathione reductase, GPx: glutathione peroxidase). Furthermore, it raised the cytoprotective peroxiredoxin 1 and 3 expression and it was assessed to be a radical scavenger through ORAC (oxygen radical absorbance capacity) and DPPH assays [160].

Urolithin, contained in pomegranate juice, proved to also be effective in a rat model of Parkinson's disease induced by rotenone. Indeed, the nutritional supplementation with pomegranate juice enhanced postural stability, impeded oxidative damage and α-synuclein aggregation, permitted neuronal survival, improved mitochondrial aldehyde dehydrogenase activity, and preserved antiapoptotic Bcl-xL protein at the control level [161].

A recent clinical trial demonstrated the beneficial effects on memory derived from the daily consumption of pomegranate juice. In fact, for a period of 12 months, the diet of two hundred and sixty-one patients (aged 50–75 y) has been implemented with pomegranate juice (8 oz (236.5 mL) per day) or a placebo drink (8 oz, matched constituents of pomegranate juice except for pomegranate polyphenols). At six and 12 months, the memory measures (Brief Visuospatial Memory Test-Revised (BVMT-R) and Buschke Selective Reminding Test (SRT)) were carried out. From the results obtained, Siddarth et al. corroborated the hypothesis that the steady consumption of pomegranate juice may stabilize the ability to learn visual information over a 12-month period [162].

Taken together, reported data corroborate the concept that pomegranate may be considered a neuroprotector and developed as an agent for treating neurodegenerative diseases.

## 3. Toxicological Aspects of Pomegranate and Potential Interaction with Drugs

Recently, some research has shown that pomegranate fruit can be considered part of a healthy diet and lifestyle without any risks or side effects [5]. Studies have shown that two doses of pomegranate extract (0.4 and 1.2 mg/kg body weight) produced no toxic reactions in rats in terms of food intake, weight changes, or behavioral or biochemical factors [163].

Heber et al. [164] carried out studies on 64 overweight subjects to assess the safety of the use of extracts in humans. Following the intake of 710 and 1420 mg capsules (containing 435 and 870 mg gallic acid equivalents, respectively), no adverse events occurred and no substantial differences in toxicity were estimated in any subject studied.

Other studies have been carried out on patients with carotid artery stenosis, showing that the intake of pomegranate juice (121 mg/L ellagic acid equivalent) for more than 3 years did not cause any toxic effect on blood parameters or liver, kidney, or heart functions [165].

Based on current research on pomegranate, considering its beneficial effects in cancer, cardiovascular diseases, etc., it was interesting to define the effects of pomegranate extracts on the cytochrome P450-3A, a liver enzyme system responsible for the metabolism of several drugs [7].

Studies in rats show that the administration of pomegranate juice possesses an inhibitory activity on the pharmacokinetics of carbamazepine, an anticonvulsant drug also substrate by the cytochrome P450-3A [7].

A single-dose randomized trial in healthy volunteers showed that treatment with pomegranate juice had no effect on the half-life or distribution time of intravenous administration of benzodiazepines with anxiolytic, hypnotic, anticonvulsant, and muscle relaxant properties, and no effect on maximum concentration or clearance after oral administration [7].

All of these findings strengthen the pomegranate value, making it an interesting and safe nutraceutical.

## 4. Conclusions

Today, it is well known that the positive effects of fruit and vegetables in preventing disorders depend on the composition of the bioactive compounds they contain (Table 1) [5,7,166–171].

**Table 1.** Bioactive compounds present in pomegranate that have been demonstrated to have biological activity.

| Structure | Activity | Bibliographic Reference |
|---|---|---|
| 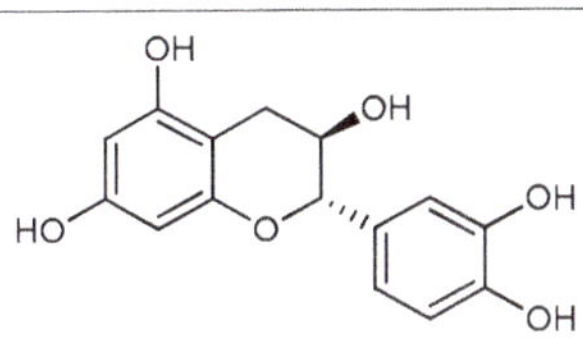<br>(+)-(2R,3S)-Catechin (1) | Antioxidant | [48] |
| 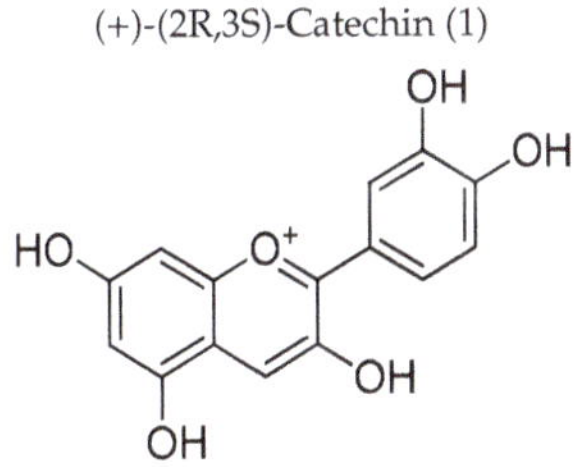<br>Cyanidin (2) | Antioxidant<br>Antitumoral<br>Anti-inflammatory | [5] |
| 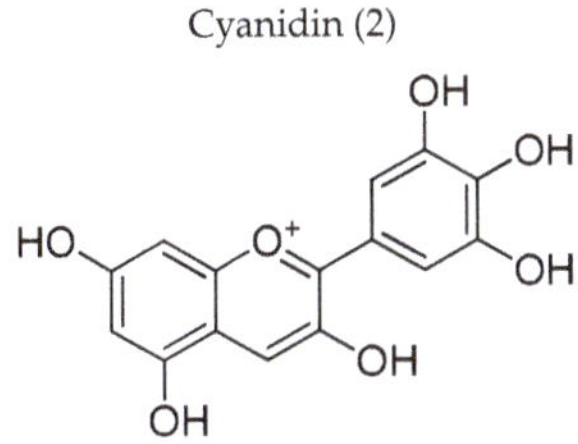<br>Delphinidin (3) | Antitumoral<br>Antioxidant<br>Anti-inflammatory | [5] |

**Table 1.** *Cont.*

| Structure | Activity | Bibliographic Reference |
|---|---|---|
| 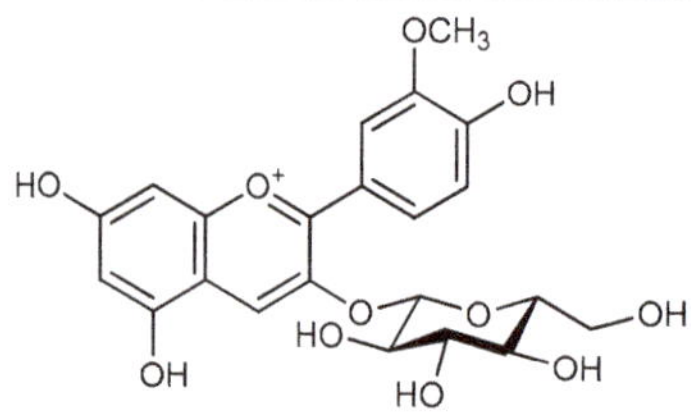 Pelargonidin-3-O-glucoside (4) | Antitumoral<br>Antioxidant<br>Anti-inflammatory | [5] |
| 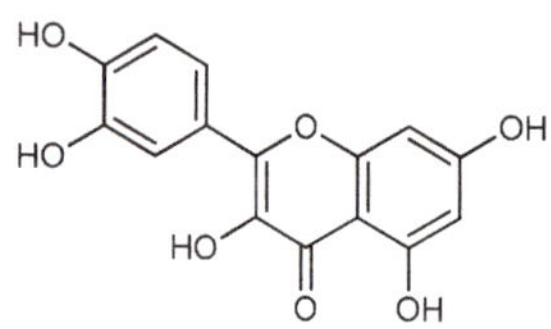 Quercetin (5) | Antiviral<br>Antioxidant Cardioprotective<br>Antitumoral | [52] |
| 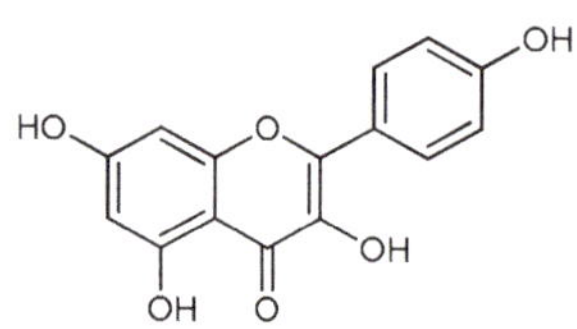 Kaempferol (6) | Anti-cancerogenic | [54] |
| 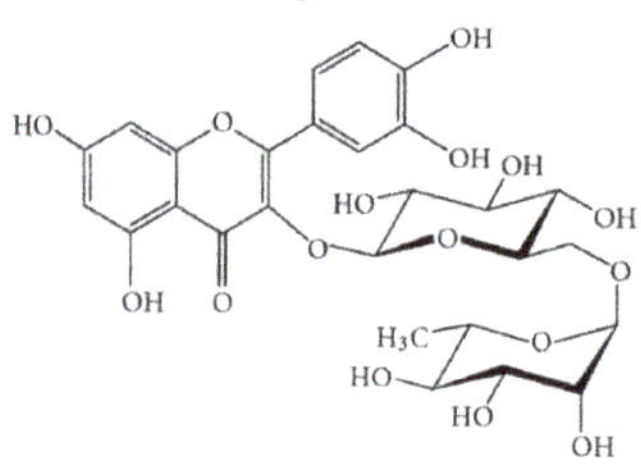 Rutin (7) | Antitumoral<br>Antimicrobial | [54] |
| 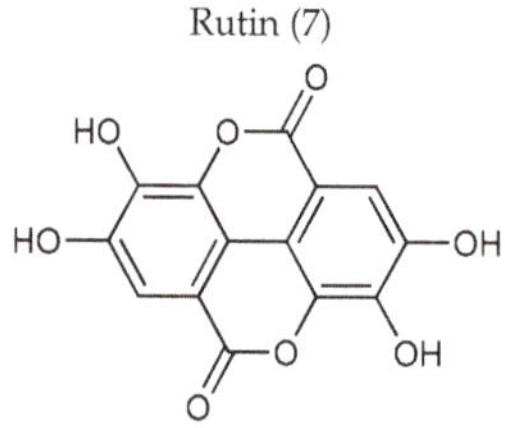 Ellagic Acid (8) | Antitumoral<br>Antiviral<br>Antimicrobial | [56] |

**Table 1.** *Cont.*

| Structure | Activity | Bibliographic Reference |
| --- | --- | --- |
| 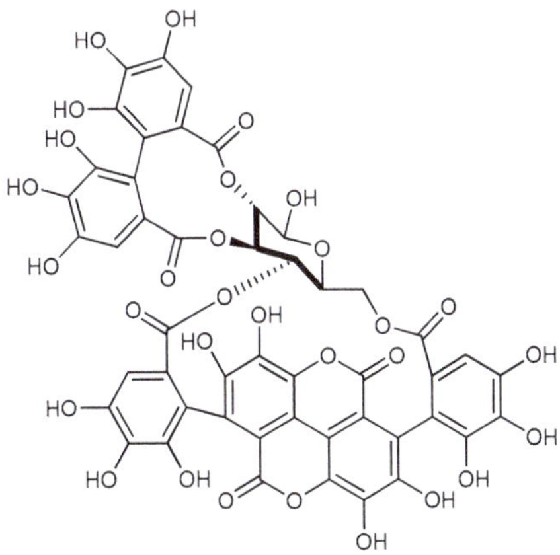 Punicalagin (9) | Anti-inflammatory | [57] |
| 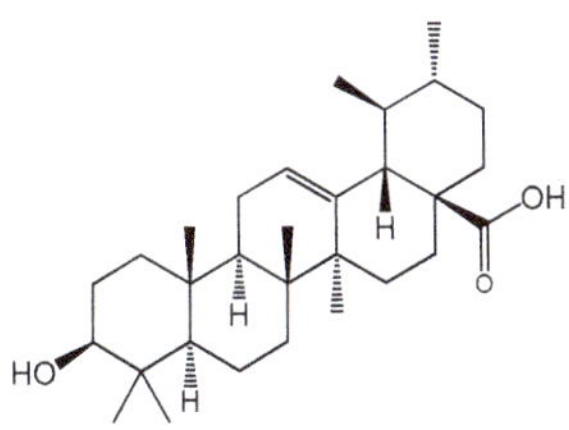 Ursolic acid (10) | Antiproliferative Pro-apoptotics Antigenotoxic Antitumoral | [65] |
| 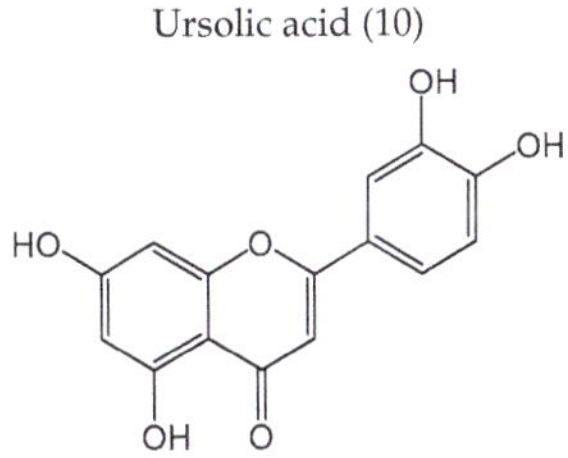 Luteolin (11) | Antitumoral | [65] |
| 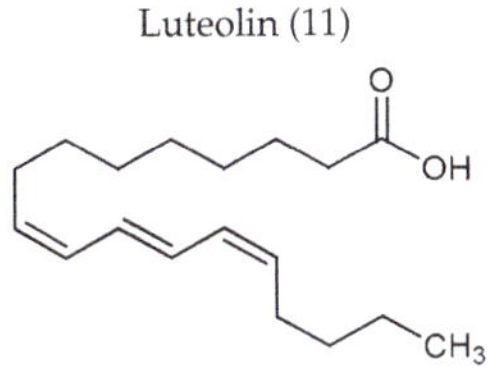 Punicic Acid (12) | Anti-inflammatory | [83] |

**Table 1.** *Cont.*

| Structure | Activity | Bibliographic Reference |
|---|---|---|
| 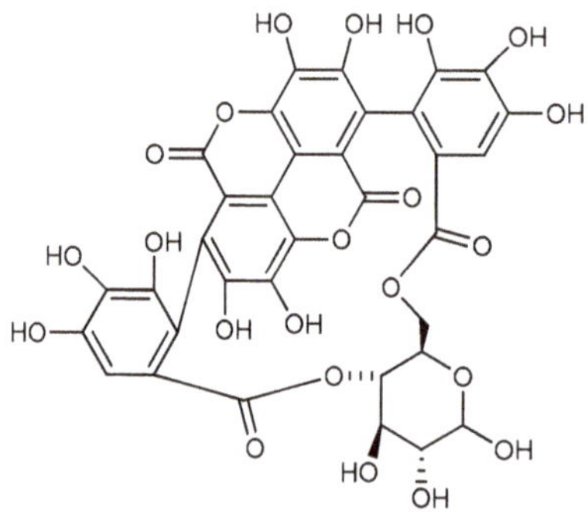 Punicalin (13) | Anti-inflammatory | [80] |
| 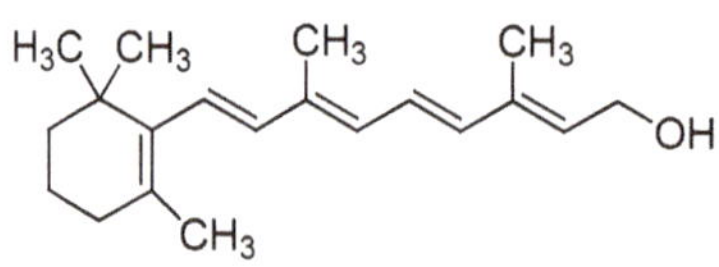 Vitamin A or retinol (14) | Antioxidant | [20] |
| 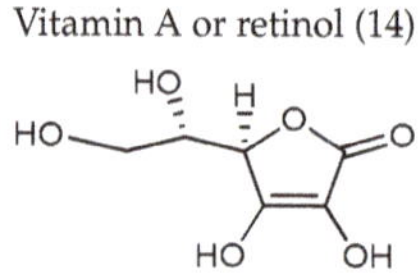 Vitamin C or ascorbic acid (15) | Antioxidant | [20] |
| 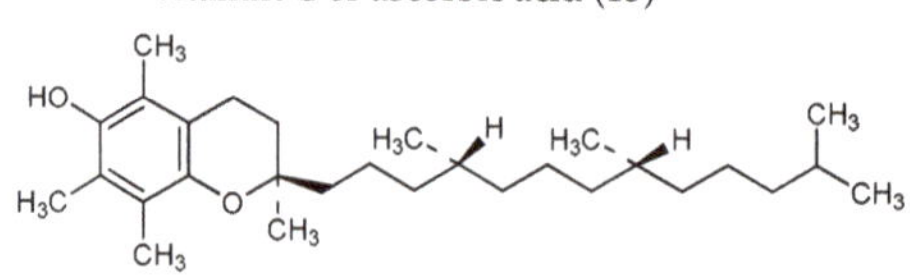 Vitamin E or tocopherol (16) | Antioxidant<br>Anti-inflammatory | [20] |
| 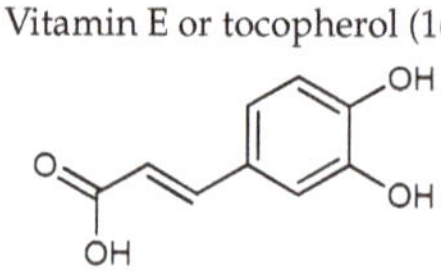 Caffeic acid (17) | Antidiabetics | [5] |
| 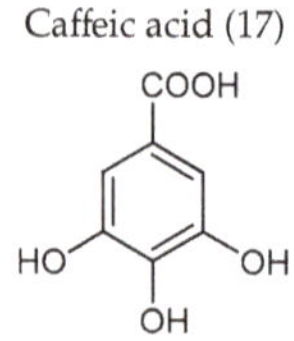 Gallic acid (18) | Antimicrobial<br>Prevention of cardiovascular diseases | [110] |

In particular, among them, pomegranate emerges, as well as nutritional resources, for its extensively documented health properties in several application fields (neoplastic, cardiovascular, viral, inflammatory, etc.) that make it a promising tool in medicinal treatments (Table 2).

**Table 2.** Biological properties of different pomegranate extracts.

| Pomegranate Extracts | Properties | References |
| --- | --- | --- |
| | **Antioxidant activity** | |
| **Juice** | ↓ ROS | [23,25,26,32] |
| **Peel (PPE)** | ↓ ROS | [25,27–31] |
| **Seed** | Radical scavenging | [24,25,32] |
| | **Anticancer properties** | |
| **Fruit (PoMx)** | Reverse tamoxifen resistant | [61] |
| | Inhibition of cell proliferation and metastasis (HSC-3) | [62] |
| **Husk** | Inhibition of cell proliferation (HepG2 and CACO) | [68] |
| **Juice** | Inhibition of cell proliferation (HT-29) | [5] |
| **Leaf** | Inhibition of cell proliferation (HeLa)<br>Necrosis<br>Apoptosis | [75] |
| **Peel** | Inhibition of cell proliferation (MCF-7 and HCT-116) | [67] |
| | Inhibition of cell proliferation (MCF-7, SKOV3, PC3 and A549) | [69] |
| | Inhibition of cell proliferation (K562)<br>Cell cycle G2/M<br>Apoptosis | [70] |
| | Inhibition of cell proliferation (HONE1) | [76] |
| **Seed** | Inhibition of cell proliferation (HepG2 and CACO) | [68] |
| **Seedless aril** | Inhibition of cell proliferation (LAPC4) | [59] |
| **Skin (PoPx)** | Inhibition of cell proliferation (MCF-7 and LAPC4)<br>Apoptosis | [59] |
| | Inhibition of cell proliferation (MCF-7)<br>Antiangiogenic<br>Inhibition of cell invasion and motility | [60] |
| | Inhibition of cell proliferation (melanocyte)<br>Free radicals block in UVA and UVB irradiated human skin | [63,64] |
| **Whole extract (PE)** | Inhibition of cell proliferation (WA-4)<br>Cell cycle arrest G0/G1<br>Apoptosis | [65] |
| | Inhibition of cell proliferation (PANC-1 and AsPC-1)<br>Cell cycle arrest G2 | [66] |
| | Immunomodulatory effects in stem cells | [71] |

**Table 2.** *Cont.*

| Pomegranate Extracts | Properties | References |
|---|---|---|
| **Anti-inflammatory properties** | | |
| **Flower (PFE)** | ↓ NO, iNOS, COX-2, PGE-2, TNF-α, IL-6, IL-1β<br>Inhibition of NF-kB activation and MAPKs phosphorylation | [89] |
| **Fruit (PoMx)** | ↓ Thrombospondin, TGF-β1 | [85] |
| **Peel (PPE)** | ↓ TNF-α, IL-1β, MCP-1, ICAM | [88] |
| **Seed** | ↓ COX, LOX | [81,82] |
| | ↓ Lipid peroxidation (LPO) | [83] |
| **Whole extract (PE)** | ↓ NO | [80,84] |
| | ↓ COX-2, PTGES | [87,91] |
| **Antidiabetic activity** | | |
| **Flower (PFE)** | Suppression of α-glucosidase activity | [95] |
| | Activation of PPAR-γ | [5] |
| | ↓ LPO | [100] |
| | Improvement of insulin sensitivity, blood lipid profile, and liver glycogen content | [102] |
| **Fruit (PoMx)** | ↓ Serum resistin | [99] |
| | ↓ FBG<br>↑ IRS-1, Akt, GLUT-2, GLUT-4 | [103] |
| **Juice** | Inhibition of DPP-4 | [104] |
| | ↓ Erythropoietin | [105] |
| **Peel (PPE)** | Inhibition of LPO<br>↓ Serum glucose concentration | [98] |
| **Seed** | Improvement of insulin sensitivity | [97] |
| **Antimicrobial activities** | | |
| **Peel (PPE)** | *S. Enteritidis, S. Kentucky* | [111] |
| | *C. albicans, C. parapsilosis, R. mucillaginosa, E. dermatitidis, S. aureus* | [112] |
| | *E.coli, P. aeruginisa, P. vulgaris, S. typhi, S. aureus, S. epidermidis, K. pneumonia* | [113] |
| **Skin (PoPx)** | *S. aureus, E. coli, Y. enterocolitica* | [110] |
| **Prevention of cardiovascular diseases** | | |
| **Juice** | ↑ Erythrocyte count, bleeding time, thrombin time<br>Activation of thromboplastin time<br>Inhibition of platelets aggregation induced by ADP | [118] |
| | Positive effects on angina pectoris | [119] |
| | Improvement of lipid profile, oxidative, and inflammatory biomarkers | [121] |
| | ↓ Blood pressure | [121,122] |
| **Peel (PPE)** | Reduction of doxorubicin cardiotoxicity | [120] |

**Table 2.** *Cont.*

| Pomegranate Extracts | Properties | References |
|---|---|---|
| | **Antiviral activities** | |
| **Peel (PPE)** | Hepatitis C virus | [127] |
| | HSV-2 | [130] |
| **Rind** | HSV-1 | [129] |
| **Whole extract (PE)** | Inhibition of H1N1, H3N2, H5N1 virus replication | [124,125] |
| | Inhibition of enterovirus 71 replication | [126] |
| | Adenovirus | [128] |
| | **Obesity** | |
| **Leaf** | Inhibition of pancreatic lipase | [137] |
| **Seed** | Inhibition of adipogenesis | [138] |
| | **Intestinal regulation** | |
| **Juice** | ↓ Intestinal motility<br>Rebalancing of intestinal microbiome | [141] |
| **Peel (PPE)** | ↓ Pathogenicity of *C. rodentium* | [142,143] |
| | **Effects on the male reproductive system** | |
| **Juice** | ↑ Sperm concentration, mobility, and density | [145] |
| | ↑ Erectile activity | [146] |
| | **Antidiarrheal effects** | |
| **Peel (PPE)** | ↓ Gastrointestinal transit | [149] |
| **Skin** | Inhibition of spontaneous ileum movements | [147] |
| | **Effects on oral health** | |
| **Whole extract (PE)** | ↓ Microorganism from the dental plaque | [151] |
| | ↓ Microorganism of gingivitis | [152] |
| | **Effects on pregnant and breastfeeding women** | |
| **Juice** | Improvement of uterine and umbilical artery function | [155] |
| | Presence of ellagic acid in breast milk | [156] |
| **Peel (PPE)** | Antibacterial effects against *S. aureus* and *E.coli* | [153] |
| **Whole extract (PE)** | ↓ bilirubin | [154] |
| | **Effects on nervous system** | |
| **Juice** | Impediment of oxidative damage and α-synuclein aggregationImprovement of mitochondrial ALDH activity | [161] |
| | Stabilization of the ability to learn visual information | [162] |
| **Whole extract (PE)** | ↓ NO, IL-6, PGE-2, TNF-α in BV-2 cells | [158,159] |
| | Protection of SH-SY5Y and BV-2 cell viability<br>Activation of Sirt-1 in BV-2 cells | [159] |
| | Improvement of mitochondrial activity, CAT, SOD, GR, and GPx activities and redox state | [160] |
| | ↑ Peroxiredoxins 1 and 3 | |

Certainly, the juice is the most widespread type of pomegranate-based product in stores for its ease of consumption, followed by capsules with a concentration of ellagic acid of 40% [3]. From what emerges from the various studies analyzed, the interest in the consumption and production of pomegranate has greatly increased in recent years [172–175].

Indeed, prior research has thoroughly supported the beneficial effects of pomegranate regarding not only the phytocomplex, but also the single components, among which the various tannins and the ellagic acid stand out. However, several studies stated that the synergistic potential of all pomegranate components (for example phenolic compounds) is powerful to that of single constituents [176–178]. Precisely, the latter has recently aroused great interest for the potential inhibitory activity against Sars-CoV-2, demonstrated through docking studies and, for this reason, it seems clear that continuing the research with the evaluation of the activity could represent an important goal for the fight against this plague [131,133].

Moreover, much research has promoted the possible employment of pomegranate together with other drugs already available (like doxorubicin), establishing its good influence in reverting their chemoresistance, thus overcoming their limits, lowering the toxicity, and increasing the action.

In light of the reasons mentioned above, it becomes evident that pomegranate turns out to be one of the most attractive nutraceuticals, and the deepening of its study could lead to the opening of new avenues for medical science domains.

**Author Contributions:** Conceptualization, A.C. (Anna Caruso) and A.B.; methodology, D.I.; software, J.C.; validation, A.F. and C.F.; resources, A.T. and G.B.; data curation, A.C. (Alessia Catalano); writing—original draft preparation, A.C. (Anna Caruso) and A.B.; writing—review and editing, A.C. (Alessia Carocci) and D.I.; supervision, M.S.S. All authors have read and agreed to the published version of the manuscript.

**Funding:** This research received no external funding

**Conflicts of Interest:** The authors declare no conflict of interest.

## References

1. Bellini, E.; Giordani, E.; La Malfa, S. Minor fruit tree species in Italy, a traditional resource for in the innavation of fruitculture: Persimmon and pomegranate as study cases. *Italus Hortus* **2010**, *17*, 75–90.
2. Sepulveda, E.L.G.; Sàenz, C.; Tapia, M. Minimal processing of pomegranate var. wonderful. In Proceedings of the Production, Processing and Marketing of Pomegranate in the Mediterranean Region: Advances in Research and Technology, Universidad Miguel Hernández (EPSO-UMH), Orihuela, Spain, 15–17 October 1998; pp. 237–242.
3. Da Silva, J.A.T.; Rana, T.S.; Narzary, D.; Verma, N.; Meshram, D.T.; Ranade, S.A. Pomegranate biology and biotechnology: A review. *Sci. Hortic.* **2013**, *160*, 85–107. [CrossRef]
4. Heber, D.; Schulman, R.N.; Seeram, N.P. *Pomegranates: Ancient Roots to Modern Medicine*; CRC Press: Boca Raton, FL, USA, 2006; p. 262.
5. Viuda-Martos, M.; Fernandez-Lopez, J.; Perez-Alvarez, J.A. Pomegranate and its Many Functional Components as Related to Human Health: A Review. *Compr. Rev. Food Sci. Food Saf.* **2010**, *9*, 635–654. [CrossRef]
6. Andreu-Sevilla, A.J.; Signes-Pastor, A.J.; Carbonell-Barrachina, A.A. La Granada Y Su Zumo. Producción, Composición Y Propiedades Beneficiosas Para La Salud. *Inf. Univ. Miguel Hernández Elchedigital* **2008**, *234*, 36–39.
7. Jurenka, J.S. Therapeutic applications of pomegranate (*Punica granatum* L.): A review. *Altern. Med. Rev.* **2008**, *13*, 128–144.
8. Ismail, T.; Sestili, P.; Akhtar, S. Pomegranate peel and fruit extracts: A review of potential anti-inflammatory and anti-infective effects. *J. Ethnopharmacol.* **2012**, *143*, 397–405. [CrossRef]
9. Cano, A.; Arnao, M.B. Hydrophilic and lipophilic antioxidant activity in different leaves of three lettuce varieties. *Int. J. Food Prop.* **2005**, *8*, 521–528. [CrossRef]

10. Plastina, P.; Apriantini, A.; Meijerink, J.; Witkamp, R.; Gabriele, B.; Fazio, A. In Vitro Anti-Inflammatory and Radical Scavenging Properties of Chinotto (*Citrus myrtifolia Raf.*) Essential Oils. *Nutrients* **2018**, *10*, 783. [CrossRef]
11. Chen, M.L.; Vigneault, C.; Raghavan, G.S.V.; Kubow, S. Importance of phytochemical content of fruits and vegetables to human health. *Stewart Postharvest Rev.* **2007**, *3*, 20–32.
12. Badria, F.A.; Zidan, O.A. Natural products for dental caries prevention. *J. Med. Food* **2004**, *7*, 381–384. [CrossRef]
13. Parisi, O.I.; Casaburi, I.; Sinicropi, M.S.; Avena, P.; Caruso, A.; Givigliano, F.; Pezzi, V.; Puoci, F. Most Relevant Polyphenols Present in the Mediterranean Diet and Their Incidence in Cancer Diseases. In *Polyphenols in Human Health and Disease*; Watson, R.R., Preedy, V.R., Zibadi, S., Eds.; Academic Press: Cambridge, MA, USA, 2014; Volume 2, pp. 1341–1351.
14. Plastina, P.; Fazio, A.; Gabriele, B. Comparison of fatty acid profile and antioxidant potential of extracts of seven Citrus rootstock seeds. *Nat. Prod. Res.* **2012**, *26*, 2182–2187. [CrossRef] [PubMed]
15. Leporini, M.; Loizzo, M.R.; Tundis, R.; La Torre, C.; Fazio, A.; Plastina, P. Non-Pungent n-3 Polyunsaturated Fatty Acid (PUFA)-Derived Capsaicin Analogues as Potential Functional Ingredients with Antioxidant and Carbohydrate-Hydrolysing Enzyme Inhibitory Activities. *Antioxidants* **2019**, *8*, 162. [CrossRef] [PubMed]
16. Benincasa, C.; La Torre, C.; Plastina, P.; Fazio, A.; Perri, E.; Caroleo, M.C.; Gallelli, L.; Cannataro, R.; Cione, E. Hydroxytyrosyl Oleate: Improved Extraction Procedure from Olive Oil and By-Products, and In Vitro Antioxidant and Skin Regenerative Properties. *Antioxidants* **2019**, *8*, 233. [CrossRef] [PubMed]
17. Gabriele, B.; Fazio, A.; Carchedi, M.; Plastina, P. In vitro antioxidant activity of extracts of Sybaris liquorice roots from Southern Italy. *Nat. Prod. Res.* **2012**, *26*, 2176–2181. [CrossRef]
18. Cannataro, R.; Caroleo, M.C.; Fazio, A.; La Torre, C.; Plastina, P.; Gallelli, L.; Lauria, G.; Cione, E. Ketogenic Diet and microRNAs Linked to Antioxidant Biochemical Homeostasis. *Antioxidants* **2019**, *8*, 269. [CrossRef]
19. Espin, J.C.; Garcia-Conesa, M.T.; Tomas-Barberan, F.A. Nutraceuticals: Facts and fiction. *Phytochemistry* **2007**, *68*, 2986–3008. [CrossRef]
20. Hounsome, N.; Hounsome, B.; Tomos, D.; Edwards-Jones, G. Plant metabolites and nutritional quality of vegetables. *J. Food Sci.* **2008**, *73*, R48–R65. [CrossRef]
21. Dorais, M.; Ehret, D.L.; Papadopoulos, A.P. Tomato *(Solanum lycopersicum)* health components: From the seed to the consumer. *Phytochem. Rev.* **2008**, *7*, 231–250. [CrossRef]
22. Amarowicz, R.; Pegg, R.B.; Rahimi-Moghaddam, P.; Barl, B.; Weil, J.A. Free-radical scavenging capacity and antioxidant activity of selected plant species from the Canadian prairies. *Food Chem.* **2004**, *84*, 551–562. [CrossRef]
23. Gil, M.I.; Tomas-Barberan, F.A.; Hess-Pierce, B.; Holcroft, D.M.; Kader, A.A. Antioxidant activity of pomegranate juice and its relationship with phenolic composition and processing. *J. Agric. Food Chem.* **2000**, *48*, 4581–4589. [CrossRef]
24. Basiri, S. Evaluation of antioxidant and antiradical properties of Pomegranate (*Punica granatum* L.) seed and defatted seed extracts. *J. Food Sci. Technol.* **2015**, *52*, 1117–1123. [CrossRef] [PubMed]
25. Derakhshan, Z.; Ferrante, M.; Tadi, M.; Ansari, F.; Heydari, A.; Hosseini, M.S.; Conti, G.O.; Sadrabad, E.K. Antioxidant activity and total phenolic content of ethanolic extract of pomegranate peels, juice and seeds. *Food Chem. Toxicol.* **2018**, *114*, 108–111. [CrossRef] [PubMed]
26. Di Stefano, V.; Pitonzo, R.; Novara, M.E.; Bongiorno, D.; Indelicato, S.; Gentile, C.; Avellone, G.; Bognanni, R.; Scandurra, S.; Melilli, M.G. Antioxidant activity and phenolic composition in pomegranate (*Punica granatum* L.) genotypes from south Italy by UHPLC-Orbitrap-MS approach. *J. Sci. Food Agric.* **2019**, *99*, 1038–1045. [CrossRef]
27. Mastrogiovanni, F.; Bernini, R.; Basirico, L.; Bernabucci, U.; Campo, M.; Romani, A.; Santi, L.; Lacetera, N. Antioxidant and anti-inflammatory effects of pomegranate peel extracts on bovine mammary epithelial cells BME-UV1. *Nat. Prod. Res.* **2020**, *34*, 1465–1469. [CrossRef] [PubMed]
28. Nur Hanani, Z.A.; Cheng Yee, F.; Nor-Khaizura, M.A.R. Effect of pomegranate (*Punica granatum* L.) peel powder on the antioxidant and antimicrobial properties of fish gelatin films as active packaging. *Food Hydrocoll.* **2019**, *89*, 253–259. [CrossRef]

29. Bertolo, M.R.V.; Martins, V.C.A.; Horn, M.M.; Brenelli, L.B.; Plepis, A.M.G. Rheological and antioxidant properties of chitosan/gelatin-based materials functionalized by pomegranate peel extract. *Carbohydr. Polym.* **2020**, *228*, 115386. [CrossRef]
30. Jalal, H.; Pal, M.A.; Ahmad, S.R.; Rather, M.; Andrabi, M.; Hamdani, S. Physico-chemical and functional properties of pomegranate peel and seed powder. *Pharm. Innov. J.* **2018**, *7*, 1127–1131.
31. Surek, E.; Nilufer-Erdil, D. Phenolic contents, antioxidant activities and potential bioaccessibilities of industrial pomegranate nectar processing wastes. *Int. J. Food Sci. Technol.* **2016**, *51*, 231–239. [CrossRef]
32. Pinilla, M.; Iglesias-Moya, J.; Jesús Campos, M.; Corpas, F.J.; Palma, J.M. Pomegranate (*Punica granatum* L.) Fruits: Characterization of the Main Enzymatic Antioxidants (Peroxisomal Catalase and SOD Isozymes) and the NADPH-Regenerating System. *Agronomy* **2019**, *9*, 338. [CrossRef]
33. Ferlay, J.; Shin, H.R.; Bray, F.; Forman, D.; Mathers, C.; Parkin, D.M. Estimates of worldwide burden of cancer in 2008: GLOBOCAN 2008. *Int. J. Cancer* **2010**, *127*, 2893–2917. [CrossRef]
34. Saturnino, C.; Caruso, A.; Iacopetta, D.; Rosano, C.; Ceramella, J.; Muia, N.; Mariconda, A.; Bonomo, M.G.; Ponassi, M.; Rosace, G.; et al. Inhibition of Human Topoisomerase II by N,N,N-Trimethylethanammonium Iodide Alkylcarbazole Derivatives. *Chem. Med. Chem.* **2018**, *13*, 2635–2643. [CrossRef] [PubMed]
35. Saturnino, C.; Caruso, A.; Longo, P.; Capasso, A.; Pingitore, A.; Caroleo, M.C.; Cione, E.; Perri, M.; Nicolo, F.; Nardo, V.M.; et al. Crystallographic study and biological evaluation of 1,4-dimethyl-N-alkylcarbazoles. *Curr. Top. Med. Chem.* **2015**, *15*, 973–979. [CrossRef]
36. Chimento, A.; Sala, M.; Gomez-Monterrey, I.M.; Musella, S.; Bertamino, A.; Caruso, A.; Sinicropi, M.S.; Sirianni, R.; Puoci, F.; Parisi, O.I.; et al. Biological activity of 3-chloro-azetidin-2-one derivatives having interesting antiproliferative activity on human breast cancer cell lines. *Bioorganic Med. Chem. Lett.* **2013**, *23*, 6401–6405. [CrossRef] [PubMed]
37. Kawanishi, S.; Hiraku, Y.; Pinlaor, S.; Ma, N. Oxidative and nitrative DNA damage in animals and patients with inflammatory diseases in relation to inflammation-related carcinogenesis. *Biol. Chem.* **2006**, *387*, 365–372. [CrossRef] [PubMed]
38. Ishikawa, K.; Takenaga, K.; Akimoto, M.; Koshikawa, N.; Yamaguchi, A.; Imanishi, H.; Nakada, K.; Honma, Y.; Hayashi, J. ROS-generating mitochondrial DNA mutations can regulate tumor cell metastasis. *Science* **2008**, *320*, 661–664. [CrossRef]
39. Caruso, A.; Iacopetta, D.; Puoci, F.; Cappello, A.R.; Saturnino, C.; Sinicropi, M.S. Carbazole derivatives: A promising scenario for breast cancer treatment. *Mini Rev. Med. Chem.* **2016**, *16*, 630–643. [CrossRef]
40. Ceramella, J.; Caruso, A.; Occhiuzzi, M.A.; Iacopetta, D.; Barbarossa, A.; Rizzuti, B.; Dallemagne, P.; Rault, S.; El-Kashef, H.; Saturnino, C.; et al. Benzothienoquinazolinones as new multi-target scaffolds: Dual inhibition of human Topoisomerase I and tubulin polymerization. *Eur. J. Med. Chem.* **2019**, *181*, 111583. [CrossRef]
41. Sinicropi, M.S.; Iacopetta, D.; Rosano, C.; Randino, R.; Caruso, A.; Saturnino, C.; Muia, N.; Ceramella, J.; Puoci, F.; Rodriquez, M.; et al. N-thioalkylcarbazoles derivatives as new anti-proliferative agents: Synthesis, characterisation and molecular mechanism evaluation. *J. Enzyme Inhib. Med. Chem.* **2018**, *33*, 434–444. [CrossRef]
42. Faienza, M.F.; Corbo, F.; Carocci, A.; Catalano, A.; Clodoveo, M.L.; Grano, M.; Wang, D.Q.-H.; D'Amato, G.; Muraglia, M.; Franchini, C.; et al. Novel insights in health-promoting properties of sweet cherries. *J. Funct. Foods* **2020**, *69*, 103945. [CrossRef]
43. Karaaslan, M.; Vardin, H.; Varliklioz, S.; Yilmaz, F.M. Antiproliferative and antioxidant activities of Turkish pomegranate (*Punica granatum* L.) accessions. *Int. J. Food Sci. Technol.* **2014**, *49*, 82–90. [CrossRef]
44. Aqil, F.; Munagala, R.; Vadhanam, M.V.; Kausar, H.; Jeyabalan, J.; Schultz, D.J.; Gupta, R.C. Anti-proliferative activity and protection against oxidative DNA damage by punicalagin isolated from pomegranate husk. *Food Res. Int.* **2012**, *49*, 345–353. [CrossRef] [PubMed]
45. Dikmen, M.; Ozturk, N.; Ozturk, Y. The antioxidant potency of *Punica granatum* L. Fruit peel reduces cell proliferation and induces apoptosis on breast cancer. *J. Med. Food* **2011**, *14*, 1638–1646. [CrossRef] [PubMed]
46. Lansky, E.P.; Jiang, W.; Mo, H.; Bravo, L.; Froom, P.; Yu, W.; Harris, N.M.; Neeman, I.; Campbell, M.J. Possible synergistic prostate cancer suppression by anatomically discrete pomegranate fractions. *Investig. New Drugs* **2005**, *23*, 11–20. [CrossRef] [PubMed]

47. Fazio, A.; Iacopetta, D.; La Torre, C.; Ceramella, J.; Muia, N.; Catalano, A.; Carocci, A.; Sinicropi, M.S. Finding solutions for agricultural wastes: Antioxidant and antitumor properties of pomegranate Akko peel extracts and beta-glucan recovery. *Food Funct.* **2018**, *9*, 6618–6631. [CrossRef] [PubMed]
48. Hong, M.Y.; Seeram, N.P.; Heber, D. Pomegranate polyphenols down-regulate expression of androgen-synthesizing genes in human prostate cancer cells overexpressing the androgen receptor. *J. Nutr. Biochem.* **2008**, *19*, 848–855. [CrossRef]
49. Di Carlo, G.; Mascolo, N.; Izzo, A.A.; Capasso, F. Flavonoids: Old and new aspects of a class of natural therapeutic drugs. *Life Sci.* **1999**, *65*, 337–353. [CrossRef]
50. Noda, Y.; Kaneyuki, T.; Mori, A.; Packer, L. Antioxidant activities of pomegranate fruit extract and its anthocyanidins: Delphinidin, cyanidin, and pelargonidin. *J. Agric. Food Chem.* **2002**, *50*, 166–171. [CrossRef]
51. Iacopetta, D.; Grande, F.; Caruso, A.; Mordocco, R.A.; Plutino, M.R.; Scrivano, L.; Ceramella, J.; Muia, N.; Saturnino, C.; Puoci, F.; et al. New insights for the use of quercetin analogs in cancer treatment. *Future Med. Chem.* **2017**, *9*, 2011–2028. [CrossRef]
52. Levin, G.M. *Pomegranate Roads: A Soviet Botanist's Exile from Eden*; Floreant Press: Forestville, CA, USA, 2006.
53. Grande, F.; Parisi, O.I.; Mordocco, R.A.; Rocca, C.; Puoci, F.; Scrivano, L.; Quintieri, A.M.; Cantafio, P.; Ferla, S.; Brancale, A.; et al. Quercetin derivatives as novel antihypertensive agents: Synthesis and physiological characterization. *Eur. J. Pharm. Sci.* **2016**, *82*, 161–170. [CrossRef]
54. Van Elswijk, D.A.; Schobel, U.P.; Lansky, E.P.; Irth, H.; van der Greef, J. Rapid dereplication of estrogenic compounds in pomegranate (*Punica granatum*) using on-line biochemical detection coupled to mass spectrometry. *Phytochemistry* **2004**, *65*, 233–241. [CrossRef]
55. Khatib, M. *Bioactive Compounds into Edible Syrian Plants: Pomegranate and Capper*; University of Florence: Florence, Italy, 2015.
56. Gonzalez-Sarrias, A.; Espin, J.C.; Tomas-Barberan, F.A.; Garcia-Conesa, M.T. Gene expression, cell cycle arrest and MAPK signalling regulation in Caco-2 cells exposed to ellagic acid and its metabolites, urolithins. *Mol. Nutr. Food Res.* **2009**, *53*, 686–698. [CrossRef] [PubMed]
57. Adams, L.S.; Seeram, N.P.; Aggarwal, B.B.; Takada, Y.; Sand, D.; Heber, D. Pomegranate juice, total pomegranate ellagitannins, and punicalagin suppress inflammatory cell signaling in colon cancer cells. *J. Agric. Food Chem.* **2006**, *54*, 980–985. [CrossRef] [PubMed]
58. Pantuck, A.J.; Leppert, J.T.; Zomorodian, N.; Aronson, W.; Hong, J.; Barnard, R.J.; Seeram, N.; Liker, H.; Wang, H.; Elashoff, R.; et al. Phase II study of pomegranate juice for men with rising prostate-specific antigen following surgery or radiation for prostate cancer. *Clin. Cancer Res.* **2006**, *12*, 4018–4026. [CrossRef] [PubMed]
59. Koyama, S.; Cobb, L.J.; Mehta, H.H.; Seeram, N.P.; Heber, D.; Pantuck, A.J.; Cohen, P. Pomegranate extract induces apoptosis in human prostate cancer cells by modulation of the IGF-IGFBP axis. *Growth Horm. IGF Res.* **2010**, *20*, 55–62. [CrossRef]
60. Khan, G.N.; Gorin, M.A.; Rosenthal, D.; Pan, Q.; Bao, L.W.; Wu, Z.F.; Newman, R.A.; Pawlus, A.D.; Yang, P.; Lansky, E.P.; et al. Pomegranate fruit extract impairs invasion and motility in human breast cancer. *Integr. Cancer Ther.* **2009**, *8*, 242–253. [CrossRef]
61. Aiyer, H.S.; Bouker, K.B.; Cook, K.L.; Facey, C.O.; Hu, R.; Schwartz, J.L.; Shajahan, A.N.; Hilakivi-Clarke, L.; Clarke, R. Interaction of dietary polyphenols with molecular signaling pathways of antiestrogen resistance: Possible role in breast cancer recurrence. *Horm. Mol. Biol. Clin. Investig.* **2012**, *9*, 127–141. [CrossRef]
62. Peng, S.Y.; Hsiao, C.C.; Lan, T.H.; Yen, C.Y.; Farooqi, A.A.; Cheng, C.M.; Tang, J.Y.; Yu, T.J.; Yeh, Y.C.; Chuang, Y.T.; et al. Pomegranate extract inhibits migration and invasion of oral cancer cells by downregulating matrix metalloproteinase-2/9 and epithelial-mesenchymal transition. *Environ. Toxicol.* **2020**, *35*, 673–682. [CrossRef]
63. Kasai, K.; Yoshimura, M.; Koga, T.; Arii, M.; Kawasaki, S. Effects of oral administration of ellagic acid-rich pomegranate extract on ultraviolet-induced pigmentation in the human skin. *J. Nutr. Sci. Vitaminol.* **2006**, *52*, 383–388. [CrossRef]
64. Manasathien, J.; Kupittayanant, S.; Indrapichate, K. Protective Efficacy of Pomegranate (*Punica Granatum* Linn., Punicaceae) Peel Ethanolic Extract on UVB-Irradiated Rat Skin. *Am. Eurasian J. Toxicol. Sci.* **2011**, *3*, 250–258.

65. Dai, Z.; Nair, V.; Khan, M.; Ciolino, H.P. Pomegranate extract inhibits the proliferation and viability of MMTV-Wnt-1 mouse mammary cancer stem cells in vitro. *Oncol. Rep.* **2010**, *24*, 1087–1091.
66. Nair, V.; Dai, Z.; Khan, M.; Ciolino, H.P. Pomegranate extract induces cell cycle arrest and alters cellular phenotype of human pancreatic cancer cells. *Anticancer. Res.* **2011**, *31*, 2699–2704.
67. Motaal, A.A.; Shaker, S. Anticancer and Antioxidant Activities of Standardized Whole Fruit, Pulp and Peel Extracts of Egyptian Pomegranate. *Open Conf. Proc. J.* **2011**, *2*, 41–45. [CrossRef]
68. El-Awady, M.A.; Awad, N.S.; El-Tarras, A.E. Evaluation of the anticancer activities of pomegranate (*Punica granatum*) and harmal (*Rhazya stricta*) plants grown in Saudi Arabia. *Int. J. Curr. Microbiol. Appl. Sci.* **2015**, *4*, 1158–1167.
69. Modaeinama, S.; Abasi, M.; Abbasi, M.M.; Jahanban-Esfahlan, R. Anti-Tumoral Properties of Punica Granatum (*Pomegranate*) Peel Extract on Different Human Cancer Cells. *Asian Pac. J. Cancer Prev.* **2015**, *16*, 5697–5701. [CrossRef] [PubMed]
70. Asmaa, M.J.; Ali, A.J.; Farid, J.M.; Azman, S. Growth inhibitory effects of crude pomegranate peel extract on chronic myeloid leukemia, K562 cells. *Int. J. Appl. Basic Med. Res.* **2015**, *5*, 100–105. [PubMed]
71. Rostami, Z.; Khorashadizadeh, M.; Ghoncheh, M.; Naseri, M. Effect of Pomegranate Extract in Mesenchymal Stem Cells by Modulation of MicroRNA-155, MicroRNA-21, MicroRNA-23b, MicroRNA-126a, and PI3K\AKT1\NF-B Expression. *DNA Cell Biol.* **2020**. [CrossRef] [PubMed]
72. Saturnino, C.; Barone, I.; Iacopetta, D.; Mariconda, A.; Sinicropi, M.S.; Rosano, C.; Campana, A.; Catalano, S.; Longo, P.; Ando, S. N-heterocyclic carbene complexes of silver and gold as novel tools against breast cancer progression. *Futur. Med. Chem.* **2016**, *8*, 2213–2229. [CrossRef]
73. Chimento, A.; De Amicis, F.; Sirianni, R.; Sinicropi, M.S.; Puoci, F.; Casaburi, I.; Saturnino, C.; Pezzi, V. Progress to Improve Oral Bioavailability and Beneficial Effects of Resveratrol. *Int. J. Mol. Sci.* **2019**, *20*, 1381. [CrossRef]
74. Parisi, O.I.; Morelli, C.; Puoci, F.; Saturnino, C.; Caruso, A.; Sisci, D.; Trombino, G.E.; Picci, N.; Sinicropi, M.S. Magnetic molecularly imprinted polymers (MMIPs) for carbazole derivative release in targeted cancer therapy. *J. Mater. Chem. B* **2014**, *2*, 6619–6625. [CrossRef]
75. Sarkar, S.; Kotteeswaran, V. Green synthesis of silver nanoparticles from aqueous leaf extract of Pomegranate (*Punica granatum*) and their anticancer activity on human cervical cancer cells. *Adv. Nat. Sci. Nanosci. Nanotechnol.* **2018**, *9*, 025014. [CrossRef]
76. Yusefi, M.; Shameli, K.; Ali, R.R.; Pang, S.W.; Teow, S.Y. Evaluating Anticancer Activity of Plant-Mediated Synthesized Iron Oxide Nanoparticles Using Punica Granatum Fruit Peel Extract. *J. Mol. Struct.* **2020**, *1204*, 127539. [CrossRef]
77. Sinicropi, M.S.; Caruso, A.; Conforti, F.; Marrelli, M.; El Kashef, H.; Lancelot, J.C.; Rault, S.; Statti, G.A.; Menichini, F. Synthesis, inhibition of NO production and antiproliferative activities of some indole derivatives. *J. Enzyme Inhib. Med. Chem.* **2009**, *24*, 1148–1153. [CrossRef] [PubMed]
78. Saturnino, C.; Popolo, A.; Ramunno, A.; Adesso, S.; Pecoraro, M.; Plutino, M.R.; Rizzato, S.; Albinati, A.; Marzocco, S.; Sala, M.; et al. Anti-Inflammatory, Antioxidant and Crystallographic Studies of N-Palmitoyl-ethanol Amine (PEA) Derivatives. *Molecules* **2017**, *22*, 616. [CrossRef] [PubMed]
79. Plastina, P.; Benincasa, C.; Perri, E.; Fazio, A.; Augimeri, G.; Poland, M.; Witkamp, R.; Meijerink, J. Identification of hydroxytyrosyl oleate, a derivative of hydroxytyrosol with anti-inflammatory properties, in olive oil by-products. *Food Chem.* **2019**, *279*, 105–113. [CrossRef]
80. Lee, C.J.; Chen, L.G.; Liang, W.L.; Wang, C.C. Anti-inflammatory effects of *Punica granatum* Linne in vitro and in vivo. *Food Chem.* **2010**, *118*, 315–322. [CrossRef]
81. Tomas-Barberan, F.A. Granada y salud: Aspectos farmacologicos y terapéuticos de la Granada. In *El Granado—I Jornadas Nacionales Sobre el Granado: Produccion, Economia, Indrustrializacion, Alimentacion y Salud*; Moreno, P.M., Garcia, F.H., Murcia, P.L., Eds.; Documentos Poscosecha: Valencia, Spain, 2010; pp. 7–22.
82. Negro, C.; Longo, L.; Vasapollo, G.; De Bellis, L.; Miceli, A. Biochemical, Antioxidant and Anti-Inflammatory Properties of Pomegranate Fruits Growing in Southern Italy (Salento, Apulia). *Acta Aliment.* **2012**, *41*, 190–199. [CrossRef]

83. Boussetta, T.; Raad, H.; Letteron, P.; Gougerot-Pocidalo, M.A.; Marie, J.C.; Driss, F.; El-Benna, J. Punicic acid a conjugated linolenic acid inhibits TNFalpha-induced neutrophil hyperactivation and protects from experimental colon inflammation in rats. *PLoS ONE* **2009**, *4*, e6458. [CrossRef]
84. Jung, K.H.; Kim, M.J.; Ha, E.; Uhm, Y.K.; Kim, H.K.; Chung, J.H.; Yim, S.V. Suppressive effect of Punica granatum on the production of tumor necrosis factor (Tnf) in BV2 microglial cells. *Biol. Pharm. Bull.* **2006**, *29*, 1258–1261. [CrossRef]
85. De Nigris, F.; Balestrieri, M.L.; Williams-Ignarro, S.; D'Armiento, F.P.; Fiorito, C.; Ignarro, L.J.; Napoli, C. The influence of pomegranate fruit extract in comparison to regular pomegranate juice and seed oil on nitric oxide and arterial function in obese Zucker rats. *Nitric Oxide Biol. Chem.* **2007**, *17*, 50–54. [CrossRef]
86. Larrosa, M.; Gonzalez-Sarrias, A.; Yanez-Gascon, M.J.; Selma, M.V.; Azorin-Ortuno, M.; Toti, S.; Tomas-Barberan, F.; Dolara, P.; Espin, J.C. Anti-inflammatory properties of a pomegranate extract and its metabolite urolithin-A in a colitis rat model and the effect of colon inflammation on phenolic metabolism. *J. Nutr. Biochem.* **2010**, *21*, 717–725. [CrossRef]
87. Ouachrif, A.; Khalki, H.; Chaib, S.; Mountassir, M.; Aboufatima, R.; Farouk, L.; Benharraf, A.; Chait, A. Comparative study of the anti-inflammatory and antinociceptive effects of two varieties of Punica granatum. *Pharm. Biol.* **2012**, *50*, 429–438. [CrossRef] [PubMed]
88. Park, S.; Seok, J.K.; Kwak, J.Y.; Suh, H.J.; Kim, Y.M.; Boo, Y.C. Anti-Inflammatory Effects of Pomegranate Peel Extract in THP-1 Cells Exposed to Particulate Matter PM10. *Evid. Based Complement. Altern. Med.* **2016**, *2016*, 1–11. [CrossRef]
89. Xu, J.; Zhao, Y.; Aisa, H.A. Anti-inflammatory effect of pomegranate flower in lipopolysaccharide (LPS)-stimulated RAW264.7 macrophages. *Pharm. Biol.* **2017**, *55*, 2095–2101. [CrossRef] [PubMed]
90. Kim, H.; Banerjee, N.; Ivanov, I.; Pfent, C.M.; Prudhomme, K.R.; Bisson, W.H.; Dashwood, R.H.; Talcott, S.T.; Mertens-Talcott, S.U. Comparison of anti-inflammatory mechanisms of mango (*Mangifera indica* L.) and pomegranate (*Punica granatum* L.) in a preclinical model of colitis. *Mol. Nutr. Food Res.* **2016**, *60*, 1912–1923. [CrossRef] [PubMed]
91. Gonzalez-Trujano, M.E.; Pellicer, F.; Mena, P.; Moreno, D.A.; Garcia-Viguera, C. Antinociceptive and anti-inflammatory activities of a pomegranate (*Punica granatum* L.) extract rich in ellagitannins. *Int. J. Food Sci. Nutr.* **2015**, *66*, 395–399. [CrossRef]
92. Mo, J.; Panichayupakaranant, P.; Kaewnopparat, N.; Songkro, S.; Reanmongkol, W. Topical anti-inflammatory potential of standardized pomegranate rind extract and ellagic acid in contact dermatitis. *Phytother. Res.* **2014**, *28*, 629–632. [CrossRef]
93. Ruffo, M.; Parisi, O.I.; Scrivano, L.; Restuccia, D.; Amone, F.; Sinicropi, M.S.; Malivindi, R.; Aiello, G.; Puoci, F. Role of Calabrian Black Rice in Metabolic Syndrome: In vitro Evaluation of *Oryza sativa* L. Indica Biological Properties. *Curr. Nutr. Food Sci.* **2018**, *14*, 121–127. [CrossRef]
94. Cheng, J.T.; Liu, I.M. Stimulatory effect of caffeic acid on alpha1A-adrenoceptors to increase glucose uptake into cultured C2C12 cells. *Naunyn Schmiedebergs Arch. Pharmacol.* **2000**, *362*, 122–127. [CrossRef]
95. Li, Y.; Wen, S.; Kota, B.P.; Peng, G.; Li, G.Q.; Yamahara, J.; Roufogalis, B.D. Punica granatum flower extract, a potent alpha-glucosidase inhibitor, improves postprandial hyperglycemia in Zucker diabetic fatty rats. *J. Ethnopharmacol.* **2005**, *99*, 239–244. [CrossRef]
96. Huang, T.H.W.; Peng, G.; Kota, B.P.; Li, G.Q.; Yamahara, J.; Roufogalis, B.D.; Li, Y.H. Anti-diabetic action of *Punica granatum* flower extract: Activation of PPAR-gamma and identification of an active component. *Toxicol. Appl. Pharmacol.* **2005**, *207*, 160–169. [CrossRef]
97. McFarlin, B.K.; Strohacker, K.A.; Kueht, M.L. Pomegranate seed oil consumption during a period of high-fat feeding reduces weight gain and reduces type 2 diabetes risk in CD-1 mice. *Br. J. Nutr.* **2009**, *102*, 54–59. [CrossRef] [PubMed]
98. Parmar, H.S.; Kar, A. Medicinal values of fruit peels from Citrus sinensis, Punica granatum, and Musa paradisiaca with respect to alterations in tissue lipid peroxidation and serum concentration of glucose, insulin, and thyroid hormones. *J. Med. Food* **2008**, *11*, 376–381. [CrossRef] [PubMed]
99. Makino-Wakagi, Y.; Yoshimura, Y.; Uzawa, Y.; Zaima, N.; Moriyama, T.; Kawamura, Y. Ellagic acid in pomegranate suppresses resistin secretion by a novel regulatory mechanism involving the degradation of intracellular resistin protein in adipocytes. *Biochem. Biophys. Res. Commun.* **2012**, *417*, 880–885. [CrossRef] [PubMed]

100. Cambay, Z.; Baydas, G.; Tuzcu, M.; Bal, R. Pomegranate (*Punica granatum* L.) flower improves learning and memory performances impaired by diabetes mellitus in rats. *Acta Physiol. Hung.* **2011**, *98*, 409–420. [CrossRef]
101. Faghihimani, Z.; Mirmiran, P.; Sohrab, G.; Iraj, B.; Faghihimani, E. Effects of Pomegranate Seed Oil on Metabolic State of Patients with Type 2 Diabetes Mellitus. *Int. J. Prev. Med.* **2016**, *7*, 124.
102. Tang, D.; Liu, L.; Ajiakber, D.; Ye, J.; Xu, J.; Xin, X.; Aisa, H.A. Anti-diabetic Effect of Punica granatum Flower Polyphenols Extract in Type 2 Diabetic Rats: Activation of Akt/GSK-3beta and Inhibition of IRE1alpha-XBP1 Pathways. *Front. Endocrinol.* **2018**, *9*, 586. [CrossRef]
103. Gharib, E.; Montasser Kouhsari, S. Study of the Antidiabetic Activity of *Punica granatum* L. Fruits Aqueous Extract on the Alloxan-Diabetic Wistar Rats. *Iran. J. Pharm. Res.* **2019**, *18*, 358–368.
104. Mustafa, S.M.; Chua, L.S.; El-Enshasy, H.A.; Abd Majid, F.A.; Hanapia, S.Z. Kinetic profile and anti-diabetic potential of fermented *Punica granatum* juice using *Lactobacillus casei*. *Process. Biochem.* **2020**, *92*, 224–231. [CrossRef]
105. Banihani, S.A.; Shuaibu, S.M.; Al-Husein, B.A.; Makahleh, S.S. Fresh Pomegranate Juice Decreases Fasting Serum Erythropoietin in Patients with Type 2 Diabetes. *Int. J. Food Sci.* **2019**, *2019*, 1269341. [CrossRef]
106. Davidson, P.M.; Branen, A.L. Food Antimicrobials—An Introduction. In *Antimicrobials in Food*; Davidson, P.M., Sofos, J.M., Branen, A.L., Eds.; CRC Press: Boca Raton, FL, USA, 2005.
107. Fawole, O.A.; Makunga, N.P.; Opara, U.L. Antibacterial, antioxidant and tyrosinase-inhibition activities of pomegranate fruit peel methanolic extract. *BMC Complement. Altern. Med.* **2012**, *12*, 200. [CrossRef]
108. Naz, S.; Siddiqi, R.; Ahmad, S.; Rasool, S.A.; Sayeed, S.A. Antibacterial activity directed isolation of compounds from Punica granatum. *J. Food Sci.* **2007**, *72*, M341–M345. [CrossRef] [PubMed]
109. Vasconcelos, L.C.; Sampaio, M.C.; Sampaio, F.C.; Higino, J.S. Use of *Punica granatum* as an antifungal agent against candidosis associated with denture stomatitis. *Mycoses* **2003**, *46*, 192–196. [CrossRef] [PubMed]
110. Al-Zoreky, N.S. Antimicrobial activity of pomegranate (*Punica granatum* L.) fruit peels. *Int. J. Food Microbiol.* **2009**, *134*, 244–248. [CrossRef] [PubMed]
111. Wafa, B.A.; Makni, M.; Ammar, S.; Khannous, L.; Hassana, A.B.; Bouaziz, M.; Es-Safi, N.E.; Gdoura, R. Antimicrobial effect of the Tunisian Nana variety *Punica granatum* L. extracts against Salmonella enterica (serovars Kentucky and Enteritidis) isolated from chicken meat and phenolic composition of its peel extract. *Int. J. Food Microbiol.* **2017**, *241*, 123–131. [CrossRef] [PubMed]
112. Visnjevec, A.M.; Ota, A.; Skrt, M.; Butinar, B.; Mozina, S.S.; Cimerman, N.G.; Necemer, M.; Arbeiter, A.B.; Hladnik, M.; Krapac, M.; et al. Genetic, Biochemical, Nutritional and Antimicrobial Characteristics of Pomegranate (*Punica granatum* L.)Grown in Istria. *Food Technol. Biotechnol.* **2017**, *55*, 151–163.
113. Devanesan, S.; AlSalhi, M.S.; Balaji, R.V.; Ranjitsingh, A.J.A.; Ahamed, A.; Alfuraydi, A.A.; AlQahtani, F.Y.; Aleanizy, F.S.; Othman, A.H. Antimicrobial and Cytotoxicity Effects of Synthesized Silver Nanoparticles from Punica granatum Peel Extract. *Nanoscale Res. Lett.* **2018**, *13*, 315. [CrossRef]
114. Esmaillzadeh, A.; Azadbakht, L. Food intake patterns may explain the high prevalence of cardiovascular risk factors among Iranian women. *J. Nutr.* **2008**, *138*, 1469–1475. [CrossRef]
115. Zirpoli, H.; Caputo, M.; Carraturo, A.; Torino, G.; Fazio, A.; Attya, M.; Rastrelli, L.; Tecce, M.F. Selective action of human sera differing in fatty acids and cholesterol content on in vitro gene expression. *J. Cell Biochem.* **2012**, *113*, 815–823. [CrossRef]
116. Heinecke, J.W. Lipoprotein oxidation in cardiovascular disease: Chief culprit or innocent bystander? *J. Exp. Med.* **2006**, *203*, 813–816. [CrossRef]
117. Khateeb, J.; Gantman, A.; Kreitenberg, A.J.; Aviram, M.; Fuhrman, B. Paraoxonase 1 (PON1) expression in hepatocytes is upregulated by pomegranate polyphenols: A role for PPAR-gamma pathway. *Atherosclerosis* **2010**, *208*, 119–125. [CrossRef]
118. Riaz, A.; Khan, R.A. Anticoagulant, antiplatelet and antianemic effects of *Punica granatum* (pomegranate) juice in rabbits. *Blood Coagul. Fibrinolysis* **2016**, *27*, 287–293. [CrossRef]
119. Razani, Z.; Dastani, M.; Kazerani, H.R. Cardioprotective Effects of Pomegranate (*Punica granatum*) Juice in Patients with Ischemic Heart Disease. *Phytother. Res.* **2017**, *31*, 1731–1738. [CrossRef]

120. Patel, I.B.; Atar, M.A.; Ali, S.A. *Punica granatum* Peel Extract Ameliorates Doxorubicin Induced Cardiotoxicity. *Anal. Chem. Lett.* **2019**, *9*, 835–844. [CrossRef]
121. Boldaji, R.B.; Akhlaghi, M.; Sagheb, M.M.; Esmaeilinezhad, Z. Pomegranate juice improves cardiometabolic risk factors, biomarkers of oxidative stress and inflammation in hemodialysis patients: A randomized crossover trial. *J. Sci. Food Agric.* **2019**. [CrossRef] [PubMed]
122. Sohrab, G.; Roshan, H.; Ebrahimof, S.; Nikpayam, O.; Sotoudeh, G.; Siasi, F. Effects of pomegranate juice consumption on blood pressure and lipid profile in patients with type 2 diabetes: A single-blind randomized clinical trial. *Clin. Nutr. Espen.* **2019**, *29*, 30–35. [CrossRef] [PubMed]
123. Caruso, A.; Ceramella, J.; Iacopetta, D.; Saturnino, C.; Mauro, M.V.; Bruno, R.; Aquaro, S.; Sinicropi, M.S. Carbazole Derivatives as Antiviral Agents: An Overview. *Molecules* **2019**, *24*, 1912. [CrossRef]
124. Sundararajan, A.; Ganapathy, R.; Huan, L.; Dunlap, J.R.; Webby, R.J.; Kotwal, G.J.; Sangster, M.Y. Influenza virus variation in susceptibility to inactivation by pomegranate polyphenols is determined by envelope glycoproteins. *Antivir. Res.* **2010**, *88*, 1–9. [CrossRef]
125. Haidari, M.; Ali, M.; Ward Casscells, S.; Madjid, M. Pomegranate (*Punica granatum*) purified polyphenol extract inhibits influenza virus and has a synergistic effect with oseltamivir. *Phytomedicine* **2009**, *16*, 1127–1136. [CrossRef]
126. Yang, Y.; Xiu, J.; Zhang, L.; Qin, C.; Liu, J. Antiviral activity of punicalagin toward human enterovirus 71 in vitro and in vivo. *Phytomedicine* **2012**, *20*, 67–70. [CrossRef]
127. Reddy, B.U.; Mullick, R.; Kumar, A.; Sudha, G.; Srinivasan, N.; Das, S. Small molecule inhibitors of HCV replication from pomegranate. *Sci. Rep.* **2014**, *4*, 5411. [CrossRef]
128. Moradi, M.; Karimi, A.; Alidadi, S.; Saedi-Marghmaleki, M.; Salehian, M. *In vitro* Anti-adenovirus activity of pomegranate (*Punica granatum* L.) peel extract. *Adv. Herb. Med.* **2016**, *2*, 1–8.
129. Houston, D.M.J.; Bugert, J.J.; Denyer, S.P.; Heard, C.M. Correction: Potentiated virucidal activity of pomegranate rind extract (PRE) and punicalagin against Herpes simplex virus (HSV) when co-administered with zinc (II) ions, and antiviral activity of PRE against HSV and aciclovir-resistant HSV. *PLoS ONE* **2017**, *12*, e0188609. [CrossRef] [PubMed]
130. Arunkumar, J.; Rajarajan, S. Study on antiviral activities, drug-likeness and molecular docking of bioactive compounds of *Punica granatum* L. to Herpes simplex virus—2 (HSV-2). *Microb. Pathog.* **2018**, *118*, 301–309. [CrossRef] [PubMed]
131. Rehman, M.T.; AlAjmi, M.F.; Hussain, A. Natural Compounds as Inhibitors of SARS-CoV-2 Main Protease (3CLpro): A Molecular Docking and Simulation Approach to Combat COVID-19. *Chem. Rxiv.* **2020**. [CrossRef]
132. Vardhan, S.; Sahoo, S.K. Searching inhibitors for three important proteins of COVID-19 through molecular docking studies. *arXiv* **2020**, arXiv:2006.1667.
133. Khalifa, I.; Zhu, W.; Nafie, M.S.; Dutta, K.; Li, C. Anti-COVID-19 Effects of Ten Structurally Different Hydrolysable Tannins through Binding with the Catalytic-Closed Sites of COVID-19 Main Protease: An In-Silico Approach. *Preprints* **2020**. [CrossRef]
134. Nettore, I.C.; Rocca, C.; Mancino, G.; Albano, L.; Amelio, D.; Grande, F.; Puoci, F.; Pasqua, T.; Desiderio, S.; Mazza, R.; et al. Quercetin and its derivative Q2 modulate chromatin dynamics in adipogenesis and Q2 prevents obesity and metabolic disorders in rats. *J. Nutr. Biochem.* **2019**, *69*, 151–162. [CrossRef]
135. Mackay, J.; Mensah, G.; Mendis, S.; Greenland, K. *The Atlas of Heart Disease and Stroke*; World Health Organization, Ed.; World Health Organization in collaboration with the USA Centers for Disease Control and Prevention; World Health Organization: Geneva, Switzerland, 2004.
136. Cerda, B.; Ceron, J.J.; Tomas-Barberan, F.A.; Espin, J.C. Repeated oral administration of high doses of the pomegranate ellagitannin punicalagin to rats for 37 days is not toxic. *J. Agric. Food. Chem.* **2003**, *51*, 3493–3501. [CrossRef]
137. Lei, F.; Zhang, X.N.; Wang, W.; Xing, D.M.; Xie, W.D.; Su, H.; Du, L.J. Evidence of anti-obesity effects of the pomegranate leaf extract in high-fat diet induced obese mice. *Int. J. Obes.* **2007**, *31*, 1023–1029. [CrossRef]
138. Trichur Khabeer, S.; Prashant, A.; Haravey Krishnan, M. Dietary fatty acids from pomegranate seeds (*Punica granatum*) inhibit adipogenesis and impact the expression of the obesity-associated mRNA transcripts in human adipose-derived mesenchymal stem cells. *J. Food Biochem.* **2019**, *43*, e12739. [CrossRef]

139. Seeram, N.P.; Henning, S.M.; Zhang, Y.; Suchard, M.; Li, Z.; Heber, D. Pomegranate juice ellagitannin metabolites are present in human plasma and some persist in urine for up to 48 hours. *J. Nutr.* **2006**, *136*, 2481–2485. [CrossRef] [PubMed]
140. Bialonska, D.; Kasimsetty, S.G.; Schrader, K.K.; Ferreira, D. The effect of pomegranate (*Punica granatum* L.) byproducts and ellagitannins on the growth of human gut bacteria. *J. Agric. Food. Chem.* **2009**, *57*, 8344–8349. [CrossRef]
141. Chaves, F.M.; Baptista, I.L.; Simabuco, F.M.; Quaresma, P.G.F.; Pena, F.L.; Bezerra, R.M.N.; Pauli, J.R.; da Cunha, D.T.; Campos-Ferraz, P.L.; Antunes, A.E.C. High-intensity-exercise-induced intestinal damage is protected by fermented milk supplemented with whey protein, probiotic and pomegranate (*Punica granatum* L.). *Br. J. Nutr.* **2018**, *119*, 896–909. [CrossRef] [PubMed]
142. George, N.S.; Cheung, L.M.; Luthria, D.L.; Santin, M.; Dawson, H.D.; Bhagwat, A.A.; Smith, A.D. Pomegranate peel extract alters the microbiome in mice and dysbiosis caused by *Citrobacter rodentium* infection. *Food Sci. Nutr.* **2019**, *7*, 2565–2576. [PubMed]
143. Smith, A.D.; George, N.S.; Cheung, L.; Bhagavathy, G.V.; Luthria, D.L.; John, K.M.; Bhagwat, A.A. Pomegranate peel extract reduced colonic damage and bacterial translocation in a mouse model of infectious colitis induced by Citrobacter rodentium. *Nutr. Res.* **2020**, *73*, 27–37. [CrossRef]
144. Turk, G.; Sonmez, M.; Aydin, M.; Yuce, A.; Gur, S.; Yuksel, M.; Aksu, E.H.; Aksoy, H. Effects of pomegranate juice consumption on sperm quality, spermatogenic cell density, antioxidant activity and testosterone level in male rats. *Clin. Nutr.* **2008**, *27*, 289–296. [CrossRef]
145. Turk, G.; Sonmez, M.; Ceribasi, A.O.; Yuce, A.; Atessahin, A. Attenuation of cyclosporine A-induced testicular and spermatozoal damages associated with oxidative stress by ellagic acid. *Int. Immunopharmacol.* **2010**, *10*, 177–182. [CrossRef]
146. Forest, C.P.; Padma-Nathan, H.; Liker, H.R. Efficacy and safety of pomegranate juice on improvement of erectile dysfunction in male patients with mild to moderate erectile dysfunction: A randomized, placebo-controlled, double-blind, crossover study. *Int. J. Impot. Res.* **2007**, *19*, 564–567. [CrossRef]
147. Qnais, E.Y.; Elokda, A.S.; Abu Ghalyun, Y.Y.; Abdulla, F.A. Antidiarrheal activity of the aqueous extract of *Punica granatum* (pomegranate) peels. *Pharm. Biol.* **2007**, *45*, 715–720. [CrossRef]
148. Olapour, S.; Mousavi, E.; Sheikhzade, M.; Hoseininezhad, O.; Najafzadeh, H. Evaluation antidiarrheal effects of pomegranate peel extract. *J. Iran. Chem. Soc.* **2009**, *6*, 115–143.
149. Zhao, S.S.; Ma, D.X.; Zhu, Y.; Zhao, J.H.; Zhang, Y.; Chen, J.Q.; Sheng, Z.L. Antidiarrheal effect of bioactivity-guided fractions and bioactive components of pomegranate (*Punica granatum* L.) peels. *Neurogastroenterol. Motil.* **2018**, *30*, e13364. [CrossRef] [PubMed]
150. Dumitrescu, A.L. Influence of periodontal disease on cardiovascular diseases. *Rom. J. Intern. Med.* **2005**, *43*, 9–21. [PubMed]
151. DiSilvestro, R.A.; DiSilvestro, D.J.; DiSilvestro, D.J. Pomegranate Extract Mouth Rinsing Effects on Saliva Measures Relevant to Gingivitis Risk. *Phytother. Res.* **2009**, *23*, 1123–1127. [CrossRef]
152. Sastravaha, G.; Gassmann, G.; Sangtherapitikul, P.; Grimm, W.D. Adjunctive periodontal treatment with Centella asiatica and Punica granatum extracts in supportive periodontal therapy. *J. Int. Acad. Periodontol.* **2005**, *7*, 70–79. [PubMed]
153. Al-Wazni, W.S.; Hadi, B.S. Antivirulence effects of pomegranate peel extracts on most common urinary tract infection pathogens in pregnant women. *J. Contemp. Med. Sci.* **2015**, *1*, 7–12. [CrossRef]
154. Manouchehrian, M.; Shakiba, M.; Shariat, M.; Lotfi, M.H.; Kamalinejad, M.; Babaeian, M. The Effect of Pomegranate Paste on Neonatal Jaundice Incidence: A Clinical Trial in Women during Pregnancy. *Int. J. Clin. Med.* **2017**, *8*, 144–151. [CrossRef]
155. Finn-Sell, S.L.; Cottrell, E.C.; Greenwood, S.L.; Dilworth, M.R.; Cowley, E.J.; Sibley, C.P.; Wareing, M. Pomegranate Juice Supplementation Alters Utero-Placental Vascular Function and Fetal Growth in the eNOS-/- Mouse Model of Fetal Growth Restriction. *Front. Physiol.* **2018**, *9*, 1145. [CrossRef]
156. Henning, S.M.; Wallenstein, M.B.; Weigel, N.; Johnson, C.; Yang, J.; Lee, R.; Korn, M.; Scala, M.; Stevenson, K.D.; Ben-Nissan, D.; et al. Appearance of Ellagic Acid Metabolites from Pomegranate Juice in Breast Milk: A Case Report. *Ann. Clin. Case Rep.* **2019**, *4*, 1738.

157. Heneka, M.T.; Carson, M.J.; El Khoury, J.; Landreth, G.E.; Brosseron, F.; Feinstein, D.L.; Jacobs, A.H.; Wyss-Coray, T.; Vitorica, J.; Ransohoff, R.M.; et al. Neuroinflammation in Alzheimer's disease. *Lancet Neurol.* **2015**, *14*, 388–405. [CrossRef]
158. DaSilva, N.A.; Nahar, P.P.; Ma, H.; Eid, A.; Wei, Z.; Meschwitz, S.; Zawia, N.H.; Slitt, A.L.; Seeram, N.P. Pomegranate ellagitannin-gut microbial-derived metabolites, urolithins, inhibit neuroinflammation in vitro. *Nutr. Neurosci.* **2019**, *22*, 185–195. [CrossRef]
159. Velagapudi, R.; Lepiarz, I.; El-Bakoush, A.; Katola, F.O.; Bhatia, H.; Fiebich, B.L.; Olajide, O.A. Induction of Autophagy and Activation of SIRT-1 Deacetylation Mechanisms Mediate Neuroprotection by the Pomegranate Metabolite Urolithin A in BV2 Microglia and Differentiated 3D Human Neural Progenitor Cells. *Mol. Nutr. Food Res.* **2019**, *63*, e1801237. [CrossRef] [PubMed]
160. Casedas, G.; Les, F.; Choya-Foces, C.; Hugo, M.; Lopez, V. The Metabolite Urolithin-A Ameliorates Oxidative Stress in Neuro-2a Cells, Becoming a Potential Neuroprotective Agent. *Antioxidants* **2020**, *9*, 177. [CrossRef] [PubMed]
161. Kujawska, M.; Jourdes, M.; Kurpik, M.; Szulc, M.; Szaefer, H.; Chmielarz, P.; Kreiner, G.; Krajka-Kuzniak, V.; Mikolajczak, P.L.; Teissedre, P.L.; et al. Neuroprotective Effects of Pomegranate Juice against Parkinson's Disease and Presence of Ellagitannins-Derived Metabolite-Urolithin A-In the Brain. *Int. J. Mol. Sci.* **2020**, *21*, 202. [CrossRef] [PubMed]
162. Siddarth, P.; Li, Z.P.; Miller, K.J.; Ercoli, L.M.; Merril, D.A.; Henning, S.M.; Heber, D.; Small, G.W. Randomized placebo-controlled study of the memory effects of pomegranate juice in middle-aged and older adults. *Am. J. Clin. Nutr.* **2020**, *111*, 170–177. [CrossRef]
163. Vidal, A.; Fallarero, A.; Pena, B.R.; Medina, M.E.; Gra, B.; Rivera, F.; Gutierrez, Y.; Vuorela, P.M. Studies on the toxicity of *Punica granatum* L. (*Punicaceae*) whole fruit extracts. *J. Ethnopharmacol.* **2003**, *89*, 295–300. [CrossRef]
164. Heber, D.; Seeram, N.P.; Wyatt, H.; Henning, S.M.; Zhang, Y.; Ogden, L.G.; Dreher, M.; Hill, J.O. Safety and antioxidant activity of a pomegranate ellagitannin-enriched polyphenol dietary supplement in overweight individuals with increased waist size. *J. Agric. Food Chem.* **2007**, *55*, 10050–10054. [CrossRef] [PubMed]
165. Aviram, M.; Rosenblat, M.; Gaitini, D.; Nitecki, S.; Hoffman, A.; Dornfeld, L.; Volkova, N.; Presser, D.; Attias, J.; Liker, H.; et al. Pomegranate juice consumption for 3 years by patients with carotid artery stenosis reduces common carotid intima-media thickness, blood pressure and LDL oxidation. *Clin. Nutr.* **2004**, *23*, 423–433. [CrossRef]
166. Chimento, A.; Sirianni, R.; Saturnino, C.; Caruso, A.; Sinicropi, M.S.; Pezzi, V. Resveratrol and Its Analogs as Antitumoral Agents for Breast Cancer Treatment. *Mini Rev. Med. Chem.* **2016**, *16*, 699–709. [CrossRef]
167. Iacopetta, D.; Rosano, C.; Puoci, F.; Parisi, O.I.; Saturnino, C.; Caruso, A.; Longo, P.; Ceramella, J.; Malzert-Freon, A.; Dallemagne, P.; et al. Multifaceted properties of 1,4-dimethylcarbazoles: Focus on trimethoxybenzamide and trimethoxyphenylurea derivatives as novel human topoisomerase II inhibitors. *Eur. J. Pharm. Sci.* **2017**, *96*, 263–272. [CrossRef]
168. Sala, M.; Chimento, A.; Saturnino, C.; Gomez-Monterrey, I.M.; Musella, S.; Bertamino, A.; Milite, C.; Sinicropi, M.S.; Caruso, A.; Sirianni, R.; et al. Synthesis and cytotoxic activity evaluation of 2,3-thiazolidin-4-one derivatives on human breast cancer cell lines. *Bioorg. Med. Chem. Lett.* **2013**, *23*, 4990–4995. [CrossRef]
169. Tundis, R.; Iacopetta, D.; Sinicropi, M.S.; Bonesi, M.; Leporini, M.; Passalacqua, N.G.; Ceramella, J.; Menichini, F.; Loizzo, M.R. Assessment of antioxidant, antitumor and pro-apoptotic effects of *Salvia fruticosa* Mill. subsp. thomasii (*Lacaita*) Brullo, Guglielmo, Pavone & Terrasi (*Lamiaceae*). *Food Chem. Toxicol.* **2017**, *106*, 155–164. [PubMed]
170. Ceramella, J.; Loizzo, M.R.; Iacopetta, D.; Bonesi, M.; Sicari, V.; Pellicano, T.M.; Saturnino, C.; Malzert-Freon, A.; Tundis, R.; Sinicropi, M.S. Anchusa azurea Mill. (*Boraginaceae*) aerial parts methanol extract interfering with cytoskeleton organization induces programmed cancer cells death. *Food Funct.* **2019**, *10*, 4280–4290. [CrossRef] [PubMed]
171. Francomano, F.; Caruso, A.; Barbarossa, A.; Fazio, A.; La Torre, C.; Ceramella, J.; Mallamaci, R.; Saturnino, C.; Iacopetta, D.; Sinicropi, M.S. β-Caryophyllene: A Sesquiterpene with Countless Biological Properties. *Appl. Sci.* **2019**, *9*, 5420. [CrossRef]

172. Khomich, L.M.; Perova, I.B.; Eller, K.I. Pomegranate juice nutritional profile. *Vopr. Pitan.* **2019**, *88*, 80–92. [PubMed]
173. Bar-Ya'akov, I.; Tian, L.; Amir, R.; Holland, D. Primary Metabolites, Anthocyanins, and Hydrolyzable Tannins in the Pomegranate Fruit. *Front. Plant Sci.* **2019**, *10*, 620. [CrossRef] [PubMed]
174. Vucic, V.; Grabez, M.; Trchounian, A.; Arsic, A. Composition and Potential Health Benefits of Pomegranate: A Review. *Curr. Pharm. Des.* **2019**, *25*, 1817–1827. [CrossRef] [PubMed]
175. Viswanath, M.; Sridevi, P.; Bhagavan, B.V.K.; Ravindra Kumar, K.; Subbaramamma, P. Toxicological, Pharmacological and Cellular properties of Pomegranate (*Punica granatum* L.): A Review. *J. Pharmacogn. Phytochem.* **2019**, *8*, 172–176.
176. Natalello, A.; Hervas, G.; Toral, P.G.; Luciano, G.; Valenti, B.; Mendoza, A.G.; Pauselli, M.; Priolo, A.; Frutos, P. Bioactive compounds from pomegranate by-products increase the in vitro ruminal accumulation of potentially health promoting fatty acids. *Anim. Feed Sci. Technol.* **2020**, *259*, 114355. [CrossRef]
177. Prakash, C.V.S.; Prakash, I. Bioactive Chemical Constituents from Pomegranate (*Punica Granatum*) Juice, Seed and Peel-A Review. *Int. J. Res. Chem. Environ.* **2011**, *1*, 1–18.
178. Sateriale, D.; Facchiano, S.; Colicchio, R.; Pagliuca, C.; Varricchio, E.; Paolucci, M.; Volpe, M.G.; Salvatore, P.; Pagliarulo, C. In vitro Synergy of Polyphenolic Extracts from Honey, Myrtle and Pomegranate Against Oral Pathogens, S. mutans and R. dentocariosa. *Front. Microbiol.* **2020**, *11*, 1465. [CrossRef]

applied sciences

MDPI

Review

# β-Caryophyllene: A Sesquiterpene with Countless Biological Properties

**Fabrizio Francomano [1,†], Anna Caruso [1,†], Alexia Barbarossa [1], Alessia Fazio [1,*], Chiara La Torre [1], Jessica Ceramella [1], Rosanna Mallamaci [2], Carmela Saturnino [3], Domenico Iacopetta [1] and Maria Stefania Sinicropi [1]**

1 Department of Pharmacy, Health and Nutritional Sciences, University of Calabria, Via P. Bucci, 87036 Arcavacata di Rende, Italy; fabriziofrancomano@libero.it (F.F.); anna.caruso@unical.it (A.C.); alexia.barbarossa@hotmail.it (A.B.); latorre.chiara@libero.it (C.L.T.); jessicaceramella@gmail.com (J.C.); domenico.iacopetta@unical.it (D.I.); s.sinicropi@unical.it (M.S.S.)

2 Department of Biosciences, Biotechnology and Biopharmaceutics, University of Bari, Via Orabona 4, 70124 Bari, Italy; rosanna.mallamaci@uniba.it

3 Department of Science, University of Basilicata, 85100 Potenza, Italy; carmela.saturnino@unibas.it

* Correspondence: alessia.fazio@unical.it; Tel.: +39-0984-493013

† The authors have contributed equally to the manuscript.

Received: 18 November 2019; Accepted: 9 December 2019; Published: 11 December 2019

**Abstract:** β-Caryophyllene (BCP), a natural bicyclic sesquiterpene, is a selective phytocannabinoid agonist of type 2 receptors (CB2-R). It isn't psychogenic due to the absence of an affinity to cannabinoid receptor type 1 (CB1). Among the various biological activities, BCP exerts anti-inflammatory action via inhibiting the main inflammatory mediators, such as inducible nitric oxide synthase (iNOS), Interleukin 1 beta (IL-1β), Interleukin-6 (IL-6), tumor necrosis factor-alfa (TNF-α), nuclear factor kapp a-light-chain-enhancer of activated B cells (NF-κB), cyclooxygenase 1 (COX-1), cyclooxygenase 2 (COX-2). Peroxisome proliferator-activated receptors alpha (PPAR-α) effects are also mediated by the activation of PPAR-α and PPAR-γ receptors. In detail, many studies, in vitro and in vivo, suggest that the treatment with β-caryophyllene improves the phenotype of animals used to model various inflammatory pathologies, such as nervous system diseases (Parkinson's disease, Alzheimer's disease, multiple sclerosis, amyotrophic lateral sclerosis, stroke), atherosclerosis, and tumours (colon, breast, pancreas, lymphoma, melanoma and glioma cancer). Furthermore, pre-clinical data have highlighted that BCP is potentially useful in *Streptococcus* infections, osteoporosis, steatohepatitis, and exerts anticonvulsant, analgesic, myorelaxing, sedative, and antidepressive effects. BCP is non-toxic in rodents, with a Lethal dose, 50% (LD50) greater than 5000 mg/kg. Nevertheless, it inhibits various cytochrome P450 isoforms (above all, CYP3A4), which metabolise xenobiotics, leading to adverse effects, due to drug levels over therapeutic window. All the reported data have highlighted that both pharmacological and toxicological aspects need to be further investigated with clinical trials.

**Keywords:** sesquiterpene; inflammatory pathologies; nutraceuticals

## 1. Introduction

Sesquiterpenoids are an extremely wide group of secondary metabolites found in the plant kingdom [1,2]. Their main carbon backbone consists of 15 atoms and several synthases action leads to a great variety of chemical structures. In particular, bicyclic sesquiterpenoids have raised attention in the bio-pharmacological field, even as essential oils. Among them, β-caryophyllene is one of the major component of essential oils extracted from spice and food plants [3–7], such as black pepper (*Piper nigrum* L.), rosemary (*Rosmarinus officinalis*), cinnamon (*Cinnamomum* spp.) oregano (*Origanum vulgare* L.) basil (*Ocimum* spp.), thyme (*Thymus vulgaris*), sage (*Salvia officinalis*), mint (*Mentha piperita*), ginger (*Zingiber officinale*), chinotto (*Citrus Myrtfolia* Raf) [8], and cloves (*Syzygium aromaticum*) [9,10]. It was also found in citronella, (*Cymbopogon*), pine tree (*Pinus*), *Chenopodium ambrosioides*, *Cannabis sativa*, in plants of the genus *Copaifera*, *Artemisia*, *Murraya*, *Cordia*, *Spiranthes*, *Ocimum*, *Croton* [9,10], and in the leaves of *Annona Cherimola* [11]. This plant compound was approved by Food and Drug Administration (FDA) and European Food Safety Authority (EFSA) and it is used as flavour enhancer [9] and in cosmetics [12]. In nature, BCP mainly occurs as *trans*-caryophyllene ((E)-BCP) (**1**) mixed with small amount of the isomers (Z)-$\beta$-caryophyllene ((Z)-BCP) (**2**) and $\alpha$-humulene ($\alpha$-caryophyllene) (**3**) [12], and its oxidation derivative, $\beta$-caryophyllene oxide (BCPO) (**4**) (Figure 1).

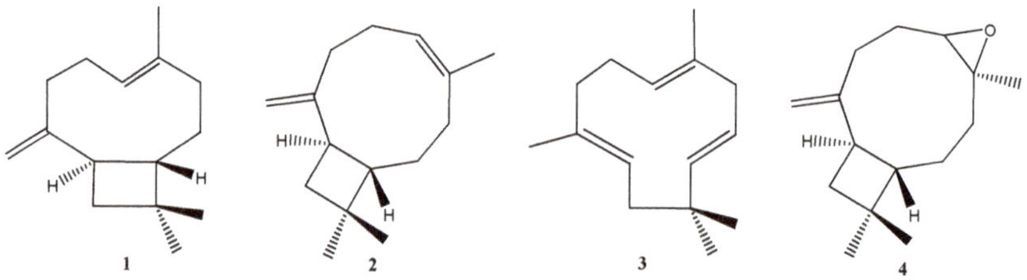


**Figure 1.** Structures of sesquiterpenes (**1–4**).

BCP belongs to cannabinoid family, which are ligands of the cannabinoid receptors present in the organism. Cannabinoid receptors CB1-R and CB2-R are metabotropic receptors are G protein (protein binding GTP)-coupled receptors, involved in the regulation of neurotransmitters responsible for maintaining an energetic balance, in the metabolism, and in the immune response. The aforementioned receptors are bound and activated by endogenous cannabinoids, derivatives of arachidonic acid, including 2-arachidonoylglycerol and *N*-arachidonoylethanolamine, better known as anandamide. Both receptors are bound by a lot of proteins in various pathways, acting as mediators of cellular responses to biological molecules.

Unlike the main traditional cannabinoids, such as $\Delta^8$-tetrahydrocannabinol (**5**), $\Delta^9$-tetrahydrocannabinol (**6**) and cannabinol (**7**) (Figure 2), able to activate both receptors, BCP has a very different chemical structure and it is a selective agonist of CB2-R. Furthermore, BCP has no side effects, not activating CB1-Rs (mainly expressed in the central nervous system, but also in the liver, lungs, heart, blood vessels and digestive tract). CB2-Rs are mainly found in peripheral tissues and in immune system cells (B and NK lymphocytes, macrophages, mast cells) and, to a lesser extent, in the central nervous system (brain, neurons, microglia) [13]. The expression of CB2-Rs in the central nervous system is increased in neurodegenerative pathologies, such as Parkinson's disease, Alzheimer's disease, amyotrophic lateral sclerosis (ALS) and in some tumours like gliomas [14].

The binding of BCP to CB2-Rs is responsible for many cellular events:

- The activation of $G_{i/o}$ $\alpha$-subunit inhibits the activity of adenylate cyclase, the enzyme which converts ATP in cAMP (cyclic AMP) and activates phospholipase C (PLC), determining calcium release from IP3 (inositol trisphosphate)-sensitive calcium channels, increasing the intracellular level of calcium;
- The activation of the complex formed by $G_{i/o}$ $\beta$- and $\gamma$-subunits, which modulate various pathways modifying the expression and the activity of many target proteins [15].

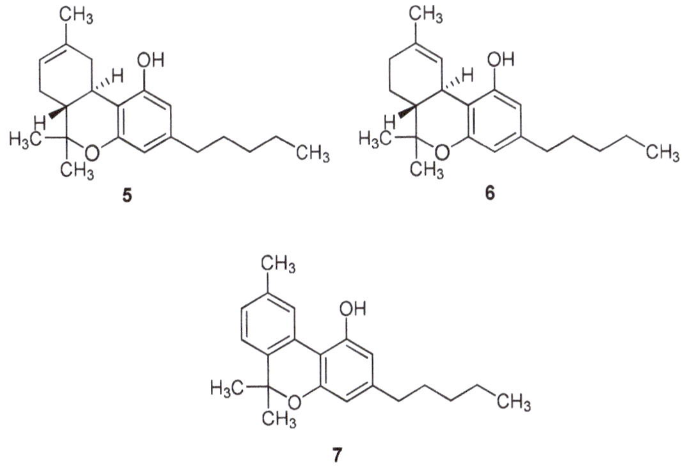


**Figure 2.** Structures of cannabinoids (**5–7**).

In addition to the involvement in physiological functions, CB2-Rs are implied in numerous pathological processes, so they can represent an interesting target in order to obtain agonist molecules for the treatment of many pathological conditions, including neuropathic pain, inflammation, neuroinflammatory and neurodegenerative pathologies (Parkinson's disease, Alzheimer's disease, multiple sclerosis, and amyotrophic lateral sclerosis), spinal and brain injuries, stroke, ischemia, anxiety disorder, depression, colitis, fibrosis and liver ischemia, atherosclerosis, osteoporosis, osteoarthritis, diabetes, obesity, and some types of cancer [9,12–14,16–21].

Molecular docking studies showed that BPC interacts with CB2-R on the binding site of CP55,940 (an agonist), precisely in a hydrophobic pocket, engaging with hydrophobic amino acid residues (valine-113, phenylalanine-117, isoleucine-198, tryptophan-258 and methionine-265) [12]. The double bond with conformation E of BCP is fundamental for receptor binding. BCPO lacks a double bond in position C4-C5 and it is not able to bind CB2-R [10].

## 2. Biological properties of $\beta$-caryophyllene

In the last few years, BCP has represented an important subject of study [9–15,22–65]. In particular, a lot of researches about the its effects in vitro and in vivo on animals have been conducted, thus experimental data about its biological properties have been verified. However, further studies are needed in order to translate the findings in animal models into promising pre-clinical and clinical trials on humans. Pre-clinical studies have revealed that BCP is a modulator of nervous system and exerts beneficial effects on numerous neurodegenerative and inflammatory pathologies. Moreover, it is able to act on the liver and bones, and has antibiotic properties.

Table 1 shows the principal studies on $\beta$-caryophyllene.

**Table 1.** Preclinical studies on $\beta$-caryophyllene.

| Experimental Conditions | Effects | Possible Mechanisms | References |
|---|---|---|---|
| 50 mg/kg (i.p.) in Wistar rats (models of Alzheimer's disease) | Neuroprotective | Decreased NO synthesis; Decreased activation of astrocytes and microglia; Reduced expression of Iba-1 e GFAP | Ojha et al. (2016) [22] |
| 10 mg/kg in rat Clone 6 (C6) microglial cells | Anti-inflammatory | Reduced production of iNOS, IL-1β, IL-6 and COX-2 | Chang et al. (2013) [24] |
| 10, 25, and 50 µM in BV2 mouse cells | Suppression of neuroinflammation due to hypoxia | NF-κB inhibition; Decreased production of NO and Prostaglandin E2 (PGE2) | Hu et al. (2017) [26] |
| 10 mg/kg in SH-SY5Y mouse cells treated with MPTP (model of Parkinson's disease) | Neuroprotective; enhancement of motor coordination | Nrf2 activation | Viveros-Paredes et al. (2017) [30] |
| 0.5 and 1 µM in rat C6 glioma cells | Antioxidant | Nrf2 activation | Assis et al. (2014) [33] |
| 25 mg/kg twice a day (p.o.) in mice treated with paclitaxel (model of peripheral neuropathy) | Attenuation of peripheral neuropathy | mitogen-activated protein kinase (MAPK) p38/NF-κB inhibition | Segat et al. (2017) [36] |
| 2.25–18 µg (s.c.) | Strengthening of analgesic effect of morphine | CB2-R activation; Stimulation of $\beta$-endorphins release and µ-opioid receptors | Katsuyama et al. (2013) [37] |
| 50 mg/kg twice a day in mice with experimental autoimmune encephalomyelitis (model of multiple sclerosis) | Decreased pain; decreased cerebral damage; antioxidant | Decreased production of IFN-γ; Increased expression of IL-10; Increased activity of catalase, superoxide dismutase and glutathione peroxidase. | Alberti et al. (2017) [38] |
| 50, 100, and 150 mg/kg of the hydroalcoholic extract and 200 mg/kg of the essential oil from basil in Syrian mice. | Sedative | | Rabbani et al. (2015) [39] |
| 50, 100 mg/kg of the essential oil from *Baccharis uncinella* (p.o.) | Sedative | | Ascari et al. (2012) [40] |
| 50, 100, 150 mg/kg of the hydroalcoholic extract and 200 mg/kg of the essential oil from basil (i.p.) in Syrian mice | Anxiolytic | | Rabbani et al. (2015) [39] |
| 50 mg/kg in (C57 black 6) C57BL/6 mice (p.o.) | Anxiolytic | Activation of CB2-Rs | Bahi et al. (2014) [41] |
| 100 and 300 µg/mL of essential oil from *Pterodon polygalaeflorus* and 30 and 100 µg/mL of BCP in rats | Muscle relaxant | | Leonhardt et al. (2010) [42] |
| 50 mg/kg in C57BL/6 mice (p.o.) | Antidepressive | Activation of CB2-Rs | Bahi et al. (2014) [41] |

**Table 1.** *Cont.*

| Experimental Conditions | Effects | Possible Mechanisms | References |
| --- | --- | --- | --- |
| 10, 30 and 100 mg/kg in C57BL/6 mice | Anticonvulsant | | de Oliveira et al. (2016) [43] |
| 30 mg/kg every four hours in mice | Anticonvulsant | | Tchekalarova et al. (2018) [44] |
| In HT29 and HTC116 (cell lines of colon cancer) | Antitumour | | Dahham et al. (2015) [45] |
| In PANC-1 (cell line of pancreas cancer) | Antitumour | | Dahham et al. (2015) [45] |
| 2.5, 5 and 10 μmol/L in B16F10 melanoma cells in high-fat diet-induced obese C57BL/6N mice | Reduction of obesity-related cancer risk | | Jung et al. (2015) [48] |
| Essential oil from *Pamburus missionis* (containing also phytolandaromadendrene oxide) in A431 and HaCaT cell lines (skin epidermoid cancer) | Antitumour | ROS production; Increased Bcl-2 Associated X protein (Bax) expression; Reduced B-cell lymphoma 2 (Bcl-2) expression; Cytochrome C release in cytoplasm; Apoptosis. | Pavithra et al. (2018) [49] |
| In BS-24-1 (mouse cell line of lymphoma) and MoFir (human lymphocytes B transformed through Epstein-Barr virus) | Antitumour | | Pavithra et al. (2018) [49] |
| In MCF-7 (breast cancer), L-929 (mouse fibroblasts), DLD-1 (colon cancer) | Increased intracellular levels of paclitaxel | Increased permeability of cell membrane; CYP3A4 inhibition | Fidyt et al. (2016) [13]; Legault et al. (2007) [47]; Nguyen et al. (2017) [64] |
| In CT26 cells transplanted in BALB/c mice and exposed to high levels of glucose (model of colorectal cancer) | Antitumour | NF-κB and arginine ADP-ribosyltransferase 1 (ART1) inhibition | Zhou et al. (2018) [51] |
| 1 or 10 μg/mL + baicalin (1, 10 μg/mL) and (+)-catechin (1, 10 μg/mL) in macrophages of RAW267.4 mouse | Reduction of macrophage proliferation | Cell cycle arrest G2/M phase; Reduction of the expression of Akt, MAPK e p65 NF-κB; Caspase-3 activation | Yamaguchi et al. (2016) [52] |
| 30 mg/kg in high-fat-fed Wistar rats (p.o.) | Antiatherogenic; Antioxidant | Activation of cannabinoid receptor 2 (CB2-R), PPAR-γ, PPAR-α, PGC1-α; Reduction of Vascular cell adhesion molecule-1 (VCAM1) expression | Youssef et al. (2019) [53] |
| 40 mg in Wistar rats (model of bilateral carotid artery occlusion and reperfusion) | Prevention of the effects induced bycarotid occlusion and reperfusion | Activation of endocannabinoid system; Increased PPAR-α expression | Poddighe et al. (2018) [55] |

**Table 1.** *Cont.*

| Experimental Conditions | Effects | Possible Mechanisms | References |
|---|---|---|---|
| 215 and 430 mg/kg/die (p.o.) in rats (model of rheumatoid arthritis) | Reduction of oxidative stress, inflammation and edema | Radical scavenging; Nrf2 induction; Increased glutathione levels; Reduction of the expression of COX-2, cytokines and NF-κB | Ames-Sibin et al. (2018) [56] |
| 30 and 300 mg/kg (p.o.) in C57BL/6J mice fed with methionine- and choline-deficient diet (model of non-alcoholic steatohepatitis, NASH) | Reduction of liver steatosis, inflammation and fibrosis | Reduction of cytokineslevels Transforming growth factor beta (TGF-β), Nox2 and collagen; Increased levels of SOD2 and GPx1 | Arizuka et al. (2017) [58] |
| In the oral cavity for about 10 min | Bactericidal against Streptococcus mutans (tooth decay) | GlucosyltransferaseB (GtfB), GtfC and GtfD Inhibition | Yoo et al. (2018) [59] |
| 0.1–100 μM in culture of mouse bone marrow cells | Stimulates osteoblast mineralization and inhibits adipogenesis and osteoclastogenesis | PPAR-γ activation in pluripotent stem cells; Inhibits TNF-α and NF-κB in osteoclasts | Yamaguchi et al. (2016) [61] |

### *2.1. β-Caryophyllene and Nervous System*

Many studies report the beneficial effects of BCP on central nervous system [22–44], in particular against neuroinflammatory and neurodegenerative pathologies.

Neuroinflammation is a process leading to nervous system degeneration, characterised by the activation of monocytes, macrophages, mast cells, lymphocytes, and the production of inflammation mediators, such as nitric oxide (NO), various cytokines (IL-1β, IL-6 and TNF-α), the protein NF-κB (nuclear factor kappa B) and the prostaglandins.

In detail, BCP, administered intraperitoneally in Wistar rats, at the dose of 50 mg/kg, has reduced the activity of inducible nitric oxide synthase (iNOS) and, consequently, nitric oxide production, thus decreasing brain oxidative stress and leading to the inhibition of lipid peroxidation and the depletion of glutathione stores [22], causing striatal and cortical neurotoxicity [9]. Moreover, nitric oxide is involved in the activation of cyclooxygenase (COX), the enzyme synthetizing prostaglandin H2 (PGH2), which is the precursor of the other prostaglandins, in particular Prostaglandin D2 (PGD2) and Prostaglandin E2 (PGE2), responsible for inflammation and pain. In this contest a frontiers of research for BCP and BCPO will be their influence on microRNA, small molecule able to regulate gene expression, considering that they are involved and proposed as biomarkers, in neuroinflammation and pain [66,67].

The essential oil of *Erymanthos erythropappus*, rich in BCP, has been shown to have anti-inflammatory in Wistar rats [23].

It has also been demonstrated that BCP, at the dose of 10 mg/kg, inhibits the transcription of iNOS, IL-1β, IL-6, and COX-2 in C6 microglia cells [9,24].

Furthermore, BCP, tested on the mouse BV2 cell line at the concentrations of 10, 25 and 50 μM, has inhibited NF-κB activation and reduced the production of nitric oxide and PGE2, thus suppressing hypoxia-induced neuroinflammatory response [25,26].

In the case of central nervous system pathologies, Iba-1, analogous protein of Aif-1, present only in macrophages and microglia, is over expressed (for example, in ischemia) [27], whereas GFAP (glial fibrillary acidic protein), which forms intermediate filaments, is expressed in a lot of central nervous system cells (including astrocytes and ependymal cells). In addition, it is involved in cell communication, in the interaction neuron-astrocytes, in the functioning of the blood–brain barrier

(BBB), particularly during mitosis, in which it modulates the filament network, and in the repair following brain injuries [9,28]. It has been demonstrated that BCP administered intraperitoneally at the dosage of 50 mg/kg, reduces the activation of astrocytes and microglia, by decreasing Iba-1 and GFAP expression, thus avoiding the death of dopaminergic neurons [22].

Amyloid plaques, formed by $\beta$-amyloid peptides, composed in turn by 36–43 amino acids and derived from amyloid precursor protein (APP), characterize Alzheimer's disease. These peptides are responsible for direct toxicity (death of neurons) and indirect toxicity (production of molecules stimulating inflammation). A considerable decrease of the $\beta$-amyloid peptide-induced overexpression of TLR4 (Toll-like receptor 4), which determines the activation of monocytes and microglia, has been noted in BV2 microglia cells treated with BCP, and, consequently, the sesquiterpene leads to a clear reduction of the biosynthesis of IL-6, IL-1$\beta$, PGE2, TNF-$\alpha$, NF-$\kappa$B and nitric oxide. In addition, also a reduced expression of COX-2 and iNOS has been observed [26].

Another common condition affecting central nervous system is Parkinson's disease, a neurodegenerative pathology characterized by neuroinflammation, oxidative stress, mitochondrial dysfunction and cell death, particularly in dopaminergic neurons. An in vitro study on SH-SY5Y cells showed that the treatment with BCP inhibits reactive oxygen species (ROS) production, restoring mitochondrial functionality and the levels of the antioxidant glutathione. BCP prevents apoptosis, by inhibiting the expression of Bax and caspase-3 and increasing Bcl-2 one. It reduces the phosphorylation of JNK (c-Jun N-terminal Kinase), which determines the increase of HO-1 (heme oxygenase-1) expression in this pathological condition. All these effects are related to CB2/Nrf2 pathway [29]. Research on a mouse model of Parkinson's disease conducted by Viveros-Paredes revealed that the treatment with BCP, at the dose of 10 mg/kg, enhances motor coordination in mice and protects the dopaminergic neurons from degeneration, reducing the production of inflammatory cytokines, in particular IL-1$\beta$, IL-6, and TNF-$\alpha$ [30].

BCP is also cytoprotective towards the central nervous system due to its modulation of the redox state and inflammation, useful during chemotherapy. The main mechanism of action regarding this aspect is the stimulation of nuclear factor (erythroid-derived 2)-like 2 (Nrf2) following the activation of cannabinoid receptors CB2. Nrf2 is a transcription factor, whose activity is stimulated by oxidative stress (in particular, reactive oxygen species or ROS) and oncogenes, such as KRAS, BRAF and MYC. Nrf2 increases the expression of the genes involved in cell survival and in the reduction of the inflammatory process and the oxidative stress in SNC. Moreover, nerve growth factor (NGF) is stimulated through Tropomyosin receptor kinase A (Trk A) pathway, which stimulates in turn PI3K-Akt pathway [68].

BCP, through Nrf2 activation, increases the expression of antiapoptotic genes (B-cell leukemia/lymphoma-2 (BCL-2), Mouse double minute 2 (mdm-2), COX-2, c-Myb) and, at the same time, reduces the one of proapoptotic genes (Bax, Bak-1, caspase-8, caspase-9) [31].

All these cellular events determine cell survival and proliferation and angiogenesis [9].

BCP also diminishes the expression of matrix metallopeptidase (MMP-9) and increases the one of occludin, claudin-5, Tight Junction Protein ZO-1 and Growth Associated Protein 43 (GAP-43) [32].

Furthermore, BCP, at the concentrations of 0.5 and 1 $\mu$M, exerts cytoprotective effect on C6 glioma cells, increasing the antioxidant activity, through CB2-R-dependent Nrf2 pathway [33]. In addition, administered orally in Wistar rats, at the dosages of 34, 102 and 306 mg/kg, it reduces oxidative stress and neuronal apoptosis, due to the increased expression of Nrf2 and OH-1.

BCP also inhibits the production of nitric oxide, hydrogen peroxide, TNF-$\alpha$, interferon gamma (IFN-$\gamma$), interleukin-17 (IL-17) reducing the extent of macrophage infiltration [9].

BCP can induce neurogenesis with a mechanism independent from the activation of CB2-R, Nrf2 and NGF, stimulating Tropomyosin receptor kinase A (TrkA) in SH-SY5Y (Cell lines Homo sapiens, human, bone marrow) and PC12 (Cell line derived from a pheochromocytoma of the rat adrenal medulla) neuroblastoma cells [34].

There are numerous studies regarding the analgesic effects of the sesquiterpene, isolated or in mixture in essential oils. In fact, the essential oil of *Erymanthos erythropappus*, containing BCP as one of the main components, administered orally at the doses of 200 and 400 mg/kg, induces analgesia in Wistar rats [23]. Further, the essential oil of *Senecio rufinervis*, containing about the 6% of BCP, is responsible for the analgesic effect in mice, when administered orally at the doses of 50 and 75 mg/kg [35].

Other studies have reported the same results in mice at the doses of 1 and 5 mg/kg intraperitoneally and at the dosages of 2.6 mg/kg/die per *os* for 2 weeks, alone or in combination with the docosahexaenoic acid (DHA) [9]. The essential oil of *Vitex agnus-castus* (BCP accounts for about 7% of the oil), has exhibited analgesic effect in Wistar rats which were subjected to an immersion test.

BCP, at the dose of 25 mg/kg, decreases the extent of peripheral neuropathyin mice, by the activation of cannabinoid receptors CB2-R and the inhibition of the MAPK p38, resulting in a reduced transcriptional activity of NF-κB, involved in phlogosis [36].

It has been demonstrated in mice that BCP actually exerts its antinociceptive action by activating CB2-Rs, in particular towards primary sensory neurons. In fact, the analgesic effect manifests itself in wild-type mice, but not in the CB2-R knockout mice. The activation of cannabinoid system indirectly leads to the modulation of opioid system. In more detail, it stimulates $\beta$-endorphin release, activating opioid receptors $\mu$ on primary afferent neurons. BCP has been shown to boost the painkilling effect of morphine, used in the treatment of severe pain, making possible the reduction of the dose of the drug and, consequently, decreasing its side effects [37].

Instead, there was no indication of synergy between BCP and the other components of the essential oils [13].

The derivative BCPO has not elicited any interest as an analgesic [13], since it does not bind to cannabinoid receptors due to the different molecular structure.

BCP could result potentially useful in multiple sclerosis management (pathology characterized by axon demyelination and neuroinflammation), as demonstrated by previous study which investigated the therapeutic potential of BCP on experimental autoimmune encephalomyelitis (EAE), a murine model of multiple sclerosis (MS).

The research has shown that the treatment with BCP reduces interferon-$\gamma$ (IFN-$\gamma$) production, the main activator of macrophages, and increases the levels of interleukin-10 (IL-10).

Moreover, it inhibits cell T migration. The oral treatment with 50 mg/kg of BCP twice a day has decreased hyperalgesia induced by the EAE and protected from brain damage, inhibiting cytokine biosynthesis and restoring the activity of catalase, superoxide dismutase and glutathione peroxidase, all enzymes involved in the detoxification of oxidant substances [38].

Two essential oils containing BCP, basil one, administered intraperitoneally (at the doses of 100, 150 and 200 mg/kg), and *Baccharis uncinella* one, administered per *os* (at the doses of 50 or 100 mg/kg), exhibit significant sedative effect [39,40].

These early results on the animal model could be attributed to BCP, being a valid alternative to the most common sedatives and anaesthetics.

The latter drugs have contraindications in patients with cardiovascular, neuromuscular and cerebrovascular diseases, and could cause many side effects, such as convulsions, insomnia, agitation, dependence and, in the most extreme cases, lead to death.

Basil essential oil has also exerted anxiolytic effect [39]. The same results have been obtained in mice, when BCP has been orally administered at the dose of 50 mg/kg [41].

Therefore, all these data suggest the possible use of the sesquiterpene as a substitute of the most common anxiolytics, like benzodiazepines and the selective serotonin reuptake inhibitors (SSRI), which cause many adverse effects, such as movement problems, sedation, dependence, muscle relaxation, anterograde amnesia, teratogenic risk, reduction of bone mineral density.

BCP also exhibits antidepressant properties, as revealed by a research conducted on albino Swiss mice receiving 50 mg/kg of the BCP per Os (oral somministration). The mechanism of this effect is due to the activation of CB2-R [41].

Various studies have shown muscle relaxant effect of BCP in animals. In this regard, the compound, alone or as a constituent of the essential oil from *Pterodon polygalaeflorus*, exerts antispasmodic activity on rat isolated ileum at variable concentrations (1 to 1000 μg/mL) [42].

BCP, orally administered at the dosages of 10, 30, and 100 mg/kg in C57BL/6 mice, exerted a dose-dependent anticonvulsant effect [9,43]. Furher research on mice has revealed that the sesquiterpene inhibits tonic-clonic seizures in MES (maximal electro shock seizure) test and reduces kainate-induced neurotoxicity, suggesting a possible use of the sesquiterpene for the treatment of epilepsy [44].

### 2.2. *β-Caryophyllene and Cancer*

Both BCP and BCPO have shown cytotoxic activity against various cancer cell lines.

In particular, BCP has significantly decreased the proliferation of two colon cancer cell lines, HT-29 and HCT-116, and a pancreas cancer cell line, PANC-1. Moreover, it has been quite successful on other types of cancer cells [45].

In the intestinal cancer cell line CaCo-2 [46], for example, BCP has not been able to exert a significant effect on cell growth, unlike the isomer α-humulene [13].

In human breast cancer cells MCF-7, BCP amplifies the cytotoxicity of the isomers isocaryophyllene and α-humulene [47]. A study realized on obese mice C57BL/6N, injected with melanoma cells, has shown that the phytocannabinoid is able to decrease the precancerous effect caused by a high-fat diet [48].

BCP, which is more than 25% of the essential oil from *Pamburus missionis*, has synergistic effect with two other important components of the essential oil, phytol (**8**) and aromadendrene oxide (**9**) (Figure 3), resulting in antitumor activity against A431 and HaCaT cell lines, by blocking the cell in phase G0/G1 or sub-G1.

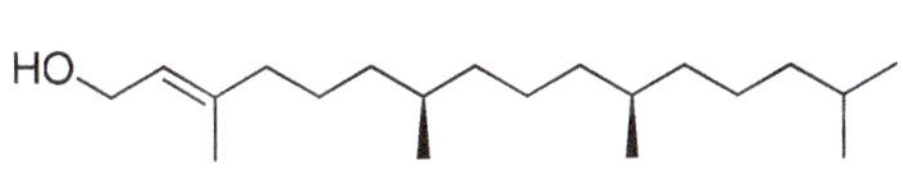


**8**

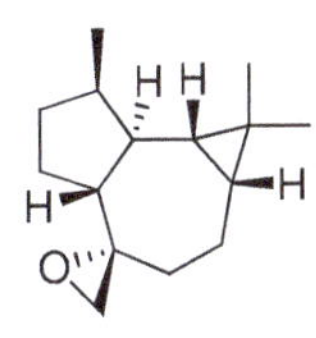


**9**

**Figure 3.** Structures of the phytol (**8**) and aromadendrene oxide (**9**).

The mechanism of action is associated to ROS production [69] and the mitochondrial membrane potential loss, by increasing Bax expression and decreasing Bcl-2 expression. Bax and Bak oligomers form pores, which increase the permeability of the external mitochondrial membrane, releasing cytochrome c in the cytoplasm, which is an apoptosis characterizing event [70–72]. The release of cytochrome c from mitochondria to cytosol leads to the formation of apoptosomes, and consequently, the activation of caspase-9, which activates the cascade of the effector caspases [49,73,74].

It has been shown that BCP causes the activation of caspase-3 and determines nucleolus fragmentation and the consequent apoptosis in two different cell lines, BS-24-1 (murine cell line of lymphoma) and MoFir (human T cell transformed through Epstein-Barr virus) [49].

BCP, one of the compounds in the essential oil, of *Commiphora gileadensis*, is responsible for antiproliferative exhibited by 3-(4,5-Dimethylthiazol-2-Yl)-2,5-Diphenyltetrazolium Bromide (MTT) assay [75–80] and proapoptotic effects (exhibited via DNA "ladder" and caspase-3 activation) in tumor cell lines, while there was no apoptosis induction in normal cell lines (FB) [13,50].

Anyhow, numerous studies have revealed that BCP enhances the effectiveness of antitumor drugs. In particular, a research has shown that it increases the activity of paclitaxel in lots of cell lines: MCF-7 (breast cancer), L-929 (mouse fibroblasts), DLD-1 (colon cancer) [47]. In detail, in the latter cell line, the BCP determines a rise of the intracellular concentrations in the drug, probably by increasing the permeability of the cell membrane [13].

BCP has been reported to exert anticancer and hypoglycemic effects in BALB/c mice transplanted with cells of the line CT26 exposed to high levels of glucose, to mimic a colorectal cancer. The sesquiterpene blocks ART1 effects, by inhibiting NF-κB. ART1 (arginine-specific mono-ADP-ribosyl transferase 1) is an enzyme, whose concentrations are higher in patients with type 2 diabetes, involved in the pathogenesis of colorectal cancer. The overexpression of ART1 probably increases glycolysis and energy metabolism, thus regulating the protein kinase B/mammalian target of rapamycin/c-Myc signaling pathway and the expression of glycolytic enzymes. This suggests that BCPO may be a potential treatment for this kind of carcinoma [51].

Further, BCPO is cytotoxic against various cell lines, including: HeLa (human cervical adenocarcinoma cells), HepG2 (human leukaemia cells), AGS (human lung cancer cells), SNU-1 and SNU-16 (human stomach cancer cells) and A-2780 (human ovarian cancer cells). It modulates many fundamental pathways in tumour pathogenesis, such as those involving MAPK, Phosphoinositide 3-kinases (PI3K)/protein kinase B (Akt)/mammalian target of rapamycin (mTOR)/S6 kinase 1 (S6K1) and Signal Transducer and Activator of Transcription (STAT3) [13].

BCPO, like BCP, shows synergy with antitumor drugs, like paclitaxel and doxorubicin. In particular, it is able to increase the concentration of the two drugs in many cancer cell lines, including CaCo-2.

Anyhow, between the two sesquiterpenes, the oxidised derivative has the greatest antitumor activity, variable according to cell types and dosages. In fact, due to the epoxide function, BCPO binds covalently to amino groups and thiol moieties of proteins and nitrogen bases, which constitute the nucleic acids. In any case, both the molecules achieve their antitumor action by causing apoptosis and blocking the cell cycle [13].

### 2.3. *β-Caryophyllene and Inflammatory Diseases*

Acute inflammation is a body defence mechanism against various factors, which may evolve in chronic inflammation, resulting in pathological disease [81]. After the acute reaction, monocytes remain in the inflammation site, where they secrete cytokines and chemokines and stimulate macrophages, amplifying the inflammatory response. As for neuroinflammation, the main phlogistic mediators are interleukins IL-1β, IL-6, and TNF-α, which increase NF-κB (nuclear factor kappa B) expression, prostaglandins and leukotrienes, synthesized by cyclooxygenases (COX) and 5-lipoxigenase (LOX) [82].

An in vitro study on macrophages of RAW267.4 [83] mice has revealed that BCP, administered in association to other two natural molecules, baicalin (**10**) and (+)-catechin (**11**) (Figure 4), at relatively low doses, suppresses the proliferation of these cells involved in inflammation [52].

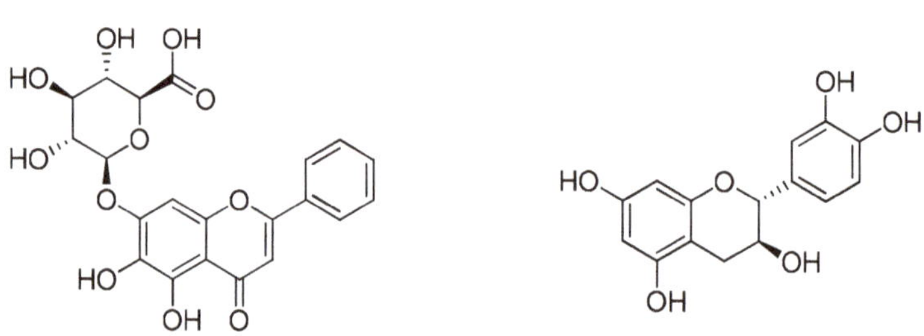


**10** **11**

**Figure 4.** Structures of the baicalin (**10**) and (+)-catechin (**11**).

The three substances act synergically, since they separately do not exert any significant activity. Their effect is achieved through the cell cycle arrest phase G2/M and the modulation of various intracellular pathways, such as PI3K/Akt, extracellular signal-regulated kinases(ERK)/MAPK, and calcium homeostasis. In particular, the decrease of the expression of Akt, MAPK p38 and p44/42, and caspase-3 activation, an important step of apoptosis, have been observed. In addition to these effects, the expression of COX-1 and COX-2 diminishes, and so the activity of the protein p65 of NF-κB family does [52].

According to a recent study, BCP exerted powerful results against the negative effects of dyslipidemia and vascular inflammation in mice [53]. In fact, it has been revealed that the treatment with the sesquiterpene, at the dose of 30 mg/kg, prevents the increase of adiposity index, glycemia and insulinemia due to a high fat-diet. It also helps dyslipidemia and reduces all atherogenic risk indexes, even if it does not modify body weight. BCP reduces oxidative stress, by decreasing the concentration of NO and malondialdehyde (MDA), a by-product of lipid peroxidation, and by increasing the level of the endogenous antioxidant glutathione [84]. The phytocannabinoid is able to suppress mediators involved both in inflammation and atherosclerosis, such as TNF-α and NF-κB. In this way, it leads to the inhibition of VCAM1 [53], a vascular cell adhesion protein, which promotes the adhesion of white cells of the vascular endothelium and favours atherosclerosis, confirming what was reported in other in vitro researches [54].

BCP normalizes the ratio between endothelial (eNOS) and inducible (iNOS) nitric oxide synthase within the aorta. The latter is activated in the phlogistic process and in the atherosclerosis following the oxidative stress-induced NF-κB activation. Further, iNOS produces a high amount of nitric oxide, which interacts with ROS, generating peroxynitrites, which amplify the oxidative stress.

Furthermore, BCP attenuates the formation of foam cells and the deposition of collagen, which plays a crucial role in the formation and progression of vulnerable atherosclerotic plaques, and protects the integrity of elastic lamina.

All these effects are attributable to direct action of BCP, as agonist of cannabinoid receptors CB2, other than of PPAR-γ receptors (receptors activated by peroxisome proliferator-activated receptors), involved in the reduction of blood levels of total cholesterol, low density lipoprotein (LDL) and very low density lipoprotein (VLDL), in the inhibition of vascular inflammation and synthesis of adhesion molecules and in the rebalancing of nitric oxide concentration and nitric oxide synthase isoforms. Moreover, it would seem that BCP is also an agonist of PPAR-α receptors, reduces fat mass and triglycerides and increases high density lipoprotein (HDL). This happens through the binding of BCP to cannabinoid receptors CB2-R, which activates PGC1-α (coactivator 1 of the peroxisome gamma proliferator) and allows the interaction among PPAR-γ and various transcriptional factors, such as PPAR-α, increasing the expression of enzymes with the function of oxidizing fat acids, above all in the liver.

The comparison with the thiazolidinedione pioglitazone (**12**) (Figure 5), a PPAR-γ agonist used for the treatment of type 2 diabetes and atherosclerosis, has shown that BCP is more effective than the drug for all the parameters, except for the glutathione levels. In addition, the sesquiterpene does not induce body weight gain, the main side effect of pioglitazone [53].

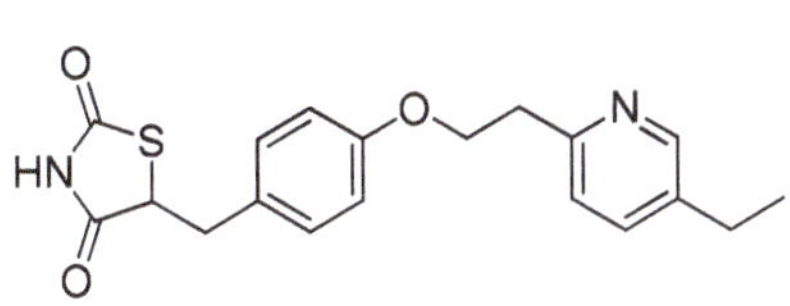


**12**

**Figure 5.** Structure of the pioglitazone.

BCP is beneficial in a model of bilateral carotid artery occlusion and reperfusion (BCCAO/R) in Wistar rats. This study has demonstrated that a single dose of BCP prevents plasmatic and tissue modifications induced by carotid obstruction and reperfusion [55]. It increases tissue levels of endocannabinoids (anandamide, 2-arachidonoylglycerol, palmitoylethanolamide, oleylethanolamide) and cannabinoid receptors CB1 and CB2, reducing anandamide blood concentrations. Furthermore, it preserves the tissue levels of the essential fatty acid docosahexaenoic acid (DHA), increases PPAR-$\alpha$ expression, and decreases lipoperoxidation damage [55].

BCP effects have been studied in a rat model of rheumatoid arthritis, characterized by inflammatory response.

BCP and copaiba oil (the sesquiterpene accounts for about 37% of the oil) have the same ability to reduce paw edema, popliteal lymph nodes weight and myeloperoxidase plasmatic activity. BCP, unlike copaiba oil, decreases also the activity of hepatic myeloperoxidase and leukocytes, both the blood ones and those present in the joints. The anti-inflammatory activity is slightly greater in copaiba oil than in BCP, probably because of the presence of other molecules with synergist effect. Neither the oil nor the single molecule are able to modify secondary injuries and body weight in rats. Both are capable, at high doses, to:

- Reduce carboxylic groups of protein and ROS in the liver, normalizing the levels;
- Increase glutathione levels and reduce lipoperoxidation [56].

Copaiba oil, because of the presence of diterpenes, like kaurenoic acid and hardwickiic acid, is hepatotoxic, reducing the liver functionality due to hepatic cholestasis. In this context, the use of isolated BCP is preferable, since it is hepatoprotective [57].

BCP exerts in vitro [45,85,86] and in vivo antioxidant capacities.

As regards the mechanism, BCP, at a dose of 430 mg / kg, exhibits antioxidant activity and acts by exerting various functions: radical scavenging ability in particular with respect to hydroxyl radicals, lipid peroxides and superoxide anions; stimulation of the endogenous antioxidant system, highlighted by the increased content of glutathione in the liver, induced by Nrf. A decrease in inflammation, characterized by reduced activity of myeloperoxidase, is a marker of infiltration of polymorphonuclear cells, and diminished expression of COX-2 and cytokines, such as TNF-α, IL-1β, IL-6, and consequently, NF-κB.

Also, BCP effects on C57BL/6J mice have been investigated. The mice, fed with an essential amino acid-deficient diet, are a model of non-alcoholic steatohepatitis (NASH), which is a hepatic inflammatory pathology associated to metabolic syndrome predisposing cardio-vascular diseases. In the case of steatohepatitis, the organ undergoes histological changes caused by oxidative stress, inflammation and fibrosis. The treatment with BCP reduced inflammation and fibrosis. Moreover, a decrease in alanine-transaminase (ALT) and cytokine expression has been observed, suggesting that the liver has been less damaged. BCP exerts antioxidant effect, by increasing the levels of the enzymes SOD2 (superoxide dismutase 2) and GPx1 (glutathione peroxidase 1), both involved in free radical detoxification. In addition, the enzyme Nox2 (reduced form of nicotinamide adenine dinucleotide phosphate (NADP) oxidase 2), TGF-$\beta$, and collagen, all elements which contribute to hepatic fibrosis, have been inhibited by the sesquiterpene [58].

### 2.4. *β-Caryophyllene and Micro-Organisms*

BCP has antimicrobial activity both against Gram-positive bacteria, such as *Staphilococcus aureus* and Gram-negative bacteria, including *Escherichia coli* [59]. It is more effective against Gram-positive and the activity of *Artemisia fed dei* extracts, containing BCP, against bacteria responsible for tooth decay and periodontitis has been reported [60]. The antimicrobial activity of the essential oil and some of its compounds was tested against 15 different genera of oral bacteria, including *Streptococcus mutans*, which produces a biofilm, constituted by a variety of extracellular polymeric substances (EPS), which protect micro-organisms of oral cavity. The antibacterial activities of the essential oil and of BCP was determined based on the minimum inhibitory concentrations (MICs) which were determined as the lowest concentration of the test samples that resulted in a complete inhibition of visible growth in the broth. BCP kills *Streptococcus mutans* with a minimum inhibiting concentration (MIC) of 0.32% by penetrating the biofilm, which constitutes a protective barrier against external substances, acting as an ion exchange resin. *Streptococcus mutans* uptakes sucrose, which is necessary for glucan synthesis, through the enzymes GtfB, GtfC and GtfD. The first two are required for the synthesis of insoluble glucans, which facilitate bacterial aggregation. GtfD, on the other hand, promotes the synthesis of water-soluble glucans, which are necessary for biofilm development. The mechanism of action of BCP is to reduce the expression of Gtf genes, thus preventing biofilm synthesis.

### 2.5. *β-Caryophyllene and Osteoporosis*

BCP exerts effects on various tissues, like bone tissue. In fact, it stimulates the differentiation of multipotent stem cells (MSC) of bone marrow (which can differentiate in many cell types, including adipocytes, osteoblasts, myoblasts, chondrocytes, etc.) in osteoblasts, limiting the formation of adipose cells. In order to modulate this process, regulated by various pathways, BCP activates PPAR-$\gamma$. It suppresses osteoclastogenesis, which is the synthesis of osteoclasts, cells derived from haematopoietic progenitors which promote bone resorption. This mechanism, TNF-$\alpha$-mediated, activates the transcription factor NF-κB in prosteoclasts and it is inhibited in vitro by BCP (in cell cultures of murine bone marrow) in the differentiation stage in prosteoclasts.

Therefore, BCP enhances osteoblast mineralization, favouring osteoblastogenesis and suppressing osteoclastogenesis and adipogenesis, which is an interesting property for the potential application in the treatment of osteoporosis, in particular the one associated to obesity and diabetes [61].

### 2.6. *β-Caryophyllene Toxicity*

BCP is classified according to OECD (Organization for Economic Co-operation and Development) guideline 423 as a category five substance (toxic at doses greater than 2000 mg/kg) [65].

Studies about acute toxicity have been carried out. The oral administration of 2000 mg/kg of the molecule in female Swiss mice induces no toxic effects [10], whereas in rats an $LD_{50}$ greater than 5000 mg/kg has been calculated [62].

Subchronic toxicity (90 days) has been assessed in studies in Wistar rats, which do not show toxicity at the dose of 700 mg/kg/die [62], and in female Swiss mice, which are not altered in terms of locomotion and muscle tone with both single and repeated doses [10].

In rodents, overall absence of neurotoxicity at the dosages used to study pharmacological effects (20 to 100 mg/kg) has been found [10,41]. Moreover, in various studies, no damage to the gastric mucosa has been observed, nor changes in other internal organs (brain, heart, liver, lungs, spleen, kidneys) or in haematological parameters have been reported both in female Swiss mice and in Wistar rats [10,62]. Moreover, an Ames test has shown no mutagenicity [63].

Body weight has resulted decreased by 5% in female Swiss mice, even if this variation is not significant [10]. Nevertheless, even though this molecule is safe, at least in animal studies, greater consideration is needed of the possible interactions of the sesquiterpene with drugs and other molecules contained in the food. In fact, a research conducted on subcellular fractions of hepatic tissue of rats and humans, demonstrates a significant inhibition by BCP and its derivatives BCPO and $\alpha$-humulene against enzymes involved in metabolism and xenobiotic detoxification. Xenobiotics undergo metabolic biotransformations, followed by degradation and subsequent elimination. It is evident that the inhibitory effects exerted by BCP on drug metabolizing enzymes considerably increases the levels of drugs in the organism, with prolonged duration of action and increased toxicity. This makes absolutely necessary a variation of the posology of drugs taken at the same time of enzymatic inhibitors. The extent of the inhibition of cytochrome P450 isoform CYP1A2 (Cytochrome P450 Family 1 Subfamily A Member 2) is greater in rat microsomes than in human ones and, above all, BCPO is able to competitively inhibit the enzyme in rats and humans through a non-competitive mechanism. However, BCP and $\alpha$-humulene result less powerful than BCPO [64].

Furthermore, they have a fair capability to inhibit CYP3A4, involved in the metabolism of many xenobiotics, such as ciclosporin, paclitaxel, ketoconazole, verapamil, nifedipine, omeprazole, statins, various sexual hormones and drugs. The inhibition causes the increase of medicine levels, which could determine severe side effects. Finally, the three terpenes exert a weak inhibition against the isoforms CYP2A6 (Cytochrome P450 Family 2 Subfamily A Member 6), CYP2B6 (Cytochrome P450 Family 2 Subfamily B Member 6), CYP2C9 (Cytochrome P450 Family 2 Subfamily C Member 9), CYP2C19 (Cytochrome P450 Family 2 Subfamily C Member 19), CYP2D6 (Cytochrome P450 Family 2 Subfamily D Member 6), CYP2E1 (Cytochrome P450 Family 2 Subfamily E Member 1) [64].

## 3. Conclusions

BCP is a plant compound which has been demonstrated to possess a great potential application for various pathological conditions, due, above all, to the selectivity towards CB2 receptors, which, in addition to making this sesquiterpene devoid of psychogenic effects typical of cannabinoids, determines its main biological effects. In fact, BCP contrast in the animals the inflammatory process, typical of various degenerative diseases, which involve central nervous system (Parkinson's disease, Alzheimer's disease, multiple sclerosis, amyotrophic lateral sclerosis, etc.), steatohepatitis, osteoporosis, but also cancer and *Streptococcus mutans* infections.

However, even if the studies on the molecule are very promising, these are only preclinical (in vitro or in vivo in animal models) and further insights and clinical trials are required for a future human application.

**Author Contributions:** Conceptualization, A.F.; methodology, D.I.; software, J.C.; validation, A.B. and C.S.; resources, R.M.; data curation, C.L.T.; writing—original draft preparation, F.F.; writing—review and editing, A.C.; supervision, M.S.S.

**Funding:** This research received no external funding.

**Conflicts of Interest:** The authors declare no conflict of interest.

## References

1. Russo, A.; Perri, M.; Cione, E.; Di Gioia, M.L.; Nardi, M.; Cristina Caroleo, M. Biochemical and chemical characterization of Cynara cardunculus L. extract and its potential use as co-adjuvant therapy of chronic myeloid leukemia. *J. Ethnopharmacol.* **2017**, *202*, 184–191. [CrossRef] [PubMed]
2. Cappello, A.R.; Dolce, V.; Iacopetta, D.; Martello, M.; Fiorillo, M.; Curcio, R.; Muto, L.; Dhanyalayam, D. Bergamot (*Citrus bergamia* Risso) Flavonoids and Their Potential Benefits in Human Hyperlipidemia and Atherosclerosis: An Overview. *Mini Rev. Med. Chem.* **2016**, *16*, 619–629. [CrossRef] [PubMed]
3. Tundis, R.; Iacopetta, D.; Sinicropi, M.S.; Bonesi, M.; Leporini, M.; Passalacqua, N.G.; Ceramella, J.; Menichini, F.; Loizzo, M.R. Assessment of antioxidant, antitumor and pro-apoptotic effects of Salvia fruticosa Mill. subsp. thomasii (Lacaita) Brullo, Guglielmo, Pavone & Terrasi (*Lamiaceae*). *Food Chem. Toxicol. Int. J. Publ. Br. Ind. Biol. Res. Assoc.* **2017**, *106*, 155–164. [CrossRef]
4. Ceramella, J.; Loizzo, M.R.; Iacopetta, D.; Bonesi, M.; Sicari, V.; Pellicano, T.M.; Saturnino, C.; Malzert-Freon, A.; Tundis, R.; Sinicropi, M.S. Anchusa azurea Mill. (*Boraginaceae*) aerial parts methanol extract interfering with cytoskeleton organization induces programmed cancer cells death. *Food Funct.* **2019**, *10*, 4280–4290. [CrossRef]
5. Chimento, A.; Sirianni, R.; Saturnino, C.; Caruso, A.; Sinicropi, M.S.; Pezzi, V. Resveratrol and Its Analogs As Antitumoral Agents For Breast Cancer Treatment. *Mini Rev. Med. Chem.* **2016**, *16*, 699–709. [CrossRef]
6. Ruffo, M.; Parisi, O.I.; Scrivano, L.; Restuccia, D.; Amone, F.; Sinicropi, M.S.; Malivindi, R.; Aiello, G.; Puoci, F. Role of Calabrian Black Rice in Metabolic Syndrome: In vitro Evaluation of Oryza sativa L. Indica Biological Properties. *Curr. Nutr. Food Sci.* **2018**, *14*, 121–127. [CrossRef]
7. Chimento, A.; De Amicis, F.; Sirianni, R.; Sinicropi, M.S.; Puoci, F.; Casaburi, I.; Saturnino, C.; Pezzi, V. Progress to Improve Oral Bioavailability and Beneficial Effects of Resveratrol. *Int. J. Mol. Sci.* **2019**, *20*, 1381. [CrossRef]

8. Plastina, P.; Apriantini, A.; Meijerink, J.; Witkamp, R.; Gabriele, B.; Fazio, A. In Vitro Anti-Inflammatory and Radical Scavenging Properties of Chinotto (*Citrus myrtifolia* Raf.) Essential Oils. *Nutrients* **2018**, *10*, 783. [CrossRef]
9. Machado, K.D.C.; Islam, M.T.; Ali, E.S.; Rouf, R.; Uddin, S.J.; Dev, S.; Shilpi, J.A.; Shill, M.C.; Reza, H.M.; Das, A.K.; et al. A systematic review on the neuroprotective perspectives of beta-caryophyllene. *Phytother. Res.* **2018**, *32*, 2376–2388. [CrossRef]
10. Oliveira, G.; Machado, K.C.; Machado, K.C.; da Silva, A.; Feitosa, C.M.; de Castro Almeida, F.R. Non-clinical toxicity of beta-caryophyllene, a dietary cannabinoid: Absence of adverse effects in female Swiss mice. *Regul. Toxicol. Pharmacol.* **2018**, *92*, 338–346. [CrossRef]
11. Ríos, M.Y.; Castrejón, F.; Robledo, N.; León, I.; Rojas, G.; Navarro, V. Chemical Composition and Antimicrobial Activity of the Essential Oils from Annona cherimola (*Annonaceae*). *Rev. De La Soc. Quím. De Méx.* **2003**, *47*, 139–142.
12. Gertsch, J.; Leonti, M.; Raduner, S.; Racz, I.; Chen, J.Z.; Xie, X.Q.; Altmann, K.H.; Karsak, M.; Zimmer, A. Beta-caryophyllene is a dietary cannabinoid. *Proc. Natl. Acad. Sci. USA* **2008**, *105*, 9099–9104. [CrossRef]
13. Fidyt, K.; Fiedorowicz, A.; Strzadala, L.; Szumny, A. beta-caryophyllene and beta-caryophyllene oxide-natural compounds of anticancer and analgesic properties. *Cancer Med.* **2016**, *5*, 3007–3017. [CrossRef]
14. Hartsel, J.A.; Eades, J.; Hickory, B.; Makriyannis, A. Cannabis sativa and Hemp. In *Nutraceuticals: Efficacy, Safety and Toxicity*; Gupta, R.C., Ed.; Academic Press: Cambridge, CA, USA, 2016; Volume 53, pp. 735–754.
15. Demuth, D.G.; Molleman, A. Cannabinoid signalling. *Life Sci.* **2006**, *78*, 549–563. [CrossRef]
16. Saturnino, C.; Iacopetta, D.; Sinicropi, M.S.; Rosano, C.; Caruso, A.; Caporale, A.; Marra, N.; Marengo, B.; Pronzato, M.A.; Parisi, O.I.; et al. N-Alkyl Carbazole Derivatives as New Tools for Alzheimer's Disease: Preliminary Studies. *Molecules* **2014**, *19*, 9307–9317. [CrossRef]
17. Catalano, A.; Carocci, A.; Corbo, F.; Franchini, C.; Muraglia, M.; Scilimati, A.; De Bellis, M.; De Luca, A.; Camerino, D.C.; Sinicropi, M.S.; et al. Constrained analogues of tocainide as potent skeletal muscle sodium channel blockers towards the development of antimyotonic agents. *Eur. J. Med. Chem.* **2008**, *43*, 2535–2540. [CrossRef]
18. Iacopetta, D.; Rosano, C.; Puoci, F.; Parisi, O.I.; Saturnino, C.; Caruso, A.; Longo, P.; Ceramella, J.; Malzert-Freon, A.; Dallemagne, P.; et al. Multifaceted properties of 1,4-dimethylcarbazoles: Focus on trimethoxybenzamide and trimethoxyphenylurea derivatives as novel human topoisomerase II inhibitors. *Eur. J. Pharm. Sci.* **2017**, *96*, 263–272. [CrossRef]
19. Ceramella, J.; Caruso, A.; Occhiuzzi, M.A.; Iacopetta, D.; Barbarossa, A.; Rizzuti, B.; Dallemagne, P.; Rault, S.; El-Kashef, H.; Saturnino, C.; et al. Benzothienoquinazolinones as new multi-target scaffolds: Dual inhibition of human Topoisomerase I and tubulin polymerization. *Eur. J. Med. Chem.* **2019**, *181*, 111583. [CrossRef]
20. Caruso, A.; Iacopetta, D.; Puoci, F.; Cappello, A.R.; Saturnino, C.; Sinicropi, M.S. Carbazole derivatives: A promising scenario for breast cancer treatment. *Mini Rev. Med. Chem.* **2016**, *16*, 630–643. [CrossRef]
21. Nettore, I.C.; Rocca, C.; Mancino, G.; Albano, L.; Amelio, D.; Grande, F.; Puoci, F.; Pasqua, T.; Desiderio, S.; Mazza, R.; et al. Quercetin and its derivative Q2 modulate chromatin dynamics in adipogenesis and Q2 prevents obesity and metabolic disorders in rats. *J. Nutr. Biochem.* **2019**, *69*, 151–162. [CrossRef]
22. Ojha, S.; Javed, H.; Azimullah, S.; Haque, M.E. beta-Caryophyllene, a phytocannabinoid attenuates oxidative stress, neuroinflammation, glial activation, and salvages dopaminergic neurons in a rat model of Parkinson disease. *Mol. Cell. Biochem.* **2016**, *418*, 59–70. [CrossRef]
23. Sousa, O.V.; Silverio, M.S.; Del-Vechio-Vieira, G.; Matheus, F.C.; Yamamoto, C.H.; Alves, M.S. Antinociceptive and anti-inflammatory effects of the essential oil from Eremanthus erythropappus leaves. *J. Pharm. Pharmacol.* **2008**, *60*, 771–777. [CrossRef]
24. Chang, H.J.; Kim, J.M.; Lee, J.C.; Kim, W.K.; Chun, H.S. Protective effect of beta-caryophyllene, a natural bicyclic sesquiterpene, against cerebral ischemic injury. *J. Med. Food* **2013**, *16*, 471–480. [CrossRef]
25. Guo, K.; Mou, X.; Huang, J.; Xiong, N.; Li, H. Trans-caryophyllene suppresses hypoxia-induced neuroinflammatory responses by inhibiting NF-kappaB activation in microglia. *J. Mol. Neurosci.* **2014**, *54*, 41–48. [CrossRef]
26. Hu, Y.; Zeng, Z.; Wang, B.; Guo, S. Trans-caryophyllene inhibits amyloid beta (Abeta) oligomer-induced neuroinflammation in BV-2 microglial cells. *Int. Immunopharmacol.* **2017**, *51*, 91–98. [CrossRef]
27. Ito, D.; Imai, Y.; Ohsawa, K.; Nakajima, K.; Fukuuchi, Y.; Kohsaka, S. Microglia-specific localisation of a novel calcium binding protein, Iba1. *Brain Res. Mol.* **1998**, *57*, 1–9. [CrossRef]

28. Bunge, M.B.; Bunge, R.P.; Ris, H. Ultrastructural study of remyelination in an experimental lesion in adult cat spinal cord. *J. Biophys. Biochem. Cytol.* **1961**, *10*, 67–94. [CrossRef]
29. Wang, G.F.; Ma, W.B.; Du, J.W. beta-Caryophyllene (BCP) ameliorates MPP plus induced cytotoxicity. *Biomed. Pharmacother.* **2018**, *103*, 1086–1091. [CrossRef]
30. Viveros-Paredes, J.M.; Gonzalez-Castaneda, R.E.; Gertsch, J.; Chaparro-Huerta, V.; Lopez-Roa, R.I.; Vazquez-Valls, E.; Beas-Zarate, C.; Camins-Espuny, A.; Flores-Soto, M.E. Neuroprotective Effects of beta-Caryophyllene against Dopaminergic Neuron Injury in a Murine Model of Parkinson's Disease Induced by MPTP. *Pharmaceuticals* **2017**, *10*, 60. [CrossRef]
31. DeNicola, G.M.; Karreth, F.A.; Humpton, T.J.; Gopinathan, A.; Wei, C.; Frese, K.; Mangal, D.; Yu, K.H.; Yeo, C.J.; Calhoun, E.S.; et al. Oncogene-induced Nrf2 transcription promotes ROS detoxification and tumorigenesis. *Nature* **2011**, *475*, 106–109. [CrossRef]
32. Tian, X.; Peng, J.; Zhong, J.; Yang, M.; Pang, J.; Lou, J.; Li, M.; An, R.; Zhang, Q.; Xu, L.; et al. beta-Caryophyllene protects in vitro neurovascular unit against oxygen-glucose deprivation and re-oxygenation-induced injury. *J. Neurochem.* **2016**, *139*, 757–768. [CrossRef]
33. Assis, L.C.; Straliotto, M.R.; Engel, D.; Hort, M.A.; Dutra, R.C.; de Bem, A.F. beta-Caryophyllene protects the C6 glioma cells against glutamate-induced excitotoxicity through the Nrf2 pathway. *Neuroscience* **2014**, *279*, 220–231. [CrossRef]
34. Santos, N.A.G.; Martins, N.M.; Sisti, F.M.; Fernandes, L.S.; Ferreira, R.S.; de Freitas, O.; Santos, A.C. The cannabinoid beta-caryophyllene (BCP) induces neuritogenesis in PC12 cells by a cannabinoid-receptor-independent mechanism. *Chem. Biol. Interact.* **2017**, *261*, 86–95. [CrossRef]
35. Mishra, D.; Bisht, G.; Mazumdar, P.M.; Sah, S.P. Chemical composition and analgesic activity of Senecio rufinervis essential oil. *Pharm. Biol.* **2010**, *48*, 1297–1301. [CrossRef]
36. Segat, G.C.; Manjavachi, M.N.; Matias, D.O.; Passos, G.F.; Freitas, C.S.; Costa, R.; Calixto, J.B. Antiallodynic effect of beta-caryophyllene on paclitaxel-induced peripheral neuropathy in mice. *Neuropharmacology* **2017**, *125*, 207–219. [CrossRef]
37. Katsuyama, S.; Mizoguchi, H.; Kuwahata, H.; Komatsu, T.; Nagaoka, K.; Nakamura, H.; Bagetta, G.; Sakurada, T.; Sakurada, S. Involvement of peripheral cannabinoid and opioid receptors in beta-caryophyllene-induced antinociception. *Eur. J. Pain* **2013**, *17*, 664–675. [CrossRef]
38. Alberti, T.B.; Barbosa, W.L.; Vieira, J.L.; Raposo, N.R.; Dutra, R.C. (-)-beta-Caryophyllene, a CB2 Receptor-Selective Phytocannabinoid, Suppresses Motor Paralysis and Neuroinflammation in a Murine Model of Multiple Sclerosis. *Int. J. Mol. Sci.* **2017**, *18*, 691. [CrossRef]
39. Rabbani, M.; Sajjadi, S.E.; Vaezi, A. Evaluation of anxiolytic and sedative effect of essential oil and hydroalcoholic extract of Ocimum basilicum L. and chemical composition of its essential oil. *Res. Pharm. Sci.* **2015**, *10*, 535–543.
40. Ascari, J.; Sens, S.L.; Nunes, D.S.; Wisniewski, A., Jr.; Arbo, M.D.; Linck, V.M.; Lunardi, P.; Leal, M.B.; Elisabetsky, E. Sedative effects of essential oils obtained from Baccharis uncinella. *Pharm. Biol.* **2012**, *50*, 113–119. [CrossRef]
41. Bahi, A.; Al Mansouri, S.; Al Memari, E.; Al Ameri, M.; Nurulain, S.M.; Ojha, S. beta-Caryophyllene, a CB2 receptor agonist produces multiple behavioral changes relevant to anxiety and depression in mice. *Physiol. Behav.* **2014**, *135*, 119–124. [CrossRef]
42. Leonhardt, V.; Leal-Cardoso, J.H.; Lahlou, S.; Albuquerque, A.A.; Porto, R.S.; Celedonio, N.R.; Oliveira, A.C.; Pereira, R.F.; Silva, L.P.; Garcia-Teofilo, T.M.; et al. Antispasmodic effects of essential oil of Pterodon polygalaeflorus and its main constituent beta-caryophyllene on rat isolated ileum. *Fundam. Clin. Pharmacol.* **2010**, *24*, 749–758. [CrossRef]
43. de Oliveira, C.C.; de Oliveira, C.V.; Grigoletto, J.; Ribeiro, L.R.; Funck, V.R.; Grauncke, A.C.; de Souza, T.L.; Souto, N.S.; Furian, A.F.; Menezes, I.R.; et al. Anticonvulsant activity of beta-caryophyllene against pentylenetetrazol-induced seizures. *Epilepsy Behav.* **2016**, *56*, 26–31. [CrossRef]
44. Tchekalarova, J.; da Conceicao Machado, K.; Gomes Junior, A.L.; de Carvalho Melo Cavalcante, A.A.; Momchilova, A.; Tzoneva, R. Pharmacological characterization of the cannabinoid receptor 2 agonist, beta-caryophyllene on seizure models in mice. *Seizure* **2018**, *57*, 22–26. [CrossRef] [PubMed]
45. Dahham, S.S.; Tabana, Y.M.; Iqbal, M.A.; Ahamed, M.B.; Ezzat, M.O.; Majid, A.S.; Majid, A.M. The Anticancer, Antioxidant and Antimicrobial Properties of the Sesquiterpene beta-Caryophyllene from the Essential Oil of Aquilaria crassna. *Molecules* **2015**, *20*, 11808–11829. [CrossRef] [PubMed]

46. Ambroz, M.; Bousova, I.; Skarka, A.; Hanusova, V.; Kralova, V.; Matouskova, P.; Szotakova, B.; Skalova, L. The Influence of Sesquiterpenes from Myrica rubra on the Antiproliferative and Pro-Oxidative Effects of Doxorubicin and Its Accumulation in Cancer Cells. *Molecules* **2015**, *20*, 15343–15358. [CrossRef] [PubMed]
47. Legault, J.; Pichette, A. Potentiating effect of beta-caryophyllene on anticancer activity of alpha-humulene, isocaryophyllene and paclitaxel. *J. Pharm. Pharmacol.* **2007**, *59*, 1643–1647. [CrossRef]
48. Jung, J.I.; Kim, E.J.; Kwon, G.T.; Jung, Y.J.; Park, T.; Kim, Y.; Yu, R.; Choi, M.S.; Chun, H.S.; Kwon, S.H.; et al. beta-Caryophyllene potently inhibits solid tumor growth and lymph node metastasis of B16F10 melanoma cells in high-fat diet-induced obese C57BL/6N mice. *Carcinogenesis* **2015**, *36*, 1028–1039. [CrossRef]
49. Pavithra, P.S.; Mehta, A.; Verma, R.S. Synergistic interaction of beta-caryophyllene with aromadendrene oxide 2 and phytol induces apoptosis on skin epidermoid cancer cells. *Phytomed. Int. J. Phytother. Phytopharm.* **2018**, *47*, 121–134. [CrossRef]
50. Amiel, E.; Ofir, R.; Dudai, N.; Soloway, E.; Rabinsky, T.; Rachmilevitch, S. β-Caryophyllene, a Compound Isolated from the Biblical Balm of Gilead (*Commiphora gileadensis*), Is a Selective Apoptosis Inducer for Tumor Cell Lines. *Evid. Based Complement. Altern.* **2012**. [CrossRef]
51. Zhou, L.; Zhan, M.L.; Tang, Y.; Xiao, M.; Li, M.; Li, Q.S.; Yang, L.; Li, X.; Chen, W.W.; Wang, Y.L. Effects of beta-caryophyllene on arginine ADP-ribosyltransferase 1-mediated regulation of glycolysis in colorectal cancer under high-glucose conditions. *Int. J. Oncol.* **2018**, *53*, 1613–1624. [CrossRef]
52. Yamaguchi, M.; Levy, R.M. The combination of beta-caryophyllene, baicalin and catechin synergistically suppresses the proliferation and promotes the death of RAW267.4 macrophages in vitro. *Int. J. Mol. Med.* **2016**, *38*, 1940–1946. [CrossRef] [PubMed]
53. Youssef, D.A.; El-Fayoumi, H.M.; Mahmoud, M.F. Beta-caryophyllene protects against diet-induced dyslipidemia and vascular inflammation in rats: Involvement of CB2 and PPAR-gamma receptors. *Chem. Biol. Interact.* **2019**, *297*, 16–24. [CrossRef] [PubMed]
54. Zhang, Z.; Yang, C.; Dai, X.; Ao, Y.; Li, Y. Inhibitory effect of trans-caryophyllene (TC) on leukocyte-endothelial attachment. *Toxicol. Appl. Pharmacol.* **2017**, *329*, 326–333. [CrossRef] [PubMed]
55. Poddighe, L.; Carta, G.; Serra, M.P.; Melis, T.; Boi, M.; Lisai, S.; Murru, E.; Muredda, L.; Collu, M.; Banni, S.; et al. Acute administration of beta-caryophyllene prevents endocannabinoid system activation during transient common carotid artery occlusion and reperfusion. *Lipids Health Dis.* **2018**, *17*, 23. [CrossRef]
56. Ames-Sibin, A.P.; Barizao, C.L.; Castro-Ghizoni, C.V.; Silva, F.M.S.; Sa-Nakanishi, A.B.; Bracht, L.; Bersani-Amado, C.A.; Marcal-Natali, M.R.; Bracht, A.; Comar, J.F. β-Caryophyllene, the major constituent of copaiba oil, reduces systemic inflammation and oxidative stress in arthritic rats. *J. Cell. Biochem.* **2018**, *119*, 10262–10277. [CrossRef]
57. Castro Ghizoni, C.V.; Arssufi Ames, A.P.; Lameira, O.A.; Bersani Amado, C.A.; Sa Nakanishi, A.B.; Bracht, L.; Marcal Natali, M.R.; Peralta, R.M.; Bracht, A.; Comar, J.F. Anti-Inflammatory and Antioxidant Actions of Copaiba Oil Are Related to Liver Cell Modifications in Arthritic Rats. *J. Cell. Biochem.* **2017**, *118*, 3409–3423. [CrossRef]
58. Arizuka, N.; Murakami, T.; Suzuki, K. The effect of beta-caryophyllene on nonalcoholic steatohepatitis. *J. Toxicol. Pathol.* **2017**, *30*, 263–273. [CrossRef]
59. Yoo, H.J.; Jwa, S.K. Inhibitory effects of beta-caryophyllene on Streptococcus mutans biofilm. *Arch. Oral Biol.* **2018**, *88*, 42–46. [CrossRef]
60. Cha, J.D.; Jung, E.K.; Kil, B.S.; Lee, K.Y. Chemical composition and antibacterial activity of essential oil from Artemisia feddei. *J. Microbiol. Biotechnol.* **2007**, *17*, 2061–2065.
61. Yamaguchi, M.; Levy, R.M. beta-Caryophyllene promotes osteoblastic mineralization, and suppresses osteoclastogenesis and adipogenesis in mouse bone marrow cultures in vitro. *Exp. Ther. Med.* **2016**, *12*, 3602–3606. [CrossRef]
62. Schmitt, D.; Levy, R.; Carroll, B. Toxicological Evaluation of beta-Caryophyllene Oil: Subchronic Toxicity in Rats. *Int. J. Toxicol.* **2016**, *35*, 558–567. [CrossRef] [PubMed]
63. Tambe, Y.; Tsujiuchi, H.; Honda, G.; Ikeshiro, Y.; Tanaka, S. Gastric cytoprotection of the non-steroidal anti-inflammatory sesquiterpene, beta-caryophyllene. *Planta Med.* **1996**, *62*, 469–470. [CrossRef] [PubMed]
64. Nguyen, L.T.; Mysliveckova, Z.; Szotakova, B.; Spicakova, A.; Lnenickova, K.; Ambroz, M.; Kubicek, V.; Krasulova, K.; Anzenbacher, P.; Skalova, L. The inhibitory effects of beta-caryophyllene, beta-caryophyllene oxide and alpha-humulene on the activities of the main drug-metabolizing enzymes in rat and human liver in vitro. *Chem. Biol. Interact.* **2017**, *278*, 123–128. [CrossRef] [PubMed]

65. (EFSA), E.F.S.A. Flavouring Group Evaluation 78 (FGE.78Rev2): Consideration of Aliphatic and alicyclic and aromatic hydrocarbons evaluated by JECFA (63rd meeting) structurally related to aliphatic and aromati hydrocarbons evaluated by EFSA in FGE.25Rev3. *EFSA J.* **2009**, *13*, 931.
66. Gallelli, L.; Cione, E.; Caroleo, M.C.; Carotenuto, M.; Lagana, P.; Siniscalchi, A.; Guidetti, V. microRNAs to Monitor Pain-migraine and Drug Treatment. *MicroRNA* **2017**, *6*, 152–156. [CrossRef] [PubMed]
67. Gallelli, L.; Cione, E.; Peltrone, F.; Siviglia, S.; Verano, A.; Chirchiglia, D.; Zampogna, S.; Guidetti, V.; Sammartino, L.; Montana, A.; et al. Hsa-miR-34a-5p and hsa-miR-375 as Biomarkers for Monitoring the Effects of Drug Treatment for Migraine Pain in Children and Adolescents: A Pilot Study. *J. Clin. Med.* **2019**, *8*, 928. [CrossRef]
68. Caroleo, M.C.; Carito, V.; Pingitore, A.; Perrotta, I.D.; Perri, M.; Mancuso, D.; Russo, A.; Cione, E. Human kidney podocyte cell population as a novel biological target of nerve growth factor. *Growth Factors* **2015**, *33*, 14–22. [CrossRef]
69. Iacopetta, D.; Grande, F.; Caruso, A.; Mordocco, R.A.; Plutino, M.R.; Scrivano, L.; Ceramella, J.; Muia, N.; Saturnino, C.; Puoci, F.; et al. New insights for the use of quercetin analogs in cancer treatment. *Future Med. Chem.* **2017**, *9*, 2011–2028. [CrossRef]
70. Rizza, P.; Pellegrino, M.; Caruso, A.; Iacopetta, D.; Sinicropi, M.S.; Rault, S.; Lancelot, J.C.; El-Kashef, H.; Lesnard, A.; Rochais, C.; et al. 3-(Dipropylamino)-5-hydroxybenzofuro[2,3-f]quinazolin-1(2H)-one (DPA-HBFQ-1) plays an inhibitory role on breast cancer cell growth and progression. *Eur. J. Med. Chem.* **2016**, *107*, 275–287. [CrossRef]
71. Iacopetta, D.; Mariconda, A.; Saturnino, C.; Caruso, A.; Palma, G.; Ceramella, J.; Muia, N.; Perri, M.; Sinicropi, M.S.; Caroleo, M.C.; et al. Novel Gold and Silver Carbene Complexes Exert Antitumor Effects Triggering the Reactive Oxygen Species Dependent Intrinsic Apoptotic Pathway. *ChemMedChem* **2017**, *12*, 2054–2065. [CrossRef]
72. Saturnino, C.; Barone, I.; Iacopetta, D.; Mariconda, A.; Sinicropi, M.S.; Rosano, C.; Campana, A.; Catalano, S.; Longo, P.; Ando, S. N-heterocyclic carbene complexes of silver and gold as novel tools against breast cancer progression. *Future Med. Chem.* **2016**, *8*, 2213–2229. [CrossRef] [PubMed]
73. Fazio, A.; Iacopetta, D.; La Torre, C.; Ceramella, J.; Muia, N.; Catalano, A.; Carocci, A.; Sinicropi, M.S. Finding solutions for agricultural wastes: Antioxidant and antitumor properties of pomegranate Akko peel extracts and beta-glucan recovery. *Food Funct.* **2018**, *9*, 6618–6631. [CrossRef] [PubMed]
74. Sinicropi, M.S.; Iacopetta, D.; Rosano, C.; Randino, R.; Caruso, A.; Saturnino, C.; Muia, N.; Ceramella, J.; Puoci, F.; Rodriquez, M.; et al. N-thioalkylcarbazoles derivatives as new anti-proliferative agents: Synthesis, characterisation and molecular mechanism evaluation. *J. Enzym. Inhib. Med. Chem.* **2018**, *33*, 434–444. [CrossRef] [PubMed]
75. Perri, M.; Pingitore, A.; Cione, E.; Vilardi, E.; Perrone, V.; Genchi, G. Proliferative and anti-proliferative effects of retinoic acid at doses similar to endogenous levels in Leydig MLTC-1/R2C/TM-3 cells. *Biochim. Biophys. Acta* **2010**, *1800*, 993–1001. [CrossRef]
76. Sirignano, E.; Saturnino, C.; Botta, A.; Sinicropi, M.S.; Caruso, A.; Pisano, A.; Lappano, R.; Maggiolini, M.; Longo, P. Synthesis, characterization and cytotoxic activity on breast cancer cells of new half-titanocene derivatives. *Bioorg. Med. Chem. Lett.* **2013**, *23*, 3458–3462. [CrossRef]
77. Chimento, A.; Saturnino, C.; Iacopetta, D.; Mazzotta, R.; Caruso, A.; Plutino, M.R.; Mariconda, A.; Ramunno, A.; Sinicropi, M.S.; Pezzi, V.; et al. Inhibition of human topoisomerase I and II and anti-proliferative effects on MCF-7 cells by new titanocene complexes. *Bioorg. Med. Chem.* **2015**, *23*, 7302–7312. [CrossRef]
78. Saturnino, C.; Caruso, A.; Iacopetta, D.; Rosano, C.; Ceramella, J.; Muia, N.; Mariconda, A.; Bonomo, M.G.; Ponassi, M.; Rosace, G.; et al. Inhibition of Human Topoisomerase II by N,N,N-Trimethylethanammonium Iodide Alkylcarbazole Derivatives. *ChemMedChem* **2018**, *13*, 2635–2643. [CrossRef]
79. Iacopetta, D.; Carocci, A.; Sinicropi, M.S.; Catalano, A.; Lentini, G.; Ceramella, J.; Curcio, R.; Caroleo, M.C. Old Drug Scaffold, New Activity: Thalidomide-Correlated Compounds Exert Different Effects on Breast Cancer Cell Growth and Progression. *ChemMedChem* **2017**, *12*, 381–389. [CrossRef]
80. Sirignano, E.; Pisano, A.; Caruso, A.; Saturnino, C.; Sinicropi, M.S.; Lappano, R.; Botta, A.; Iacopetta, D.; Maggiolini, M.; Longo, P. Different 6-Aryl-Fulvenes Exert Anti-proliferative effects on Cancer Cells. *Anti Cancer Agents Med. Chem.* **2015**, *15*, 468–474. [CrossRef]

81. Saturnino, C.; Popolo, A.; Ramunno, A.; Adesso, S.; Pecoraro, M.; Plutino, M.R.; Rizzato, S.; Albinati, A.; Marzocco, S.; Sala, M.; et al. Anti-Inflammatory, Antioxidant and Crystallographic Studies of N-Palmitoyl-ethanol Amine (PEA) Derivatives. *Molecules* **2017**, *22*, 616. [CrossRef]
82. Augimeri, G.; Plastina, P.; Gionfriddo, G.; Rovito, D.; Giordano, C.; Fazio, A.; Barone, I.; Catalano, S.; Ando, S.; Bonofiglio, D.; et al. N-Eicosapentaenoyl Dopamine, A Conjugate of Dopamine and Eicosapentaenoic Acid (EPA), Exerts Anti-inflammatory Properties in Mouse and Human Macrophages. *Nutrients* **2019**, *11*, 2247. [CrossRef] [PubMed]
83. Plastina, P.; Benincasa, C.; Perri, E.; Fazio, A.; Augimeri, G.; Poland, M.; Witkamp, R.; Meijerink, J. Identification of hydroxytyrosyl oleate, a derivative of hydroxytyrosol with anti-inflammatory properties, in olive oil by-products. *Food Chem.* **2019**, *279*, 105–113. [CrossRef] [PubMed]
84. Sinicropi, M.S.; Caruso, A.; Conforti, F.; Marrelli, M.; El Kashef, H.; Lancelot, J.C.; Rault, S.; Statti, G.A.; Menichini, F. Synthesis, inhibition of NO production and antiproliferative activities of some indole derivatives. *J. Enzym. Inhib. Med. Chem.* **2009**, *24*, 1148–1153. [CrossRef] [PubMed]
85. Gabriele, B.; Fazio, A.; Carchedi, M.; Plastina, P. In vitro antioxidant activity of extracts of Sybaris liquorice roots from Southern Italy. *Nat. Prod. Res.* **2012**, *26*, 2176–2181. [CrossRef] [PubMed]
86. Plastina, P.; Fazio, A.; Gabriele, B. Comparison of fatty acid profile and antioxidant potential of extracts of seven Citrus rootstock seeds. *Nat. Prod. Res.* **2012**, *26*, 2182–2187. [CrossRef] [PubMed]

applied sciences

MDPI

Review

# Effect of *Tinospora cordifolia*-Derived Phytocomponents on Cancer: A Systematic Review

**Babji Deepa [1], Harsha V. Babaji [2], Jagadish V. Hosmani [3], Abdul Wahab H. Alamir [4], Shazia Mushtaq [5], A. Thirumal Raj [6] and Shankargouda Patil [7,*]**

1 Department of Oral Pathology and Microbiology, Maratha Mandal's NATHAJIRAO G. HALGEKAR Institute of Dental Sciences and Research Centre, Belgaum 590010, India; drdeepababji80@gmail.com
2 Department of Maxillofacial Surgery, College of Dental Sciences, Rajiv Gandhi University of Health Sciences, Davanagere 577004, India; drharshavb81@yahoo.co.in
3 Department of Dental Sciences, Oral Pathology Division, College of Dentistry, King Khalid University, Abha 62529, Saudi Arabia; jhosmani@gmail.com
4 Department of Maxillofacial Surgery and Diagnostic Sciences, College of Dentistry, Jazan University, Jazan 45142, Saudi Arabia; dr.abdulwahab@hotmail.com
5 Department of Dental Health, College of Applied Medical Sciences, King Saud University, Riyadh 11451, Saudi Arabia; smushtaqdr@gmail.com
6 Department of Oral Pathology and Microbiology, Sri Venkateswara Dental College and Hospital, Chennai 600130, India; thirumalraj666@gmail.com
7 Department of Maxillofacial Surgery and Diagnostic Sciences, Division of Oral Pathology, College of Dentistry, Jazan University, Jazan 45142, Saudi Arabia
* Correspondence: dr.ravipatil@gmail.com

Received: 28 October 2019; Accepted: 26 November 2019; Published: 28 November 2019

**Abstract:** The major cancer therapeutic modalities include surgery, chemotherapy and radiotherapy. Although these treatment regimens have played a significant role in effectively inhibiting cancer, their associated morbidity reduces the overall quality of life. Thus, researchers are striving to identify any alternate therapeutic approach capable of inhibiting cancer without eliciting the added morbidity. Among the alternate cancer therapeutics being researched, much importance is being given to the use of plants due to the presence of a wide variety of anti-carcinogenic compounds. *Tinospora cordifolia (Tc)* is one such plant and has shown to exhibit anti-carcinogenic properties. The present review aimed to systematically analyze published data on the effect of *Tinospora cordifolia*-derived phytocomponents on cancer. PubMed, Scopus, Web of Science, Embase and Cochrane library were searched using the keywords *Tinospora cordifolia*; anticancer; phytocomponents until March 20, 2019. In vivo and in vitro original studies in the English language were included. Of the 342 articles identified, only 25 articles met the selection criteria and were included in the review. Significant anti-carcinogenic properties were exhibited by *Tinospora cordifolia*-derived phytocompounds including palmative, berberine, new clerodane furanoditerene glycoside, arabinogalactan, phenolic compounds and epoxy cleodane diterpene. No significant side effects have been elicited with its use. Based on the data from the included studies, *Tinospora cordifolia* could be a natural therapeutic agent for cancer, provided its anti-carcinogenic properties can be elicited consistently at a large scale in clinical trials.

**Keywords:** cancer; phytocomponents; *Tiospora cardifolia*

## 1. Introduction

Cancer is a growing economic burden worldwide. The Globocan 2018 database estimates 18.1 million new cancer cases and 9.6 million cancer-related deaths [1]. This frightening and dreadful statistic has motivated the hunt for efficacious anticancer agents. Current therapeutic modalities in cancer, including surgery, chemotherapy, and radiotherapy have shown to be associated with significant morbidity [2]. Thus, there is much interest shown in the advent of alternative therapeutics including medicinal plants. Although several studies have explored the therapeutic benefits of natural plant-based products, there is a relative deficit in the number of systematic reviews available, especially with respect to plants present in remote areas such as the *Tinospora* genus. In the *Tinospora genus*, only *Tinospora cordifolia (Tc)*, has shown to exhibit anti-carcinogenic properties. Thus, *Tc* was selected as the subject of the review. *Tc* belongs to Menispermaceae family (universally named as "Guduchi"/ Giloy in Sanskrit). It is a deciduous climbing shrub with typical greenish-yellow flowers, found at higher altitudes like India, Myanmar and Sri Lanka. *Tc* has shown to exhibit several unique features, distinguishing them from other closely related species in the *Tinospora* genus, including *Tinospora malabarica* (*Tm*). *Tc* has an ash-colored cork, higher lenticels, nodes and internode, and a lower mucilage content than *Tm*. A variety of active components are derived from *Tc* including alkaloids, steroids, diterpenoid lactones, aliphatics, and glycoside. Apart from the anti-neoplastic properties, *Tc* is also shown to exhibit anti-diabetic, anti-periodic, anti-spasmodic, anti-inflammatory, anti-arthritic, antioxidant, anti-allergic, anti-stress, anti-leprotic, anti-malarial, hepatoprotective and immunomodulatory properties [3,4]. The present review systematically analyzes the published data assessing the effects of *Tc*-derived phytocomponents on cancer.

## 2. Methods

### 2.1. Inclusion Criteria

In vitro cell line and in vivo animal model studies in the English language assessing the effect of the *Tc*-derived phytocomponents on cancer.

### 2.2. Exclusion Criteria

Narrative reviews, systematic reviews, meta-analysis, letters, editorials, conference abstracts, articles in languages other than English.

### 2.3. Information Sources and Search Strategies

PubMed, Scopus, Web of Science, Embase and Cochrane library were searched using the keywords *Tinospora cordifolia*; anticancer; and phytocomponents until March, 2019.

### 2.4. Study Selection

The review was conducted in two steps. In the first step, the titles and abstracts of the articles identified from the databases were screened to remove any duplicates and irrelevant articles. The full texts of the articles selected from the first step were assessed in the second step using the inclusion criteria. Both steps one and two were conducted by two reviewers (B.D and H.V.B) independently. The kappa coefficient was assessed to determine the inter-observer reliability between the two reviewers for steps one and two.

## 3. Results

Selected Studies: A total of 342 articles (PubMed—14, Scopus—36, Web of Science—20, Embase—203, Cochrane—69) were identified using the keywords. Titles and the abstract of the identified articles were screened to remove any duplicates or irrelevant articles in step one. Only 50 articles were included in step two. Based on the full-text assessment of these 50 articles, only 25 articles were found to satisfy the inclusion criteria and were included in the systematic review. Figure 1 illustrates the search strategy employed. The kappa coefficient for steps one and two was found to be 0.9 and 1, respectively. Table 1; Table 2 respectively summarize the overall in vitro and in vivo effect of *Tc*-derived phytocomponents on cancer. The various phytocomponents present in the different parts of *Tc*, along with their anti-carcinogenic effects on cancer (both in vitro and in vivo), are elaborated in detail in the discussion.

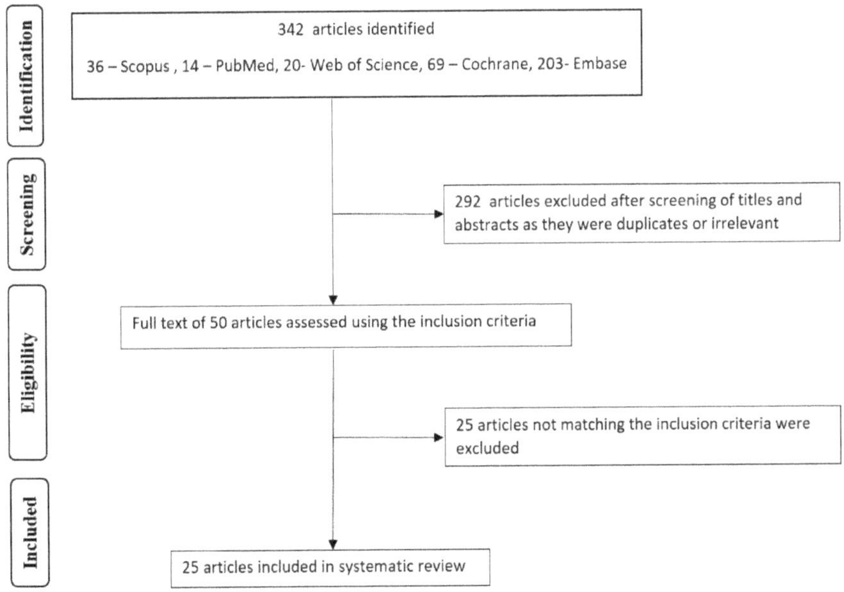


**Figure 1.** Summary of the search strategy.

**Table 1.** Summary of the data extracted from in vitro studies.

| S. No. | Phytocomponents | Part in Use | Cancer Cell Lines | Anticancer Drugs Used with *Tc* | Effect on the Cancer Cells |
|---|---|---|---|---|---|
| 1 | Berberine and isoquinolone [5] | Information not provided | HEP2 human laryngeal cancer cell line (in vitro) | 5-fluorouracil and cisplatin | Decreased gene expression of cell cycle, differentiation, and epithelial–mesenchymal transition |
| 2 | A new clerodane furano diterpene glycoside [6] | Fresh stems aqueous alcoholic extract | (in vitro) Human lung carcinoma cell line (A549), Prostate (PC-3), SF-269 (CNS), MDA-MB-435 (Melanoma), HCT-116 (Colon) and Breast (MCF-7) | Paclitaxel | Induction of mitochondria-mediated apoptosis and autophagy in HCT-116 cells |
| 5 | Information was not provided [7] | stem part aqueous and hydroalcohol extracts | Human breast carcinoma cell line MCF-7 | Doxorubicin | Induction of apoptosis |
| 6 | Phenol (ellagic acid and kaempferol) [8] | Leaf and stem Phenolic extract | CHO (Chinese Hamster Ovary) cell line | Doxorubicin | Mild cytotoxic effect noted at a high concentration of the extract |

**Table 1.** *Cont.*

| S. No. | Phytocomponents | Part in Use | Cancer Cell Lines | Anticancer Drugs Used with *Tc* | Effect on the Cancer Cells |
|---|---|---|---|---|---|
| 7 | N-formylannonain, magnoflorine, jatrorrhizine palmatine, 11-hydroxymustakone, cordifolioside A, tinocordiside and yangambin [9] | Stem Ethanol extract | Human cancer cell lines, KB (human oral squamous carcinoma), CHOK-1 (hamster ovary), HT-29 (human colon cancer) and SiHa (human cervical cancer) and murine primary cells | Doxorubicin | Cytotoxicity of cells |
| 8 | Ethanol phytofraction [10] | Powdered plant samples hexane, ethanol, and water extract | Human cancer cell lines HeLa-B75, HL-60, HEP-3B, PN-15, and normal liver cell lines | Suramin | Cytotoxicity of cells and induction of apoptosis |
| 9 | Phytochemicals of Ethanolic extract [11] | Whole plant and stem ethanolic extracts aqueous extracts | on human breast cancer cells (MCF7 and MDA MB 231) | Doxorubicin | Cytotoxicity of cells, induction of apoptosis, cell cycle arrest in the G2/M phase |
| 11 | Anthraquinones, terpenoids, and saponins and phenol [12] | Stem solvents like petroleum, ether, chloroform, ethyl acetate, acetone, and water extract | prostrate (DU-145), ovary (IGROV-1), and breast (MCF-7) cell lines | mitomycin-C (DU-145), paclitaxel against breast (MCF-7), and adriamycin (against ovary (IGR-OV-1) | Cell growth inhibition |
| 15 | Pyrrole-based small molecule [13] | Leaves ethyl acetate and aqueous extract | MDA-MB-231 breast cancer cells | Doxorubicin | Induction of apoptosis |
| 17 | Information was not provided [14] | Stem alcohol extract | human IMR-32 cell line | None | Upregulation of senescence and apoptosis |
| 18 | Phenolics contents quercetin and rutin [15] | Stems methanol extract | human breast cancer MDA-MB-231 cells | None | Anti-proliferative activity |
| 19 | Information was not provided [16] | ethanolic extract | Rat C6 glioma, U87MG human glioma, PC3 prostate cancer cell line, and HeLa cell line | None | Anti-proliferative, anti-migratory/ anti-metastatic potential activity, and induction of apoptosis. |
| 20 | Ready product from standard ayurvedic pharmacy [17] | Information was not provided | KB cancer cell lines | Methotrexate | Cell cycle arrest at G0/G1 phase |

**Table 2.** Summary of the data extracted from in vivo studies.

| S. No. | Phytocomponents | Part in Use | Animals | Outcome |
|---|---|---|---|---|
| 3 | Arabinogalactan a polysaccharide [18] | Stem Aqueous extract | Male BALB/c mice (25–30 g) benzo(a) pyrene-induced pulmonary tumor | Reduced tumor incidence and multiplicity, induction of apoptosis |
| 4 | Phenolic component [19] | Plant semiautomated capsule | DABA-induced mammary carcinogenesis in female Sprague-Dawley rats (breast cancer) | Tumor inhibition |

**Table 2.** *Cont.*

| S. No. | Phytocomponents | Part in Use | Animals | Outcome |
|---|---|---|---|---|
| 6 | Phenol (ellagic acid and kaempferol) [8] | Stem phenolic extract | Freshwater air-breathing fish *Channa punctatus* with DNA damage induction by nonionic surfactant nonylphenol | Non-cytotoxic, non-mutagenic, significant antioxidant activity, genoprotective effect |
| 10 | Alkaloid palmatine [20] | Stems methanol and aqueous extracts | Swiss albino mice injected with DMBA | Tumor inhibition |
| 12 | alkaloids including berberine [21] | stems dichloromethane alcoholic Extract | Swiss albino mice injected with Ehrlich ascites carcinoma (EAC) | Cytotoxicity of the cells |
| 13 | Antarth [22] | Plants aqueous extract | male Swiss albino mice injected with Ehrlich ascites carcinoma (EAC). | Reduces the cardiotoxicity associated with doxorubicin, but independently has no anti-carcinogenic effect |
| 14 | Triterpenoids and alkaloids [23] | stem methanolic extract | BALB RC and Swiss albino mice injected with Ehrlich ascites tumor cells | Tumor inhibition |
| 15 | Pyrrole-based small molecule [13] | Leaves methanolic extract | female Swiss albino mice injected with Ehrlich ascites tumor cells | Reduced tumor burden and two-fold increase in survival |
| 16 | Hexane fraction [24] | Stems solvents like hexane, benzene and chloroform | Swiss albino female mice injected intraperitoneally with Ehrlich ascites tumor (EAT) cells | Cell growth inhibition and induction of apoptosis |
| 20 | Polysaccharide [25] | Stem methanolic extract | C57BL/6 MICE injected by B16F-10 melanoma cell lines | Tumor inhibition |
| 21 | Clerodane-derived diterpenoids [26] | Stems alcoholic extraction | Male Wistar albino strain rats with diethylnitrosamine-induced hepatocellular carcinoma | Inhibiting tumor growth by blocking carcinogen metabolic activation and enhancing carcinogen detoxification. |
| 22 | Crude powder [27] | hydroethanolic (1:1) extract | Dalton lymphoma ascites (DLA) tumor model in Swiss albino mice | Reduced clonogenicity |
| 23 | Information was not provided [28] | Plant alcoholic extract | Inbred BALB/c mice tumor-associated macrophage (TAM)-derived dendritic cell to Dalton's lymphoma-bearing mice | Enhances the differentiation of dendritic cells |

## 4. Discussion

Phytocomponents are natural components capable of exerting a therapeutic effect on disease entities, including cancer. Unlike conventional cytotoxic chemotherapeutic agents, phytocomponents have shown to inhibit cancer cells without eliciting systemic toxicity. *Tc* is one such plant, whose phytocomponents have shown therapeutic value against several diseases, including cancer [3]. Thus, the present manuscript systematically reviewed the published literature to provide comprehensive data on the effect of *Tc* on cancer based on both in vitro and in vivo experimental models. Assessment of the in vitro studies has shown *Tc* to have a potent anti-carcinogenic property based on induction of apoptosis, cell cycle arrest, anti-migratory, anti-metastatic effect, upregulation of cellular senescence and cell growth inhibition in cancer cells (Table 1). Assessment of in vivo studies has shown anti-carcinogenic properties in the form of apoptosis induction, decrease in the expression of proliferative markers, an overall reduction in disease burden, and recurrence rate, with a significant increase in the survival rate (Table 2).

### 4.1. Parts of the Tc with Anti-Carcinogenic Effect

Phytocompounds have been isolated from all parts of *Tc* including the body of the plant, leaves and stem [4]. The stem of *Tc* has shown to contain most of the high phytocomponents, hence, most of the included studies have preferred the use of the stem. Rashmi KC et al. used *Tc* leaves, due to the presence of bis(2-ethylhexyl)-1H-pyrrole-3,4-dicarboxylate. Some studies have even used the whole plant [13].

### 4.2. Tc-Based Extracts

The various parts of *Tc* were subjected to different types of extracts for isolating the bioactive phytocomponents, including the alcoholic extract and aqueous extract. For alcoholic extracts, the parts were dried, ground in an electrical grinder and dissolved in either ethanol or methanol. Soxhlet apparatus was used for the extracts. For aqueous extracts, either distilled water or double-distilled water was used. Phytocomponents isolation was also accomplished using solvents like petroleum ether, chloroform, ethyl acetate, and acetone with alcohol [8,10]. Some authors preferred using both aqueous as well as alcoholic extracts [7,10]. Hexane extracts utilized solvents including hexane, benzene, chloroform and ethyl acetate [24] Singh B et al. used phenol to extract the dried leaves and stems of *Tc* [18]. Overall, most of the included studies preferred alcoholic extracts.

### 4.3. Dose Effect of Phytocomponents

According to the US National Cancer Institute, an $IC_{50}$ value (drug concentration required for 50% inhibition in vitro) of less than 100 µg/mL from a medicinal plant is sufficient to be considered as an anticancer agent [29]. Components of such plants are isolated and characterized to delineate their bioactive molecules. The methanolic extracts of *Tc* have shown to exhibit an $IC_{50}$ value of less than 100 µg/mL [29]. Sharma N used 1.5 kg of dried, crushed *Tc* soaked in 4.5 liters and found an $IC_{50}$ value of less than 100 µg/mL [6]. Priya M S et al. used varying dosage (200, 400 and 600 µg/mL) of *Tc*, revealing dose-dependent inhibition [7]. Bala M et al. used 2 kg of dried stem in 80% ethanol extract. They extracted three phytocomponents and elicited an $IC_{50}$ value of less than 100 µg/ml [9]. Maliakkal et al. used 2 kg of the dried crushed stem through alcohol extract and observed a dose which depended on cytotoxicity, with an ideal $IC_{50}$ value of less than 100 µg/mL. They also showed that combining different phytocomponents resulted in a profound anti-carcinogenic effect [11]. Ansari et al. used 10 kg of 50% methanolic extract of *Tc*. The extract showed the anti-carcinogenic effect, while its rutin concentration was found to be higher than quercetin [15]. Ali H et al. found the phytocomponent palmatine showed anticancer effect against environmentally induced carcinogenesis. Jagetia et al. observed that combining the various alkaloid with berberine increased the antineoplastic effect. Thus, in addition to the dose-dependent effect, the overall anti-carcinogenic effect also depended on the type, number and dosage of the used phytocomponents [7,14,15,22].

### 4.4. Phytocomponents and Its Mechanism of Action against Cancer Cells

The active phytocomponents of *Tc* include alkaloids, glycosides, steroids, aliphatic compounds, essential oils, a mixture of fatty acid, calcium, phosphorous, protein and polysaccharides [4]. The various phytocomponents identified from the studies analyzed in the systematic review included berberine, new clerodane furanodiptherineglycosidae, elligic acid, kaempferol, N-formylannonain, magnoflorine, jatrorrhizine palmatine, 11-hydroxymustakone, cordifolioside A, tinocordiside, yangambin, anthraquinones, terpenoids, saponins and phenol, pyrrole-based small molecules, quercetin and rutin, arabinogalectian, palmatine, clerodane-derived diterpenoids and hexane fractions. These phytocomponents induced anticancer effect via mitochondrial-mediated apoptosis, cytotoxic activity, mutagenic activity, reduction in tumor size, triggering reactive oxygen species, decreased gene expression of the cell cycle, effectively inhibiting cancer proliferation [5–23]. The mechanism of action of *Tc* depends on the phytocomponents used. The new clerodane furanodiptherineglycosidae exhibits anticancer activity through induction of mitochondrial-mediated apoptosis by triggering reactive oxygen species and autophagy [6]. Phenolic compounds have genoprotective and antioxidant effects on cancer cells [8]. *Tc* ethanolic extracts induced apoptosis via. increased sub G0 phase without altering cell cycle [11]. Arabinogalactans present in aqueous extracts of *Tc* shown to produce immunological activity and cytotoxic activity. Phenols have shown antimutagenic and anti-malignant effects. Flavanoids have a chemopreventive role in cancer. Pyrrole-based molecules induced apoptosis and cytotoxic effects [13]. Palmatine showed enhanced antioxidant activity by the increase in the level of antioxidant enzymes and also showed inhibition of lipid peroxidation showing role in detoxification pathway [20]. Berberine has shown to inhibit tumor cell growth by a reduction in the secretion of growth factors [5]. Hexane fractions have induced apoptosis via caspase 3-activated DNase [24]. Leyo et al. reported the polysaccharides in *Tc* to show antineoplastic effect by reducing the protein levels [25]. Epoxy-clerodane-diterpene blocks the carcinogen metabolic activation and enhances carcinogen detoxification. Singh N et al. reported *Tc* extracts have shown anticancer effect by direct tumoricidal actions [28]. Thus, *Tc* extracts have shown anti-carcinogenic properties through several mechanisms including induction of DNA damage, apoptosis, inhibiting topoisomerase II, clonogenicity, antioxidant activity, glutathione S transferase activity and increasing lipid peroxidase activity.

### *4.5. Anti-Carcinogenic Effect of Tc Phytocomponents*

In vitro studies: In the included studies, the *Tc*-extracted phytocompounds were used in combination with conventional chemotherapeutics including: fluorouracil, cisplatin, paclitaxel, suramin, doxorubicin, mitomycin, adriamycin, and methotrexate. The *Tc* extract berberine (an alkaloid) has shown to inhibit cell cycle, differentiation, and epithelial–mesenchymal transition on HEP2 human laryngeal cancer cell lines [5]. New clerodane-furano-diterpene-glycoside obtained as an aqueous-alcoholic extract through bioassay-guided fractionation exhibited significant cytotoxic effect and induced apoptosis in human lung carcinoma (A549), prostate (PC-3), SF-269(CNS), melanoma (MDA-MB-435), colon cancer (HCT-116) and breast cancer (MCF-7) cell lines. It was observed the induction of apoptosis was Reactive oxygen species-mediated through mitochondria by activation of the caspase pathway [6]. Singh B et al. identified phenolic compounds from a fungal extract of endophytic fungus *Cladosporium velox* TN-9S isolated from the stem of *Tc*. Total phenol content was 730 μg gallic equivalents/mL as determined by Folin Ciocalteu reagent. The $IC_{50}$ value was less than 100 μg/mL. These phenolic compounds have shown to exhibit a mild genoprotective potential against DNA damage on Chinese hamster ovary cell lines after the treatment with non-ionic surfactant nonylphenol. It was also noted that the endophyte's capability to synthesize phytocomponent was similar to the host plant. Their non-mutagenic and non-cytotoxic nature was suggested to enhance the antioxidant and genoprotective potential [8]. Bala et al. identified the phytocompounds from *Tc* extracts such as N-formylannonain, magnoflorine, jatrorrhizine palmatine, 11-hydroxymustakone, cordifolioside A, tinocordiside and yangambin through spectroscopic analysis. These phytocompounds were shown to exhibit anti-cancer properties on several human cancer cell lines including KB (human oral squamous carcinoma), CHOK-1 (hamster ovary), HT-29 (human colon cancer) and SiHa (human cervical cancer). Bala et al. compared the anticancer activity for different fractions of the *Tc* extract. It was noted that combining the phytocompounds increased the anti-carcinogenic properties through a synergistic effect [9]. The ethanol phytofraction obtained from plant samples of *Tc* by Mishra R et al. were cytotoxic to IMR- 32 human neuroblastoma cancer cell lines. Analysis of the cellular and nuclear morphology through immunostaining revealed Tc induced apoptosis, increased expression of senescence markers. Anti-metastatic activity in the form of a reduced cell migration capacity was also observed. Protein assays result expressed the arrest of cells in the Go/G1 phase. Mishra R et al. extracted phytocomponents from Tc plant including anthraquinones, terpenoids, saponins and phenol. This component effectively inhibited the growth of prostate, ovary and breast cancer cell lines. *Tc* extracts were also shown to exhibit antiproliferative, apoptotic-inducing, anti-migratory and antimetastatic potential on glioma cells. [12]. Butanoic fractions (pyrrole-based small molecules) were shown to induce apoptosis on breast cancer cells. Rashmi KC et al. determined apoptotic induction by evaluation by various apoptotic markers, ROS generation, caspase activity, and cell cycle analyzing. They found phytocomponents of *Tc* extract having anticancer activity and also observed inhibition of tumor proliferation. Despite promising results, the major limitation of most of the abovementioned studies is that several key aspects, including the complete mechanism of action, signaling, and pharmacological actions were nor clearly delineated [13]. Quercetin and rutin belong to phenolic phytocomponents extracted from Tc showed antiproliferative activity and was confirmed on human breast cancer MDA-MB-231 cells through induction of apoptosis, expression of altered genes and checking for the levels of intracellular ROS. The pharmacokinetics profiles, pharmacodynamic profiles and preclinical evaluation are some examples of the ongoing research by the authors [15]. Table 1 provides a summary of the overall effects elicited by *Tc* against the various cell lines.

In vivo studies: Animals that are used for the study in this review include male or female Swiss albino mice injected with Ehrlich ascites cells; male BALB/c mice with benzopyrene-induced pulmonary tumor; DABA-induced mammary carcinogenesis female Sprague Dawley rats; freshwater air-breathing fish *C. punctatus*; Swiss albino with DABA-induced carcinoma, C57BL/6 mice injected by B16F-10 melanoma cell lines; and male Wistar albino-strain rats with hepatocellular carcinoma. Arabinogalactan, a polysaccharide, was shown to inhibit cancer in male BALB/c mice. The arabinogalactan and the stem extract of *Tc* were shown to have a higher anticancer effect than only *Tc* [18]. Mishra A et al. conducted a scientific evaluation of phenolic components such as ellagic acid and kaempferol obtained from *Tc* extracts. These components were shown to have a genoprotective effect on fresh-air-breathing fish as elicited by the observations made on the morphology of the nucleus. Injecting the extract of fungus of *Tc* plant and nonylphenol caused a drastic reduction in the nuclear abnormalities. [19] Hall et al. extracted alkaloid phytocomponent palmative from *Tc* and studied anticancer property against DMBA induced skin cancer. Palmative caused a gradual decrease in the bodyweight of tumor size. Palmatine phytocomponent was shown to enhance the antioxidant enzyme levels and also inhibit carcinogenesis when administrated orally. [20] Another alkaloid phytocomponent, berberine, showed tumor remission on Swiss albino mice. The study was conducted by Jagetia G C et al. The anticancer effect was dose dependent. The exact mechanism was unknown, and authors concluded that the combinational effect of the alkaloids caused a higher anticancer effect [21]. Phytocomponents including triterpenoids, alkaloids, pyrrole-based small molecules, hexane fraction and clerodane-derived diterpenoids were also shown to exhibit significant anti-carcinogenic effect in different cancer induced in animals by showing reduction in solid tumor growth [22–24,26–28]. Leyon P V and Kuttan G extracted polysaccharide from *Tc* to observe the metastatic effect on C55BL/6 mice and the highly metastatic melanoma cell line B16F-10. Significant inhibition was noted, and although the exact mechanism of action was not known, an antimitogenic effect through natural killer cell-mediated immune modulation was suggested as a possible pathway [25]. Table 2 provides a summary of the overall effects elicited by *Tc* against the various in vivo animal models.

## 5. Conclusions

*Tc* has been shown to contain several phytocomponents with significant anti-carcinogenic properties as elicited by the included in vitro and in vivo studies. Despite promising results in laboratory settings, the future scope of *Tc* application in cancer therapy depends primarily on the success of translating the in vitro and in vivo results on to the clinical trials. Thus, large-scale multicenter prospective studies are required to elicit the potential application of *Tc* in cancer therapy.

**Author Contributions:** Conceptualization, search strategy B.D., H.V.B., J.V.H., and A.W.H.A. Extraction of data from the included studies, analysis of the data- S.M., A.T.R., S.P. All authors were involved in drafting and editing the final manuscript.

**Funding:** This research received no external funding.

**Conflicts of Interest:** The authors declare no conflict of interest.

## References

1. Bray, F.; Ferlay, J.; Soerjomataram, I.; Siegel, R.L.; Torre, L.A.; Jemal, A. Global cancer statistics 2018: GLOBOCAN estimates of incidence and mortality worldwide for 36 cancers in 185 countries. *CA Cancer J. Clin.* **2018**, *68*, 394–424. [CrossRef] [PubMed]
2. Anastyuk, S.D.; Shevchenko, N.M.; Ermakova, S.P.; Vishchuk, O.S.; Nazarenko, E.L.; Dmitrenok, P.S.; Zvyagintseva, T.N. Anticancer activity in vitro of a fucoidan from the brown alga Fucusevanescens and its low-molecular fragments, structurally characterized by tandem mass-spectrometry. *Carbohydr. Polym.* **2012**, *87*, 186–194. [CrossRef]
3. Saha, S.; Ghosh, S. Tinosporacordifolia: One plant, many roles. *Anc. Sci. Life* **2012**, *31*, 151–159.
4. Tiwari, P.; Nayak, P.; Prusty, S.K.; Sahu, P. Phytochemistry and pharmacology of Tinosporacordifolia: A Review. *Syst. Rev. Pharm.* **2018**, *9*, 70–78. [CrossRef]
5. Palmieri, A.; Iapichino, A.; Cura, F.; Scapoli, L.; Carinci, F.; Mandrone, M.; Martinelli, M. Pre-treatment with berberine enhances effect of 5-fluorouracil and cisplatin in HEP2 laryngeal cancer cell line. *J. Biol. Regul. Homeost. Agents* **2018**, *32*, 167–177.
6. Sharma, N.; Kumar, A.; Sharma, P.R.; Qayum, A.; Singh, S.K.; Dutt, P.; Paul, S.; Gupta, V.; Verma, M.K.; Satti, N.K.; et al. A new clerodane furano diterpene glycoside from Tinosporacordifolia triggers autophagy and apoptosis in HCT-116 colon cancer cells. *J. Ethnopharmacol.* **2018**, *211*, 295–310. [CrossRef]
7. Priya, M.S.; Venkateswaran, K.V.; Vijayanand, T. Determination of apoptosis by flow cytometric analysis in MCF-7 cells treated with TinosporaCordifolia. *Ind. Vet. J.* **2017**, *94*, 73–75.
8. Singh, B.; Sharma, P.; Kumar, A.; Chadha, P.; Kaur, R.; Kaur, A. Antioxidant and in vivo genoprotective effects of phenolic compounds identified from an endophytic Cladosporium velox and their relationship with its host plant Tinosporacordifolia. *J. Ethnopharmacol.* **2016**, *194*, 450–456. [CrossRef]
9. Bala, M.; Pratap, K.; Verma, P.K.; Singh, B.; Padwad, Y. Validation of ethnomedicinal potential of Tinosporacordifolia for anticancer and immunomodulatory activities and quantification of bioactive molecules by HPTLC. *J. Ethnopharmacol.* **2015**, *175*, 131–137. [CrossRef]
10. Shaikh, R.; Pund, M.; Dawane, A.; Iliyas, S. Evaluation of anticancer, antioxidant, and possible anti-inflammatory properties of selected medicinal plants used in indian traditional medication. *J. Tradit. Complement. Med.* **2014**, *4*, 253–257. [CrossRef]
11. Maliyakkal, N.; Udupa, N.; Pai, K.S.R.; Rangarajan, A. Cytotoxic and apoptotic activities of extracts of withaniasomnifera and tinosporacordifolia in human breast cancer cells. *Int. J. Appl. Res. Nat. Prod.* **2013**, *6*, 1–10.
12. Mishra, A.; Kumar, S.; Pandey, A.K. Scientific validation of the medicinal efficacy of tinosporacordifolia. *Sci. World J.* **2013**, *2013*, 1–8.
13. Rashmi, K.C.; Raj, M.H.; Paul, M.; Girish, K.S.; Salimath, B.P.; Aparna, H.S. A new pyrrole based small molecule from Tinosporacordifolia induces apoptosis in MDA-MB-231 breast cancer cells via ROS mediated mitochondrial damage and restoration of p53 activity. *Chem. Biol. Interact.* **2019**, *299*, 120–130. [CrossRef] [PubMed]
14. Mishra, R.; Kaur, G. Tinosporacordifolia Induces Differentiation and Senescence Pathways in Neuroblastoma cells. *Mol. Neurobiol.* **2015**, *52*, 719–733. [CrossRef] [PubMed]
15. Ansari, J.A.; Rastogi, N.; Ahmad, M.K.; Mahdi, A.A.; Khan, A.R.; Thakur, R.; Srivastava, V.K.; Mishra, D.P.; Fatima, N.; Khan, H.J.; et al. ROS mediated pro-apoptotic effects of Tinosporacordifolia on breast cancer cells. *Front. Biosci.* **2017**, *9*, 89–100.
16. Mishra, R.; Kaur, G. Aqueous ethanolic extract of Tinosporacordifolia as a potential candidate for differentiation based therapy of glioblastomas. *PLoS ONE* **2013**, *8*, 1–13. [CrossRef] [PubMed]
17. Bansal, P.; Malik, M.A.; Das, S.N.; Kaur, J. TinosporaCordifolia Induces Cell Cycle Arrest in Human Oral Squamous Cell Carcinoma Cells. *Gulf J. Oncol.* **2017**, *1*, 10–14.

18. Mohan, V.; Koul, A. Anticancer potential of Tinosporacordifolia and arabinogalactan against benzo(a) pyrene induced pulmonary tumorigenesis: A study in relevance to various biomarkers. *J. Herbmed. Pharmacol.* **2018**, *7*, 225–235. [CrossRef]
19. Bakrania, A.K.; Nakka, S.; Variya, B.C.; Shah, P.V.; Patel, S.S. Antitumor potential of herbomineral formulation against breast cancer: Involvement of inflammation and oxidative stress. *Indian J. Exp. Biol.* **2017**, *55*, 680–687.
20. Ali, H.; Dixit, S. Extraction optimization of Tinosporacordifolia and assessment of the anticancer activity of its alkaloid palmatine. *Sci. World J.* **2013**, *2013*, 1–10. [CrossRef]
21. Jagetia, G.C.; Rao, S.K. Evaluation of the antineoplastic activity of guduchi (Tinosporacordifolia) in Ehrlich ascites carcinoma bearing mice. *Biol. Pharm. Bull.* **2006**, *29*, 460–466. [CrossRef] [PubMed]
22. Jagetia, G.C.; Reddy, TK.; Malagi, K.J.; Nayak, B.S.; Naidu, M.B.; Ravikiran, P.B.; Kamath, S.U.; Shetty, P.C.; Reddy, D.S. Antarth, a polyherbal preparation protects against the doxorubicin-induced toxicity without compromising its Antineoplastic activity. *Phytother. Res.* **2005**, *19*, 772–778.
23. Mathew, S.; Kuttan, G. Immunomodulatory and antitumour activities of Tinosporacordifolia. *Fitoterapia* **1999**, *70*, 35–43. [CrossRef]
24. Thippeswamy, G.; Salimath, B.P. Induction of caspase-3 activated DNase mediated apoptosis by hexane fraction of Tinosporacordifolia in EAT cells. *Environ. Toxicol. Pharmacol.* **2007**, *23*, 212–220. [CrossRef]
25. Leyon, P.V.; Kuttan, G. Inhibitory effect of a polysaccharide from Tinosporacordifolia on experimental metastasis. *J. Ethnopharmacol.* **2004**, *90*, 233–237. [CrossRef]
26. Dhanasekaran, M.; Baskar, A.A.; Ignacimuthu, S.; Agastian, P.; Duraipandiyan, V. Chemopreventive potential of Epoxy clerodane diterpene from Tinosporacordifolia against diethylnitrosamine-induced hepatocellular carcinoma. *Investig. New Drugs* **2009**, *27*, 347–355. [CrossRef]
27. Adhvaryu, M.R.; Reddy, N.; Parabia, M.H. Anti-tumor activity of four Ayurvedic herbs in Dalton lymphoma ascites bearing mice and their short-term in vitro cytotoxicity on DLA-cell-line. *Afr. J. Tradit. Complement. Altern. Med.* **2008**, *5*, 409–418. [CrossRef]
28. Singh, N.; Singh, S.M.; Shrivastava, P. Effect of Tinosporacordifolia on the antitumor activity of tumor-associated macrophages-derived dendritic cells. *Immunopharmacol. Immunotoxicol.* **2005**, *27*, 1–14.
29. Ahamad, R.; Shrivastava, A.N.; Khan, M.A. Evaluation of in vitro anticancer activity of stem of Tinosporacordiflia against human breast cancer Vero cell lines. *J. Med. Plants Stud.* **2015**, *3*, 33–37.

*applied sciences*

MDPI

*Review*

# The Clinical and Economic Value of Triclosan-Coated Surgical Sutures in Abdominal Surgery

**Marco Ceresoli [1], Francesca Carissimi [1], Alessandra Piemontese [2], Vito Paragò [2], Thibaut Galvain [3], Giovanni A. Tommaselli [2] and Luca Gianotti [1,*]**

1 School of Medicine and Surgery, Milano-Bicocca University and Department of Surgery, San Gerardo Hospital, 20900 Monza, Italy; m.ceresoli4@campus.unimib.it (M.C.); f.carissimi@campus.unimib.it (F.C.)
2 Johnson & Johnson Medical SpA, 00071 Pomezia, Italy; apiemont@its.jnj.com (A.P.); vparago@its.jnj.com (V.P.); gtommase@its.jnj.com (G.A.T.)
3 Johnson & Johnson Medical SAS, 97132 Issy-les-Moulineaux, France; tgalvain@its.jnj.com
* Correspondence: luca.gianotti@unimib.it; Tel.: +39-039-2332391

Received: 11 December 2019; Accepted: 30 January 2020; Published: 6 February 2020

**Abstract:** Surgical site infection (SSI) is a frequent complication of surgical procedures. The aim of this study was to analyze the clinical evidence for SSI prevention with triclosan-coated sutures (TCS) in abdominal surgery and to investigate the economic impact of TCS in this type of procedure compared with conventional absorbable sutures (CS). A literature review was carried out to identify meta-analyses that were published between 1990 and 2019 that assessed the use of TCS in abdominal surgery. A budget impact analysis was performed from an Italian hospital perspective based on the most recently published evidence to simulate the financial impact of TCS in a general surgery unit. Uncertainty was explored through scenario analysis, as well as deterministic and probabilistic sensitivity analyses. Nine meta-analyses and two additional randomized clinical trials were retrieved. All meta-analyses described a reduction (range 19%–44%) in the risk of SSI when TCS were used. The use of TCS was associated with an overall annual net saving for the general surgery unit of €14,785 and a reduction of 3.2 SSIs compared with CS. Sensitivity analyses resulted in a positive annual saving associated with TCS in 98% of scenarios. TCS are a valuable, cost-saving SSI prevention strategy. TCS additional costs would be offset by the reduction in SSIs.

**Keywords:** triclosan; surgical site infection; suture; abdominal surgery; cost-analysis; budget impact analysis

---

## 1. Introduction

A surgical site infection (SSI) is defined as an infection that occurs within 30 days of surgery if no implant is left in place or within 90 days if an implant is left in place [1]. According to the World Health Organization (WHO) definition, SSIs may involve skin and subcutaneous tissue (superficial incisional), the deep soft tissue of the incision such as fascia or muscle (deep incisional), and/or any part of the body other than the incision that was opened or manipulated during an operation (organ/space) [1].

Despite established preventive measures [2], SSI remains the most frequent complication following abdominal surgery—defined as any surgical procedure on the abdominal cavity followed by abdominal wall closure—with an incidence rate of 10%–20% in contaminated and dirty surgery [3]. SSI is not only a surgical complication that affects the length of stay, it also has a significant impact on the patient's perceived quality of life. Moreover, SSI is an important risk factor for readmission, reintervention, and the development of incisional hernia [4,5], resulting in increased healthcare-related costs [6]. Postoperative infectious complications also negatively affect long-term outcomes [7,8].

The pathogenesis of SSI is multifactorial. The main patient-related risk factors for SSIs include active smoking, impaired nutritional status, obesity, immune deficits, and diabetes, while surgery-related risk factors include the degree of field contamination, the duration of surgery, hypothermia in the operating room, and the colonization of bacterial biofilm on sutures [9–13]. To help reduce the risk of SSI, sutures coated with antimicrobial materials have been introduced in clinical practice. The most common antibacterial agent that is used to coat sutures is triclosan, which offers broad-spectrum bactericidal action against Gram-positive and Gram-negative bacteria [14]. In both in vitro and in vivo studies, microorganisms that are responsible for SSIs were prevented from binding to and colonizing triclosan-coated sutures (TCS), with the action lasting roughly 30 days from the implantation of these sutures [15,16].

Guidelines, including those issued by the US Center for Disease Control (CDC), WHO, the American College of Surgeons (ACS), and National Institute for Health and clinical Excellence (NICE), recommend the use of TCS for the prevention of SSIs [16–19].

Several randomized clinical trials have investigated the efficacy of TCS in reducing SSI following abdominal surgery; some have established a positive effect of TCS, whereas others have failed to demonstrate statistically significant differences versus conventional absorbable sutures (CS) [20–27]. Recent meta-analyses reported a significant effect, or at least a positive trend, in favor of TCS over CS in reducing SSI rates following abdominal surgery [28,29]. The partial disagreement between the meta-analyses results is likely attributable to differences in the number of studies included, the degree of field contamination (CDC classification of the wounds), the heterogeneity of surgical procedures considered, and the inclusion of trials with low-level evidence, all of which introduced uncertainty in the overall interpretation of results.

The purpose of this study was to critically analyze and appraise the latest available evidence on the role of TCS on SSI prevention in a more homogeneous population represented by patients undergoing abdominal surgery. A literature review of meta-analyses that considered TCS in abdominal surgery was conducted, and a cost analysis was subsequently performed to investigate the economic impact of TCS to provide a more comprehensive representation of the value of this technology in clinical practice.

## 2. Materials and Methods

### 2.1. Literature Review

#### 2.1.1. Criteria for Considering Studies for Literature Review

Types of studies: Meta-analyses including randomized controlled trials (RCTs), RCTs not included in the meta-analyses, and systematic reviews with the target intervention (TCS) in one arm and in abdominal surgery were included. The endpoints included measures of TCS efficacy on SSI. Meta-analyses not including abdominal procedures or not performing a sub-analysis for abdominal procedures were excluded.

Types of participants: The review included adult patients undergoing abdominal surgery where TCS was considered to be an option to approximate any of the soft tissue layer of the surgical wound.

Types of interventions: The target intervention was the use of TCS to approximate any soft tissue layer(s) in abdominal procedures. The comparator intervention was the use of CS.

Types of outcome measures: Outcomes had to be quantitative and reported either on an intention-to-treat or per protocol basis in the respective studies. Compilations and comments were excluded. The primary endpoint was defined as overall SSI rates. SSI was defined according to the CDC criteria [1].

#### 2.1.2. Search Methods for Identification of Studies

The electronic databases PubMed (MEDLINE), Scopus, Ovid, Institute for Scientific Information Web of Science, and the Cochrane Central Register of Controlled Trials and Cochrane Library database (CENTRAL, CDRS, NHSEED and HTA) were searched from January 1990 to June 2019.

The search strategy used the following syntax: (plus OR antimicrobial OR triclosan) AND (abdominal OR intestinal OR bowel OR gastrointestinal) AND (suture OR sutures) AND ([literature AND review] OR meta-analysis OR meta-analysis OR [meta AND analysis]) AND (infection) AND "humans" [MeSH Terms]. No geographic limits were applied, and only articles in English were considered.

The reference list for each literature review was examined for missing studies within the scope of this review.

#### 2.1.3. Study Selection

Following the removal of duplicate references, two reviewers (M.C. and F.C.) independently assessed the eligibility of each of the remaining pre-selected references by screening the titles, keywords and abstracts according to the systematic review eligibility criteria. The reviewers subsequently and independently assessed the full texts of all potentially relevant references for their eligibility. Following selection, discrepancies between the reviewers were resolved by discussion, and, when a consensus could not be reached, a final decision was taken by majority after consultation with the senior reviewer (L.G.). Multiple publications reporting on the same study were jointly analyzed as reporting one study. Reference lists of full-text articles were hand-searched to identify relevant publications that were missed.

The inclusion criteria for study selection were systematic reviews and meta-analyses including only RCTs investigating the role of TCS in abdominal surgery. We also searched for additional RCTs that were not included in the meta-analyses published between January 2013 and June 2019. The exclusion criteria were meta-analyses of studies that investigated operations other than, or not only, abdominal surgery, as well as observational studies. The most recently available meta-analysis that fulfilled the selection criteria was then critically appraised. The Appendix A table reports the list of all other published meta-analyses on TCS that were excluded from this study since they were not in line with the above-mentioned inclusion criteria.

#### 2.1.4. Data Extraction and Management

Data extraction was performed in a standard electronic spreadsheet, with the reviewers independently extracting the data from all eligible articles. Extractions were compared and disagreements resolved by discussion or by consultation with a third reviewer. The following data and information were extracted: year of publication, number of studies included, number of patients included in the two groups, and outcome measurements (overall SSI risk including relative risks (RR), odds ratios (OR), 95% confidence intervals (CI), and all possible sources of bias).

### 2.2. Budget Impact Model

The economic model was developed in Microsoft® Excel and used to perform a budget impact analysis in accordance with the International Society for Pharmacoeconomics and Outcomes Research (ISPOR) guidelines "Principles of Good Practice for Budget Impact Analysis" [30,31]. This analysis evaluated the impact of introducing TCS as an alternative to conventional sutures (CS), for the prevention of SSIs from an Italian hospital perspective, and more specifically from a general surgery unit. The model compared the incremental cost of adopting the technology with the potential savings derived from the reduction of SSI incidence on the total number of performed abdominal procedures. The model output consisted of a net potential saving for the surgery that was achievable by a total conversion from CS to TCS and a reduction of SSI episodes.

#### 2.2.1. Model Inputs

The model included the following inputs: baseline SSI rate, SSI reduction, the additional cost of TCS over CS, the number of sutures used per procedure, and the cost of SSI. The baseline SSI rate was taken from an Italian trial that reported the SSI rate in 281 patients undergoing colorectal resection as equal to 10.6% [23]. SSI reduction, attributable to implementing TCS into clinical practice, was selected from the reviewed meta-analysis as previously described. The differential cost of sutures was assumed to be an additional €1 for TCS compared with CS. According to expert opinion, six absorbable sutures are required per abdominal procedure, with 600 sutures used when assuming that 100 procedures are performed annually. The cost of a single SSI episode was taken from an Italian study, published in 2000, that evaluated the hospital resources that are consumed to treat postoperative infections among patients with gastrointestinal cancer who underwent abdominal procedures [32]. The cost that was identified for each wound infection was inflated to 2019 costs according to the Italian Institute of Statistics Consumer Price Index of 1.296. An SSI cost—specific for an abdominal surgery and referred to the Italian environment—of €4,838 was therefore considered in this economic analysis. The cost breakdown was attributed to additional resource use (14%) and prolonged length of stay (LOS) (86%), as previously described [32].

2.2.2. Scenario and Sensitivity Analyses

To assess the robustness of the primary analysis, scenario and sensitivity analyses were conducted from the Italian hospital perspective (general surgery unit). In the scenario analyses, the model assumed different scenarios from the base case according to the SSI risk reduction: 1) The SSI reduction rate was set at the lower bound of the CI, and 2) the SSI reduction rate was set at the upper bound of the CI. Uncertainties surrounding model parameters were also assessed through one-way sensitivity analyses that varied key parameters by using the 95% CIs where possible or by applying a ± 25% range of variation. The considered parameters included the baseline SSI rate, the cost of a single SSI, the number of sutures used per surgical procedure, and the SSI reduction rate. Furthermore, a probabilistic sensitivity analysis (PSA) was performed in which the above parameters were simultaneously varied by randomly sampling these values from the distribution assigned for each parameter for 1000 iterations. Distributions used for each parameter are presented in Table 1.

**Table 1.** Parameters and respective distributions considered in the probabilistic sensitivity analysis.

| PSA Parameters | Base Case Value | Lower Range | Upper Range | Distribution |
|---|---|---|---|---|
| Baseline SSI rate | 10.6% | 6.1% | 16.2% | Beta |
| Cost of SSI | €4838 | €4355 | €5321 | Gamma |
| Number of sutures per procedure | 6 | 4.5 | 7.5 | Gamma |
| SSI reduction rate (relative risk) | 0.70 | 0.49 | 0.98 | Log-normal |

Abbreviations: SSI, surgical site infection.

## 3. Results

### 3.1. Clinical Results

Nine meta-analyses were identified from database searching [28,29,33–39] (Table 2). After the critical appraisal of all publications, Henriksen et al. was selected as the latest meta-analysis that best matched the inclusion and exclusion criteria [28]. This meta-analysis included eight randomized trials that investigated the role of TCS in preventing SSIs in adult patients undergoing gastrointestinal surgery [20–27]. In the pooled analysis, the authors found a significant reduction of SSIs in the experimental group (SSI rate 10.1% for TCS versus 13.5% for CS; OR: 0.67; 95% CI: 0.46–0.98; $p = 0.04$). Another meta-analysis, with similar characteristics that included 10 RCTs, published in 2018, was not considered due to an input mistake in the forest plot analysis of the TCS effect in gastrointestinal surgery [29]. In particular, for Diener et al. [21], the authors considered 334 patients in the TCS arm versus 331 in the control arm instead of 587 and 598 patients, respectively, as reported in the original publication, thus resulting in an unreliable overall treatment OR.

The characteristics of the studies included in the Henriksen et al. meta-analysis are summarized in Table 3. The heterogeneity among the RCTs was high: four trials [20,23,26,27] encompassed only elective colorectal surgery, one [21] comprised all types of elective procedures through a midline laparotomy, another [22] included both elective and emergency laparotomies, another [24] evaluated only emergency surgery for fecal peritonitis, and the final study [25] analyzed only patients undergoing open appendectomy. The study sample size varied from 100 to 1185 patients; the role of laparoscopy was not clearly reported in the majority of studies.

Following publication of the above mentioned meta-analysis, two additional RCTs that investigated the role of TCS in abdominal surgery were identified [40,41]. Ichida et al. [41] evaluated 1013 adult patients undergoing both laparoscopic and open gastrointestinal operations; the authors declared full adherence to the WHO recommendations for SSI prevention. After a follow-up of 30 days, the study results showed no difference between the groups (6.9% in the TCS group versus 5.9% in the control group; $p = 0.609$). Furthermore, no significant differences within all subgroups were identified when comparing interventions for different target abdominal organs.

Olmez et al. [40] reported a RCT that investigated the effect of triclosan-coated monofilament polydioxanone (PDS) compared to standard PDS on SSI incidence after laparotomy for any type of gastrointestinal disease in 890 patients. The Consolidated Standards of Reporting Trials diagram of the randomization process and the concealment assignment were not available. The authors reported antibiotic prophylaxis as the only standard strategy for SSI prevention. Likewise, patients in the two study groups had significant differences in baseline characteristics (Body mass index, smoking habit, anemia, hypertension, and diabetes mellitus) that are potentially involved in SSI risk. The overall effect of TCS was a significant but untrustworthy reduction in the rate of SSI (19.1% vs. 25.8% in the TCS and CS groups, respectively; $p = 0.016$).

**Table 2.** Meta-analyses reviewed.

| Study | N. studies | N. pts | Surgeries | Population | Publication Bias | Results for Abdominal Surgery and RCTs Only |
|---|---|---|---|---|---|---|
| Wang et al. [42] | 17 RCTs | 3720<br>TCS:1726; CS:1994 | Abdominal<br>Breast<br>Cardiovascular | Adult (15 studies)<br>Pediatric (2 studies) | None | RR = 0.69<br>95%CI: 0.50–0.97<br>$p = 0.03$; $I^2 = 34\%$ |
| Daoud et al. [34] | 15 RCTs | 4800<br>TCS: 2323; CS = 2477 | Neurosurgery<br>Abdominal/CR<br>Breast<br>Cardiovascular<br>Various | Adult (14)<br>Pediatric (3) | None | RR = 0.67<br>95%CI: 0.54–0.84<br>$p = 0.00053$; $I^2$ = N/A |
| Apisarnthanarak et al. [35] | 29<br>22 RCTs<br>7 non-RCTs | 11,942<br>TCS: 5,802; CS: 6,140 | Neurosurgery<br>Abdominal/CR<br>Breast<br>Cardiovascular<br>Head/neck<br>Pelvic surgery<br>Various | NR | None | RR: 0.56<br>95% CI: 0.41–0.77<br>$p < 0.05$; $I^2 = 64.6\%$ |
| Guo et al. [36] | 13 RCTs | 5256<br>TCS: 2,592; CS: ,2592 | Abdominal<br>Breast<br>Cardiovascular<br>Vascular<br>Various | Adult | Low | RR = 0.70<br>95% CI: 0.50–0.99<br>$p = 0.04$; $I^2 = 52\%$ |
| Sandini et al. [37] | 6 RCTs | 2168<br>TCS: 1,102; CS: 1,066 | Colorectal | Adult | None | OR = 0.81<br>95% CI: 0.58–1.13<br>$p = 0.220$; $I^2 = 44.9\%$ |
| Elsolh et al. [38] | 5 RCTs | 3117 | Abdominal | Adult | None | OR = 0.79<br>95%CI: 0.57–1.09<br>$p = 0.15$; $I^2 = 44\%$ |
| Konstantelias et al. [39] | 30<br>RCTs: 19<br>non-RCTs: 11 | 15,385 | Neurosurgery<br>Abdominal/CR<br>Breast<br>Cardiovascular<br>Head and neck<br>Spine<br>Various | NR | None | RCTs: RR = 0.74<br>95%CI: 0.44–1.27<br>$p$ = NS; $I^2 = 54\%$ |
| Henriksen et al. [28] | 8 RCTs | 3502<br>TCS: 1,797; CS: 1,705 | Abdominal/CR | Adult | NR | OR = 0.67<br>95%CI: 0.46–0.98<br>$p = 0.04$; $I^2 = 56\%$ |
| Uchino et al. [29] | 15<br>RCTs: 10<br>non-RCTs: 5 | 5703<br>TCS: 2,889; CS: 2,814 | Abdominal/CR | Adult | Minimal | RCTs: RR = 0.67<br>95% CI: 0.48–0.94<br>$p = 0.02$; $I^2 = 55\%$ |

Abbreviations: CR, colorectal; CS, conventional sutures; NR, not reported; RCT, randomized controlled trial; TCS, Triclosan-coated sutures; RR, risk ratio; and OR, odds ratio. Surgeries were reported if analyzed in sub-analysis (for meta-analyses performing sub-analyses).

**Table 3.** Details of the randomized trials included in the Henriksen et al. meta-analysis.

| First Author | N° of pts (TCS/CS) | Type of Surgery | Additional Preventive Strategies for SSI | Follow-Up | SSI Rates | | |
|---|---|---|---|---|---|---|---|
| | | | | | TCS | CS | *p*-Value |
| Baracs et al. [20] | 385 (188/197) | Elective open colorectal surgery | antibiotic prophylaxis | Clinical examination during hospital stay, telephonic follow-up at 30 d | 12.2% | 12.2% | 0.98 |
| Diener et al. [21] | 1185 (587/598) | Elective midline laparotomies | antibiotic prophylaxis routine scrub site preparation | Clinical examination at 10 and 30 d from discharge | 14.8% | 16.1% | 0.64 |
| Justinger et al. [22] | 856 (485/371) | Emergency or elective laparotomies | bowel preparation iodine shower site preparation antibiotic prophylaxis | Clinical examination during hospital stay and at 14 d | 6.4% | 11.3% | <0.05 |
| Mattavelli et al. [23] | 281 (140/141) | Elective open colorectal surgery | Hair removal, skin disinfection, antibiotic prophylaxis, prevention of hypothermia | Weekly examination until 30 d from discharge | 12.9% | 10.6% | 0.564 |
| Mingmalairak et al. [25] | 100 (50/50) | Open appendectomy | Antibiotic prophylaxis | Clinical examination at 1,3,7,14, 30 d | 10% | 8% | 0.727 |
| Nakamura et al. [26] | 410 (206/204) | Elective colorectal surgery | Antibiotic prophylaxis and wound protector | Daily during hospital stay, weekly until 30 d after discharge | 4.3% | 9.3% | 0.047 |
| Rasic et al. [27] | 184 (91/93) | Elective colorectal surgery | Antibiotic prophylaxis | Clinical evaluation during hospital stay | 4.3% | 13.2% | 0.039 |
| Ruiz-Tovar et al. [24] | 101 (50/51) | Emergency laparotomies for fecal peritonitis | Antibiotic prophylaxis wound irrigation; sterile-drape | Clinical examination at 5, 30 and 60 d | 10% | 35.3% | 0.004 |

Abbreviations: ASA, American Society of Anesthesiologists; HIV, human immunodeficiency virus; and SSI, surgical site infection.

## 3.2. Economic Results

### 3.2.1. Base Case Analyses

A 33% reduction from the Henriksen et al. meta-analysis was used in the budget impact analysis (OR: 0.67; 0.46–0.98; $p = 0.04$), as established from the results of the literature review [28]. We considered that when the incidence of an outcome of interest was common (>10%), the adjusted OR that was derived from the logistic regression could no longer approximate the risk ratio. We therefore approximated a risk ratio from an adjusted OR ratio and derived an estimate of an SSI reduction that better represented the true RR [43]. The obtained RR (95% CI) was 0.70 (0.49–0.98), and, accordingly, a 30% SSI reduction for TCS introduction was used in the model.

The budget impact analysis estimated an additional annual cost of €600 for a complete conversion of CS to TCS. TCS reduced the number of SSIs by 30%, which led to the avoidance of 3.2 SSIs—over 100 surgeries performed—and associated hospital resources totaling €15,385 (€13,230 for additional LOS and €2155 for additional resource use). The use of TCS resulted in an overall annual net saving of €14,785. The additional €600 associated with using TCS compared with CS were offset by the SSI reduction. The model also calculated the minimal SSI reduction required to cancel out the incremental cost of TCS: The additional cost of €600 for advancing wound closure technology was offset by an SSI reduction rate of approximately 1.2%, equaling less than a single avoided SSI (0.1 episodes).

### 3.2.2. Scenario and Sensitivity Analyses

When assuming an SSI reduction rate of 2%, the economic model resulted in an overall annual net saving of €426. In this scenario, 0.2 SSI episodes were avoided, thus accounting for a saving of €1026. The incremental expenditure of €600 was offset by the SSI reduction risk. Conversely, when the SSI reduction was set at 51%, the annual net saving for the hospital was €25,554 with 5.4 SSI avoided episodes accounting for €26,154 of hospital resources.

The results of the one-way sensitivity analysis, represented as a tornado chart, are presented in Figure 1. The variables with the greatest influence on the annual savings in the model were related to efficacy parameters (SSI risk reduction and baseline SSI rate), with other parameters that were related to cost events found to demonstrate a minor influence on the annual savings. The annual savings remained positive across a broad variation in the ranges for each parameter.

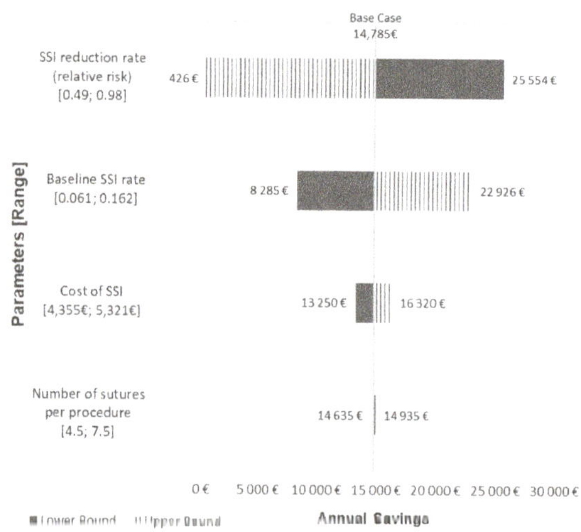


**Figure 1.** Tornado diagram with one-way sensitivity analysis results. Abbreviations: SSI, surgical site infection.

The results of 1000 Monte-Carlo simulations of the annual savings are presented in Figure 2, and the number of avoided SSIs, based on PSA, are presented in Figure 3. The analysis demonstrated that the reduction of SSI episodes due to TCS ranged between 1 and 6 cases in 90% of the simulations. The mean (interquartile range (IQR)) annual savings that were obtained from the PSA was €13,935 (9068–18,665). The whiskers extended the IQR by 1.5 to a maximum of €32,369 and a minimum of −€4987. The analysis demonstrated a 98% probability of TCS being cost saving, based on the assumptions made.

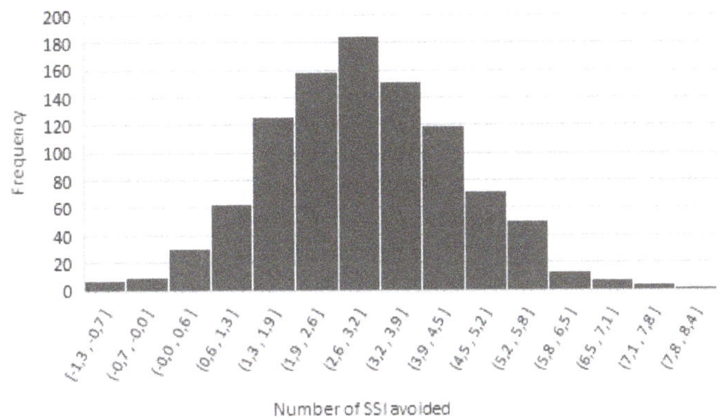


**Figure 2.** Histogram of the number of avoided SSI episodes according the probabilistic sensitivity analysis. Abbreviations: SSI, surgical site infection.

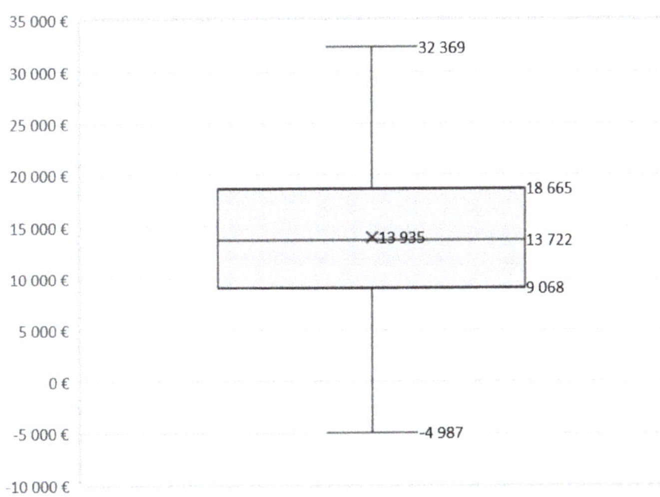


**Figure 3.** Boxplot of the annual budget impact (savings) according to the probabilistic sensitivity analysis.

## 4. Discussion

SSIs primarily result from the disruption of the equilibrium between the quantity of inoculated microorganisms, their virulence, and the ability of the immune system to clear them. TCS, which are coated in a broad-spectrum antimicrobial agent, reduce the bacterial load at the wound site, thereby potentially helping to reduce the risk of SSIs.

Meta-analysis is a useful tool in that it allows for data pooling from multiple trials and thus provides a more comprehensive estimation of treatment effects. However, this approach only partially restricted the confounding effect of the differences and heterogeneity among the included studies and populations. Among the available meta-analyses and in accordance with the defined inclusion criteria, we assessed and estimated the results of the 2017 meta-analysis reported by Henriksen et al. [28] as the most recently available evidence that fulfilled our clinical focus. The authors included all RCTs that evaluated the effect of TCS in patients who underwent abdominal surgery. The effect of TCS was evaluated in several studies that enrolled subjects with distinct patient-related and operation-related risk factors. There was heterogeneity between trials in terms of surgery type which included elective colorectal surgeries, elective procedures through midline laparotomy, elective and emergency laparotomies for any type of target organ, emergency surgery for fecal peritonitis, and open appendectomy. The pooled results yielded a significant 33% reduction in the odds of having an SSI after surgery for patients who received a wound closure with TCS versus CS. However, it should be noted that intervention type (elective versus emergency), target organs, and the lack of implementation of recognized peri-operative strategies to reduce SSI likely limit the ability to define the true effect of TCS, adding high heterogeneity and possible confounders. Two additional RCTs [40,41] that were published afterwards did not contribute to solving these uncertainties.

The effect of TCS should, ideally, be evaluated in a population with a fixed risk of SSI and, importantly, in a setting where all the WHO recommendations [16] (Table 4) are fulfilled to assess the true "over the top" protective effect of this strategy. The heterogeneity of studies' settings in terms of adherence to SSI prevention guidelines can therefore be considered as an additional bias; however, these limitations notwithstanding, all nine published meta-analyses reported a consistent decrease in the risk of having an SSI with the use of the used TCS (Table 2). Despite the fact that some of the available meta-analyses did not demonstrate significant differences compared to the control arms, the decrease in the SSI rate can still be considered important as it may result in a net financial saving even if the cost of TCS is higher than CS.

**Table 4.** World Health Organization recommendations for SSI prevention, 2016 version.

| **Preoperative Measures** |
|---|
| Shower before surgery with either plain or antimicrobial soap |
| Mupirocin 2% decolonization in *S. aureus* nasal carrier in cardiac and orthopedic surgery |
| Not remove patient hair, or if necessary, prefer clipper to shaver |
| Antibiotic prophylaxis in 120 min preceding surgical incision |
| Adequate surgical hand scrubbing |
| Oral administration of multiple nutrient-enhanced formula before surgery |
| Not interrupt immunosuppressive treatment |
| **Intraoperative Measures** |
| Skin preparation with alcohol-based chlorhexidine gluconate solution |
| Use wound protector devices |
| Consider prophylactic negative pressure therapy in high-risk wound |
| Not use plastic adhesive incise drapes |
| Not use antimicrobial sealants after surgical site skin preparation |
| Administration of oxygen with 80% FiO2 |
| Use warming device |
| Use protocol for intensive blood glucose control |
| Wound irrigation with aqueous povidone iodine solution before closure |
| Not perform antibiotic wound irrigation |
| Use of triclosan-coated sutures for the purpose of reducing the risk of SSI, independent of the type of surgery |
| **Postoperative Measures** |
| Not prolong antibiotic prophylaxis in postoperative period |
| Not continue antibiotic prophylaxis due to presence of drain |
| Administration of oxygen with 80% FiO2 for 2–6 h post-op |
| Appropriate wound evaluation and management |
| Not use advanced dressing of any sort, prefer standard dressing |

Abbreviations: SSI, surgical site infection.

A budget impact analysis was performed to evaluate the economic impact of adopting TCS into clinical practice from the perspective of a general surgery unit in an Italian hospital. The model demonstrated that the incremental cost of adopting TCS was offset by the cost of SSI episodes that were avoided. Despite an upfront incremental acquisition cost of €600, the use of TCS resulted in an overall annual net saving of €14,785 for the institution. Furthermore, sensitivity analyses demonstrated that, across a broad variation in the ranges for each parameter, the annual savings remained positive with a 98% probability that TCS was cost-saving.

Leaper et al. [44], by using a deterministic decision-tree, a stochastic cost model, and the National Health Service (NHS England)-based cost of inpatient admissions for infections and differential costs of TCS versus CS, estimated a significant mean saving of £91.25 per surgical procedure from using antimicrobial sutures across all surgical wound types. This is aligned with the findings in this analysis, where the overall saving was split for single hospitalization with a saving of €147.85. This was further supported by three trials that were published between 2007 and 2013 that evaluated the clinical efficacy and economic impact of TCS and which demonstrated a net economic saving resulting from adopting TCS rather than CS [26,45,46].

The relevance of this study to the Italian healthcare system lies in the use of Italian-specific input data for SSI risk and SSI cost. This analysis could therefore be valuable to support decision-making processes at the hospital level in Italy. The robustness of the model is increased by the performed sensitivity analyses; however, the limitations of the model lie in some of the inputs being extrapolated from literature research, not being real-world data, or being inflated to current values from outdated data like the SSI cost. To alleviate this, additional analyses, based on recent real-world data from one or more Italian hospitals could be performed to demonstrate the real budget impact of adopting TCS into Italian clinical practice.

Hence, in the decision-making process to adopt a new technology or device, a robust economic analysis should always couple the clinical results to gather additional and essential information on the potential dominant cost-effectiveness ratio.

## 5. Conclusions

This critical review provides an overview of the available evidence on the role of TCS in preventing SSI in patients undergoing abdominal surgery, and it reports the economic impact of adopting TCS into clinical practice. The results did not completely rule out and solve the conflicting results in the literature on the clinical benefit of TCS on the prevention of wound infection. However, the economic cost model that was used in this study found that the use of TCS results in significant and robust cost saving from the perspective of an Italian hospital that performs abdominal operations. The results might be transferred to many other different hypothetical settings. Triclosan antimicrobial surgical sutures should be considered for wound closure as a cost-saving measure to prevent SSIs.

**Author Contributions:** Conceptualization, A.P., T.G., V.P., L.G., G.A.T., F.C. and M.C.; methodology, A.P., T.G., L.G., V.P., G.A.T., F.C. and M.C.; software, A.P., T.G.; validation, A.P., T.G., V.P., L.G., G.A.T., F.C. and M.C.; formal analysis, A.P., T.G., V.P. and L.G.; writing, A.P., T.G., V.P., L.G., G.A.T., F.C. and M.C.; visualization, T.G.; supervision, G.A.T. and L.G. All authors have read and agreed to the published version of the manuscript.

**Funding:** Johnson and Johnson funded medical writing services for this research. The authors received no financial support for the research, authorship, and publication of this article.

**Acknowledgments:** We thank James Woolnough and Roderick Walker (Mtech) for medical writing support.

**Conflicts of Interest:** Alessandra Piemontese, Giovanni Tommaselli, Thibaut Galvain and Vito Parago all declare to be employees of Johnson & Johnson.

## Appendix A

**Table A1.** Meta-analyses published between January 1990 and June 2019 not included in the study results.

| Study | Publication Year | N. Studies/ N. Pts | Results for all Surgery and Study Type | Reason for Exclusion |
|---|---|---|---|---|
| Chang et al. | 2012 | 7 RCTs<br>836 pts | OR = 0.77<br>95% CI: 0.40–1.51<br>$p = 0.45$ | All surgery types<br>No abdominal surgery results |
| Sajid et al. | 2013 | 7 RCTs<br>1631 pts | RR = 0.61<br>95% CI: 0.37–0.99<br>$p = 0.04$ | All surgery types<br>No abdominal surgery results |
| Edmiston et al. | 2013 | 13 RCTs<br>3568 pts | RR = 0.734<br>95% CI: 0.590–0.913<br>$p = 0.005$ | All surgery types<br>No abdominal surgery results |
| Wu et al. | 2017 | 18<br>13 RCTs<br>5 non-RCTs<br>7458 pts | RCTs: OR = 0.72<br>95% CI: 0.59–0.88<br>$p = 0.001$<br>Non-RCTs: OR = 0.58<br>95% CI: 0.40–0.83<br>$p = 0.003$ | All surgery types<br>No abdominal surgery results |
| de Jonge et al. | 2017 | 21 RCTs<br>6462 pts | RR = 0.72<br>95% CI: 0.60–0.86<br>$P < 0.001$ | All surgery types<br>No abdominal surgery results |
| Leaper et al. | 2017 | 34<br>20 RCTs<br>14 non-RCTs<br>NR pts | OR = 0.61<br>95% CI: 0.52–0.73<br>$P < 0.001$ | All surgery types<br>No abdominal surgery results<br>No RCTs results |
| Hunger et al. | 2018 | 6<br>3 RCTs<br>3 non-RCTs<br>5188 pts | RCTs: RR = 0.67<br>95% CI: 0.48–0.94<br>$p = 0.02$<br>Non-RCTs: OR = 0.4<br>95% CI: 0.3–0.54<br>$P < 0.001$ | All surgery types<br>No abdominal surgery results |

Abbreviations: NR, not reported; RCT, randomized controlled trial.

## References


1. European Centre for Disease Prevention and Control. *Surveillance of Surgical Site Infections and Prevention Indicators in European Hospitals: HAI-Net SSI Protocol*; Version 2.2; Tommi Kärki: Carl Suetens, Stockholm, May 2017; ISBN 978-92-9498-060-1.
2. Preventing Surgical Site Infections: Implementation Approaches for Evidence-Based Recommendations. World Health Organization: Geneva, 2018. Licence: CC BYNC-SA 3.0 IGO. Available online: https://apps.who.int/iris/bitstream/handle/10665/273154/9789241514385-eng.pdf (accessed on 1 August 2019).
3. Aga, E.; Keinan-Boker, L.; Eithan, A.; Mais, T.; Rabinovich, A.; Nassar, F. Surgical site infections after abdominal surgery: Incidence and risk factors. A prospective cohort study. *Infect. Dis.* **2015**, *47*, 761–767. [CrossRef] [PubMed]
4. Veljkovic, R.; Protic, M.; Gluhovic, A.; Potic, Z.; Milosevic, Z.; Stojadinovic, A. Prospective clinical trial of factors predicting the early development of incisional hernia after midline laparotomy. *J. Am. Coll. Surg.* **2010**, *210*, 210–219. [CrossRef]
5. Murray, B.W.; Cipher, D.J.; Pham, T.; Anthony, T. The impact of surgical site infection on the development of incisional hernia and small bowel obstruction in colorectal surgery. *Am. J. Surg.* **2011**, *202*, 558–560. [CrossRef] [PubMed]

6. Badia, J.M.; Casey, A.L.; Petrosillo, N.; Hudson, P.M.; Mitchell, S.A.; Crosby, C. Impact of surgical site infection on healthcare costs and patient outcomes: A systematic review in six European countries. *J. Hosp. Infect.* **2017**, *96*, 1–15. [CrossRef] [PubMed]
7. O'Brien, W.J.; Gupta, K.; Itani, K.M.F. Association of Postoprative infection With Risk of Long-term Infection and Mortality. *JAMA Surg.* **2020**, *155*, 61–68. [CrossRef] [PubMed]
8. Nespoli, A.; Gianotti, L.; Totis, M.; Bovo, G.; Nespoli, L.; Chiodini, P.; Brivio, F. Correlation between postoperative infections and long-term survival after colorectal resection for cancer. *Tumori* **2004**, *90*, 485–490. [CrossRef] [PubMed]
9. Chuah, L.L.; Papamargaritis, D.; Pillai, D.; Krishnamoorthy, A.; le Roux, C.W. Morbidity and mortality of diabetes with surgery. *Nutr. Hosp.* **2013**, *28*, 47–52.
10. Sørensen, L.T.; Hemmingsen, U.; Kallehave, F.; Wille-Jørgensen, P.; Kjærgaard, J.; Møller, L.N.; Jørgensen, T. Risk Factors for Tissue and Wound Complications in Gastrointestinal Surgery. *Ann. Surg.* **2005**, *241*, 654–658. [CrossRef]
11. Dobner, J.; Kaser, S. Body mass index and the risk of infection - from underweight to obesity. *Clin. Microbiol. Infect.* **2018**, *24*, 24–28. [CrossRef]
12. Thelwall, S.; Harrington, P.; Sheridan, E.; Lamagni, T. Impact of obesity on the risk of wound infection following surgery: Results from a nationwide prospective multicentre cohort study in England. *Clin. Microbiol. Infect.* **2015**, *21*, 1008.e1–1008.e8. [CrossRef]
13. Alexander, J.W.; Kaplan, J.Z.; Altemeier, W.A. Role of suture materials in the development of wound infection. *Ann. Surg.* **1967**, *165*, 192–199. [CrossRef]
14. Kampf, G.; Kramer, A. Epidemiologic background of hand hygiene and evaluation of the most important agents for scrubs and rubs. *Clin. Microbiol. Rev.* **2004**, *17*, 863–893. [CrossRef] [PubMed]
15. Leaper, D.; Assadian, O.; Hubner, N.-O.; McBain, A.; Barbolt, T.; Rothenburger, S.; Wilson, P. Antimicrobial sutures and prevention of surgical site infection: Assessment of the safety of the antiseptic triclosan. *Int. Wound J.* **2011**, *8*, 556–566. [CrossRef] [PubMed]
16. Global Guidelines for the Prevention of Surgical Site Infection, second edition. World Health Organization: Geneva, 2018. Licence: CC BY-NC-SA 3.0 IGO. Available online: https://apps.who.int/iris/bitstream/handle/10665/277399/9789241550475-eng.pdf?ua=1 (accessed on 1 August 2019).
17. Ban, K.A.; Minei, J.P.; Laronga, C.; Harbrecht, B.G.; Jensen, E.H.; Fry, D.E.; Itani, K.M.F.; Dellinger, E.P.; Ko, C.Y.; Duane, T.M. American College of Surgeons and Surgical Infection Society: Surgical Site Infection Guidelines, 2016 Update. *J. Am. Coll. Surg.* **2017**, *224*, 59–74. [CrossRef] [PubMed]
18. Berríos-Torres, S.I.; Umscheid, C.A.; Bratzler, D.W.; Leas, B.; Stone, E.C.; Kelz, R.R.; Reinke, C.E.; Morgan, S.; Solomkin, J.S.; Mazuski, J.E.; et al. Centers for Disease Control and Prevention Guideline for the Prevention of Surgical Site Infection, 2017. *JAMA Surg.* **2017**, *152*, 784–791. [CrossRef] [PubMed]
19. Surgical Site Infections: Prevention and Treatment NICE Guideline [NG125]. Available online: https://www.nice.org.uk/guidance/ng125 (accessed on 1 August 2019).
20. Baracs, J.; Huszár, O.; Sajjadi, S.G.; Horváth, O.P. Surgical site infections after abdominal closure in colorectal surgery using triclosan-coated absorbable suture (PDS Plus) vs. uncoated sutures (PDS II): A randomized multicenter study. *Surg. Infect.* **2011**, *12*, 483–489. [CrossRef] [PubMed]
21. Diener, M.K.; Knebel, P.; Kieser, M.; Schüler, P.; Schiergens, T.S.; Atanassov, V.; Neudecker, J.; Stein, E.; Thielemann, H.; Kunz, R.; et al. Effectiveness of triclosan-coated PDS Plus versus uncoated PDS II sutures for prevention of surgical site infection after abdominal wall closure: The randomised controlled PROUD trial. *Lancet* **2014**, *384*, 142–152. [CrossRef]
22. Justinger, C.; Slotta, J.E.; Ningel, S.; Gräber, S.; Kollmar, O.; Schilling, M.K. Surgical-site infection after abdominal wall closure with triclosan-impregnated polydioxanone sutures: Results of a randomized clinical pathway facilitated trial (NCT00998907). *Surgery* **2013**, *154*, 589–595. [CrossRef]
23. Mattavelli, I.; Rebora, P.; Doglietto, G.; Dionigi, P.; Dominioni, L.; Luperto, M.; La Porta, A.; Garancini, M.; Nespoli, L.; Alfieri, S.; et al. Multi-Center Randomized Controlled Trial on the Effect of Triclosan-Coated Sutures on Surgical Site Infection after Colorectal Surgery. *Surg. Infect.* **2015**, *16*, 226–235. [CrossRef]
24. Ruiz-Tovar, J.; Alonso, N.; Morales, V.; Llavero, C. Association between Triclosan-Coated Sutures for Abdominal Wall Closure and Incisional Surgical Site Infection after Open Surgery in Patients Presenting with Fecal Peritonitis: A Randomized Clinical Trial. *Surg. Infect.* **2015**, *16*, 588–594. [CrossRef]

25. Mingmalairak, C.; Ungbhakorn, P.; Paocharoen, V. Efficacy of antimicrobial coating suture coated polyglactin 910 with tricosan (Vicryl plus) compared with polyglactin 910 (Vicryl) in reduced surgical site infection of appendicitis, double blind randomized control trial, preliminary safety report. *J. Med. Assoc. Thai.* **2009**, *92*, 770–775. [PubMed]
26. Nakamura, T.; Kashimura, N.; Noji, T.; Suzuki, O.; Ambo, Y.; Nakamura, F.; Kishida, A. Triclosan-coated sutures reduce the incidence of wound infections and the costs after colorectal surgery: A randomized controlled trial. *Surgery* **2013**, *153*, 576–583. [CrossRef]
27. Rasić, Z.; Schwarz, D.; Adam, V.N.; Sever, M.; Lojo, N.; Rasić, D.; Matejić, T. Efficacy of antimicrobial triclosan-coated polyglactin 910 (Vicryl* Plus) suture for closure of the abdominal wall after colorectal surgery. *Coll. Antropol.* **2011**, *35*, 439–443. [PubMed]
28. Henriksen, N.A.; Deerenberg, E.B.; Venclauskas, L.; Fortelny, R.H.; Garcia-Alamino, J.M.; Miserez, M.; Muysoms, F.E. Triclosan-coated sutures and surgical site infection in abdominal surgery: The TRISTAN review, meta-analysis and trial sequential analysis. *Hernia* **2017**, *21*, 833–841. [CrossRef] [PubMed]
29. Uchino, M.; Mizuguchi, T.; Ohge, H.; Haji, S.; Shimizu, J.; Mohri, Y.; Yamashita, C.; Kitagawa, Y.; Suzuki, K.; Kobayashi, M.; et al. The Efficacy of Antimicrobial-Coated Sutures for Preventing Incisional Surgical Site Infections in Digestive Surgery: A Systematic Review and Meta-analysis. *J. Gastrointest. Surg.* **2018**, *22*, 1832–1841. [CrossRef] [PubMed]
30. Mauskopf, J.A.; Sullivan, S.D.; Annemans, L.; Caro, J.; Mullins, C.D.; Nuijten, M.; Orlewska, E.; Watkins, J.; Trueman, P. Principles of good practice for budget impact analysis: Report of the ISPOR Task Force on good research practices–budget impact analysis. *Value Health* **2007**, *10*, 336–347. [CrossRef] [PubMed]
31. Sullivan, S.D.; Mauskopf, J.A.; Augustovski, F.; Jaime Caro, J.; Lee, K.M.; Minchin, M.; Orlewska, E.; Penna, P.; Rodriguez Barrios, J.-M.; Shau, W.-Y. Budget impact analysis-principles of good practice: Report of the ISPOR 2012 Budget Impact Analysis Good Practice II Task Force. *Value Health* **2014**, *17*, 5–14. [CrossRef]
32. Gianotti, L.; Braga, M.; Frei, A.; Greiner, R.; Di Carlo, V. Health care resources consumed to treat postoperative infections: Cost saving by perioperative immunonutrition. *Shock* **2000**, *14*, 325–330. [CrossRef]
33. Wang, G.J.; Hungerford, D.S.; Savory, C.G.; Rosenberg, A.G.; Mont, M.A.; Burks, S.G.; Mayers, S.L.; Spotnitz, W.D. Use of fibrin sealant to reduce bloody drainage and hemoglobin loss after total knee arthroplasty: A brief note on a randomized prospective trial. *J. Bone Jt. Surg. Am.* **2001**, *83*, 1503–1505. [CrossRef]
34. Daoud, F.C.; Edmiston, C.E.; Leaper, D. Meta-analysis of prevention of surgical site infections following incision closure with triclosan-coated sutures: Robustness to new evidence. *Surg. Infect.* **2014**, *15*, 165–181. [CrossRef]
35. Apisarnthanarak, A.; Singh, N.; Bandong, A.N.; Madriaga, G. Triclosan-coated sutures reduce the risk of surgical site infections: A systematic review and meta-analysis. *Infect. Control. Hosp. Epidemiol.* **2015**, *36*, 169–179. [CrossRef]
36. Guo, J.; Pan, L.-H.; Li, Y.-X.; Yang, X.-D.; Li, L.-Q.; Zhang, C.-Y.; Zhong, J.-H. Efficacy of triclosan-coated sutures for reducing risk of surgical site infection in adults: A meta-analysis of randomized clinical trials. *J. Surg. Res.* **2016**, *201*, 105–117. [CrossRef]
37. Sandini, M.; Mattavelli, I.; Nespoli, L.; Uggeri, F.; Gianotti, L. Systematic review and meta-analysis of sutures coated with triclosan for the prevention of surgical site infection after elective colorectal surgery according to the PRISMA statement. *Medicine* **2016**, *95*, e4057. [CrossRef] [PubMed]
38. Elsolh, B.; Zhang, L.; Patel, S.V. The Effect of Antibiotic-Coated Sutures on the Incidence of Surgical Site Infections in Abdominal Closures: A Meta-Analysis. *J. Gastrointest. Surg.* **2017**, *21*, 896–903. [CrossRef] [PubMed]
39. Konstantelias, A.A.; Andriakopoulou, C.S.I.; Mourgela, S. Triclosan-coated sutures for the prevention of surgical-site infections: A meta-analysis. *Acta Chir. Belg.* **2017**, *117*, 137–148. [CrossRef] [PubMed]
40. Olmez, T.; Berkesoglu, M.; Turkmenoglu, O.; Colak, T. Effect of Triclosan-Coated Suture on Surgical Site Infection of Abdominal Fascial Closures. *Surg. Infect.* **2019**, *20*, 658–664. [CrossRef] [PubMed]
41. Ichida, K.; Noda, H.; Kikugawa, R.; Hasegawa, F.; Obitsu, T.; Ishioka, D.; Fukuda, R.; Yoshizawa, A.; Tsujinaka, S.; Rikiyama, T. Effect of triclosan coated sutures on the incidence of surgical site infection after abdominal wall closure in gastroenterological surgery: A double-blind, randomized controlled trial in a single center. *Surgery* **2018**. [CrossRef]

42. Wang, Z.X.; Jiang, C.P.; Cao, Y.; Ding, Y.T. Systematic review and meta-analysis of triclosan-coated sutures for the prevention of surgical-site infection. *Br. J. Surg.* **2013**, *100*, 465–473. [CrossRef]
43. Zhang, J.; Yu, K.F. What's the relative risk? A method of correcting the odds ratio in cohort studies of common outcomes. *JAMA* **1998**, *280*, 1690–1691. [CrossRef]
44. Leaper, D.J.; Edmiston, C.E.; Holy, C.E. Meta-analysis of the potential economic impact following introduction of absorbable antimicrobial sutures. *Br. J. Surg.* **2017**, *104*, e134–e144. [CrossRef]
45. Galal, I.; El-Hindawy, K. Impact of using triclosan-antibacterial sutures on incidence of surgical site infection. *Am. J. Surg.* **2011**, *202*, 133–138. [CrossRef] [PubMed]
46. Fleck, T.; Moidl, R.; Blacky, A.; Fleck, M.; Wolner, E.; Grabenwoger, M.; Wisser, W. Triclosan-coated sutures for the reduction of sternal wound infections: Economic considerations. *Ann. Thorac. Surg.* **2007**, *84*, 232–236. [CrossRef] [PubMed]

applied sciences

MDPI

Project Report

# New Local Drug Delivery with Antibiotic in the Nonsurgical Treatment of Periodontitis—Pilot Study

**Aleksandra Sender-Janeczek, Jacek Zborowski *, Małgorzata Szulc and Tomasz Konopka**

Department and Division of Periodontology Wroclaw Medical University ul. Krakowska 26, 50-425 Wrocław, Poland; aleksandra.sender-janeczek@umed.wroc.pl (A.S.-J.); malgorzata.szulc@umed.wroc.pl (M.S.); tomasz.konopka@umed.wroc.pl (T.K.)

* Correspondence: jacek.zborowski@umed.wroc.pl

Received: 24 October 2019; Accepted: 19 November 2019; Published: 25 November 2019

**Abstract:** Combination of the classical subgingival instrumentation (scaling and root planing procedure, SRP) with an antibiotic administered to periodontal pockets in a suitable medium is a promising alternative protocol of nonsurgical periodontal treatment. It enables obtaining the long-term minimum drug concentration inhibiting the development of periopathogens. Objectives: Clinical and microbiological evaluation of periodontal pockets two months after single application of a gel containing piperacillin and tazobactam (Gelcide)®in relation to the nonsurgical treatment procedure (SRP). Materials and methods: Ten patients aged 24–56 years (mean 39.5) with chronic periodontitis, nonsmokers with acceptable oral hygiene and no classical exclusion criteria were qualified for treatment. In the maxilla area, SRP was performed and the assessed gel was inserted to two randomly selected adjacent periodontal pockets. Clinical evaluation included the assessment of bleeding on probing (BoP), pocket depth (PD), and clinical attachment loss (CAL) at six measurement points. A microbiological examination with the use of PET deluxe diagnostic kit in the drug-administered pockets and symmetrically in two pockets on the other side of the dental arch was performed. The examination was conducted before the treatment and two months later. Results: Two months after the treatment, a significant improvement in all analyzed clinical parameters was observed. However, the extent of this improvement did not differ significantly between the compared treatment methods. No statistically significant differences were found in the number of bacteria before and after the treatment, except for a significant decrease in the number of Micromonas micros (2957 vs. 589, $p = 0.028$) and a higher number of the green complex bacteria Capnocytophaga gingivalis (5439 vs. 2050, $p = 0.041$) after antibiotic had been used. Conclusion: No significant clinical and microbiological differences were found after additional administration of gel with piperacillin and tazobactam in relation to SRP in the preliminary study.

**Keywords:** periodontitis; SRP; LDD; gelcide

## 1. Introduction

In the light of the latest classification adopted during the World Panel for Periodontal Classification in 2017, periodontitis is a chronic, multifactorial, inflammatory disease associated with dysbiotic biofilm of the plaque, and is characterized by progressive destruction of dental attachment apparatus [1]. The most characteristic features of the disease include clinical attachment loss (CAL), pocket deepening (PD), bleeding on probing, and loss of bone tissue which can be radiologically observed [2].

Periodontitis constitutes a significant social problem due to the incidence of the disease, and the consequences associated with failure to take up treatment. Loss of teeth that takes place over the course of periodontitis results in the loss of prosthetic support areas, and significantly affects the quality of life of patients as well as the way they function in the society.

Epidemiological studies from 2013 show that 26% of Warsaw residents aged 35–44 require specialist periodontal treatment [3]. In the nationwide studies, the ratio was 16% in the same age group [4] and 19.7% in the group of people aged 65–74. In a 2012 study conducted by Konopka et al. [5], according to the definition of Page and Eke, periodontitis was diagnosed in 23.1% of the patients; in 1.6% of them the disease was severe. According to the definition by Offenbacher et al., 41.2% of the examined patients were diagnosed with periodontitis (0.7% of p1, 22.7% of p2, and 17.8% of p3). In the studies conducted in 2017 [6], the incidence of periodontitis according to DCD/AAP among the residents of Wrocław and Oława aged 65–74 was 30.7% in the cpi4 group, i.e., 47.9, out of which the disease was severe in 21.9 cases [6].

The scaling and root planing procedure (SRP) [7] is considered to be the gold standard in nonsurgical treatment of periodontitis. However, subgingival instrumentation does not eliminate all pathogens from the pocket, therefore reinfections and relapses are very common, especially in the case of deep pockets. One of the promising alternative procedural protocols is a combination of classical subgingival instrumentation (SRP) and intrapocket antibiotic administration in a suitable carrier, which makes it possible to obtain long-term minimum concentration of the drug inhibiting periopathogen development.

Such procedure is known as local drug delivery (LDD). There are many limitations to general antibiotic therapy on periodontal grounds, and it is accompanied by many adverse effects. For this reason, in 1979 Goodson postulated local application of tetracycline [8] administrated into the pockets in the form of vinyl acetate copolymer fibers (Actisite). Local application reduces or eliminates some adverse effects associated with general antibiotic therapy, especially bacterial resistance and allergic reactions. Moreover, concentrations of the active substance in LDD in the area of administration considerably exceed minimal inhibitory concentration (MIC) and enable administration to the particular area that requires treatment [9]. Since 1979, many preparations that release antibiotic active substances and can be applied into the pockets have been introduced to the market and to clinical use. They differ in terms of the carrier, the active substance and the form of administration. Currently, preparations with tetracycline (2 mg), minocycline in 2% concentration and doxycycline in 10% concentration have been in use. These preparations consist of one or two components and can be easily administered to the periodontal pockets.

One of the new forms of LDD for intrapocket application that has been recently introduced is Gelcide. It is a patented mixture that consists of piperacillin and tazobactam combined by a carrier. The preparation is packaged into two vials. After mixing the contents of the vials, the preparation gelifies, and forms a coating that seals the periodontal pocket. Thanks to these properties, the active substances of the product should be retained in the area to which the preparation was administered. The components of Gelcide are slowly released in concentrations exceeding MIC over the period of 8 to 10 days. The coating is permeable, but it is insoluble to fluids—it protects the pocket from exposure to bacteria of the oral cavity and from further irritation and infection.

## 2. Aim

The aim of the paper was to assess the effectiveness of the activity of piperacillin and tazobactam combination in the form of an intrapocket-administered chemotherapy. Its influence on the course of treatment, microbiome of the pocket, healing rate, adverse effects or lack of thereof, is in comparison with the gold standard of periodontitis treatment—the SRP procedure.

## 3. Materials and Methods

Ten patients aged 24 to 56 (the average age was 39.5, 5 males, 5 females) with general periodontitis (loss of CAL on interproximal surfaces, CAL in >30% of teeth, with moderately symmetric bone loss in pantomographic image) were qualified for the study. The patients were healthy nonsmokers who had not used antibiotics within the period of 6 months prior to the study, were not allergic to penicillin, and had at least 10 teeth in the jaw. The excluded criteria were also diabetes mellitus and hypertension.

The second stage of periodontitis was diagnosed in 1 case, the third stage in 6 cases, and the fourth in 3 cases. Degree distribution was 7 from B and 3 from C. All these individuals provided written consent for the proposed treatment protocol. The study was granted approval KB-600/2019 by the Bioethical Committee of Wrocław Medical University.

The procedure would begin with providing instructions on oral hygiene (the obtained average API value was below 20%). Subsequently, machine SRP (ultrasonic scaling with subgingival tips) under nerve block anesthesia in the mandible was performed during two visits. Examination of clinical parameters was conducted by one standardized periodontist (TK). The examinations determined parameters such as bleeding on probing (BoP), clinical attachment location (CAL), and periodontal depth (PD) in six measurement points around each tooth, except for the third molars. Then, microbiological material was collected from two symmetric pockets with similar PD, located on both sides of the dental arch; sterile filter paper was placed in the pockets for 30 s, and the samples were then transferred to transport test-tubes. During the same visit, the periodontist (ASJ) conducted analogous SRP under anesthesia in the jaw. At the end of the visit, a third periodontist (JZ) introduced Gelcide into the periodontal pockets, in accordance with the instructions of the manufacturer, applying randomization to odd-numbered patients in the order of their applications on the left, and even-numbered patients on the right. The randomization was known to only one person (JZ). The patients were informed about the possibility of attending an additional follow-up visit, if any adverse effects associated with the conducted treatment were noticed. If no such effects occurred, the follow-up visits took place one, four, and eight weeks after the initial examination. Clinical examination was conducted again at the last follow-up visit, and the material for microbiological tests (TK) was collected from the same pockets.

During microbiological testing, the PET deluxe test was applied. The test uses the PCR method in real time that makes it possible to quantify the total bacterial count (TBC) in the pocket as well as the red-complex bacteria (*Porphyromonas gingivalis*—P.g., *Treponema denticola*—T.d., and oraz *Tannerella forsythia*—T.f.), *Aggregatibacter actinomycetemcomitans*—A.a.), orange-complex bacteria (*Fusobacterium nucleatum*—F.n., *Prevotella intermedia*—P.i., and *Micromonas micros*—P.m., *Eubacterium nodatum-E.n.*), and green-complex bacteria (*Capnocytophaga gingivalis*—C.g.).

Due to lack of normality of distribution of the analyzed variables, nonparametric tests were used in statistical analysis: the Mann–Whitney U test for nonrelated variables and the Wilcoxon T-test for related variables and for groups below 25 participants. $p \leq 0.05$ was adopted as the significance level. The Statistica package, version 13.1, was used.

## 4. Results

In the vast majority of the examined patients, no symptoms of adverse effects related to the treatment were observed during the scheduled follow-up visits. The protective layer which, according to the manufacturer, was supposed to protect the patients from rapid decrease in drug concentration in the pocket within 8–10 days was unnoticeable. At the last follow-up visit, purulent effusion in the area of drug application was found in one of the patients; inflammatory erythema and exfoliative gingivitis were found in another patient, also in the area of drug application.

Table 1 presents a summary of clinical examination results for the jaw and, separately, for the pockets from which microbiological material was collected and into which the assessed gel was administered, as well as the pockets which were treated using solely SRP (the control group). In the jaw, the treatment resulted in significant decrease in the BoP index (−33%), shallowing of the average PD (−0.45 mm) and PD on the interproximal surfaces (−0.55 mm), as well as decrease in the number of areas with CAL loss ≥5 mm (improvement of the median by 11.5). In the pockets treated using the compared methods, the researchers found similar, considerable reduction in BoP (−49.2% vs. 47.5% for SRP) as well as shallowing of the average PD (0.82 vs. 0.83 mm for SRP) and, additionally, considerable improvement of attachment location in the control pockets (−0.81 mm). Two months after the treatment, no differences between the pockets treated with the use of the evaluated methods were found in terms of the assessed clinical parameters.

**Table 1.** Results of the clinical studies.

| Place of the Test | BoP | PD śr | PD | Mediana PD > 5 mm | Mediana CAL1-2 | Mediana CAL3-4 | Mediana CAL ≥ 5 |
|---|---|---|---|---|---|---|---|
| Upper Jaw | 63.3 ± 22.3<br>30.3 ± 12.1<br>$p < 0.000$ | 4.23 ± 0.92<br>3.78 ± 0.97<br>$p = 0.049$ | 4.86 ± 1.15<br>4.31 ± 1.12<br>$p = 0.047$ | 10<br>5.5<br>$p = 0.59$ | 7.5<br>11<br>$p = 0.17$ | 13<br>7.5<br>$p = 0.44$ | 26<br>14.5<br>$p = 0.012$ |
| Pockets SRP+A | 87.5 ± 20.9<br>38.3 ± 12.5<br>$p < 0.000$ | 5.0 ± 1.04<br>4.18 ± 0.92<br>$p = 0.008$ | CAL śr<br>3.96 ± 1.36<br>3.1 ± 2.02<br>$p = 0.053$ | | | | |
| Pockets SRP | 85.8 ± 14.2<br>38.3 ± 23.9<br>$p < 0.000$ | 4.83 ± 1.18<br>4.0 ± 0.99<br>$p = 0.005$ | 3.41 ± 1.63<br>2.6 ± 1.34<br>$p = 0.037$ | | | | |

Lack of significant differences in the final bleeding on probing (BoP) values ($p = 0.73$), average PD (0.53), and average clinical attachment loss (CAL) (0.63) between the pockets treated with SRP+A and SRP alone.

The results of microbiological observations are presented in Table 2. In general, no statistically significant differences in the assessment of the related variables (the same pockets before and after treatment), and the nonrelated variables (symmetric pockets before and after treatment using both methods) were found. The only exceptions concerned the pockets additionally treated with antibiotic gel. Two months post-treatment, a significant increase of *Treponema denticola* and decrease of *Micromonas micros* as well as a significantly higher amount of *Capnocytophaga gingivalis* in relation to the control pockets were found. For the latter, reduction in the amount of *Fusobacterium nucletaum* and *Micromonas micros* was considerably similar to the significant decrease after SRP.

**Table 2.** Results of microbiological tests.

| Type of Bacteria | SRP+A Treatment | SRP Treatment | P for Rows |
|---|---|---|---|
| TBC before after | 455,603,800 ± 774,832,582<br>225,700,000 ± 160,305,299<br>T26, $p = 0.88$ | 138,810,000 ± 184,812,292<br>400,090,000 ± 710,668,522<br>T16, $p = 0.24$ | U = 50 ns<br>U = 50 ns |
| P.g. before after | 55,050 ± 105,157<br>94,000 ± 239,040<br>T8, $p = 0.6$ | 54,328 ± 146,867<br>182,514 ± 365,513<br>T5, $p = 0.25$ | U = 43 ns<br>U = 47 ns |
| T.d. before after | 4802 ± 5793<br>36,188 ± 79,754<br>T3, $p = 0.036$ | 6456 ± 11,483<br>35,460 ± 50,421<br>T9, $p = 0.11$ | U = 42 ns<br>U = 48.5 ns |
| T.f. before after | 11,372 ± 23,587<br>7996 ± 10,964<br>T11, $p = 0.33$ | 40,227 ± 106,011<br>16,657 ± 29,545<br>T21, $p = 0.86$ | U = 49.5 ns<br>U = 47 ns |
| A.a before after | Liczba wykryć-3<br>Liczba wykryć-2 | Liczba wykryć-2<br>Liczba wykryć-1 | |
| F.n. before after | 6800 ± 14,499<br>466 ± 615<br>T10, $p = 0.26$ | 8257 ± 18,822<br>716 ± 1320<br>T6, $p = 0.051$ | U = 35 ns<br>U = 43 ns |
| P.i. before after | 43,404 ± 106,163<br>199,610 ± 497,956<br>T8, $p = 0.6$ | 61,153 ± 144,493<br>23,765 ± 69,086<br>T5, $p = 0.25$ | U = 46 ns<br>U = 40 ns |
| P.m. before after | 2877 ± 2644<br>631 ± 767<br>T3, $p = 0.036$ | 5914 ± 11,514<br>589 ± 616<br>T6, $p = 0.051$ | U = 49.5 ns<br>U = 48.5 ns |
| E.n. before after | Liczba wykryć-1<br>Liczba wykryć-1 | Liczba wykryć-1<br>Liczba wykryć-2 | |
| C.g. before after | 14,636 ± 37,211<br>5439 ± 4962<br>T25, $p = 0.8$ | 5878 ± 6165<br>2050 ± 1796<br>T8, $p = 0.086$ | U = 41.5 ns<br>U = 22.5 |

The influence of the current periodontitis classification on the assessed amounts of bacteria is presented in Table 3. A small number of analyzed samples and lack of detection of particular pathogens in the pockets influenced considerable values of standard deviations and the significance of the intergroup differences. Quantitative differentiation of the studied compounds of the pocket microbiome was influenced to a larger extent by the degrees of periodontitis before the treatment (considerably higher number of TBC in degree B, as well as higher—though marginally—numbers of *Porphyromonas gingivalis*, *Treponema denticola*, *Prewvotella intermedia*, and *Capnocytophaga* in relation to C).

**Table 3.** Results of microbiological tests depending on the stadium and the degree of periodontitis.

| Type of Bacteria | Stage II and III vs. IV | p | Grade B vs. C | p |
|---|---|---|---|---|
| TBC before after | 404,295,571 ± 660,136,448 vs. 473,333,333 ± 493,301,180<br>386,071,429 ± 595,735,933 vs. 142,150,000 ± 114,175,369 | 0.23<br>0.16 | **410,078,571 ± 656,851,542 vs. 33,839,667 ± 42,535,311**<br>369,428,571 ± 602,988,291 vs. 180,983,333 ± 102,863,996 | **0.048**<br>0.74 |
| P.g. before after | 64,141 ± 145,840 vs. 32,633 ± 50,957<br>73,439 ± 230,011 vs. 289,500 ± 417,614 | 0.59<br>0.59 | 68,641 ± 144,276 vs. 22,133 ± 52,856<br>79,653 ± 229,047 vs. 275,000 ± 427,446 | 0.3<br>0.71 |
| T.d. before after | 5731 ± 9219 vs. 5392 ± 8920<br>40,956 ± 74,652 vs. 23,850 ± 35,928 | 0.9<br>0.48 | 7438 ± 10,099 vs. 1408 ± 2223<br>41,263 ± 74,510 vs. 23,133 ± 36,272 | 0.3<br>0.54 |
| T.f. before after | 32,595 ± 90,773 vs. 9943 ± 14,466<br>11,238 ± 21,684 vs. 14,867 ± 25,083 | 0.62<br>0.77 | 34,431 ± 90,504 vs. 5660 ± 8455<br>11,224 ± 21,730 vs.15,133 ± 24,856 | 0.71<br>0.97 |
| A.a before after | Liczba wykryć-3 vs. 2<br>Liczba wykryć-2 vs. 1 | | Liczba wykryć-3 vs. 2<br>Liczba wykryć-2 vs. 1 | |
| F.n. before after | 9221 ± 19,245 vs. 3578 ± 5346<br>565 ± 856 vs. 652 ± 1405 | 0.59<br>0.65 | 9264 ± 19,221 vs. 3478 ± 5442<br>457 ± 873 vs. 903 ± 1318 | 0.97<br>0.26 |
| P.i. before after | 40,398 ± 92,291 vs. 80,000 ± 186,333<br>148,125 ± 424,800 vs. 26,667 ± 65,320 | 0.53<br>0.2 | 66,669 ± 14,447 vs. 18,700 ± 44,736<br>158,075 ± 423,223 vs. 3450 ± 7209 | 0.34<br>0.33 |
| P.m. before after | 2948 ± 2751 vs. 7773 ± 14,874<br>695 ± 676 vs. 412 ± 702 | 0.77<br>0.2 | 5547 ± 9698 vs. 1708 ± 1891<br>544 ± 564 vs. 765 ± 937 | 0.36<br>0.77 |
| E.n. before after | Liczba wykryć-2 vs. 0<br>Liczba wykryć-3 vs. 0 | | Liczba wykryć-2 vs. 0<br>Liczba wykryć-3 vs. 0 | |
| C.g. before after | 12,261 ± 31,319 vs. 5580 ± 7049<br>3264 ± 2426 vs. 4867 ± 6646 | 0.45<br>0.87 | 13,679 ± 31,124 vs. 2272 ± 3107<br>3336 ± 2366 vs. 4698 ± 6746 | 0.09<br>0.65 |

## 5. Discussion

Currently, only one study assessing clinical and microbiological effectiveness of locally applied piperacillin with tazobactam in patients with periodontitis is available.

Lack of placebo application may constitute a limitation of the study; both the person who collected the material and the laboratory staff were subjected to a blind test.

In our own studies on clinical parameters, such as bleeding on probing (BoP) index, periodontal pocket depth (PPD), and location of the connective tissue attachment (CAL), the following results were obtained after 2 months: BoP was reduced (−49.2% vs. 47.5% for SRP), the average PPD was shallowed (0.82 vs. 0.83 mm for SRP), and additionally, attachment location in the control pockets was improved (−0.81 mm). On the other hand, after 2 months of treatment, no differences in the assessed clinical parameters between the two groups were observed. In the studies conducted by Lauenstein et al. [9,10], in which piperacillin with tazobactam were applied locally, similar reduction in the values of clinical parameters was obtained. The difference in PPD measurements between the initial examination and the examination conducted in the control group after 26 weeks was 1.8 mm (SE ± 0.3; 95% CI 1.2, 2.3; $p < 0.001$). The average difference in PPD between the initial value and the 26th week in the study group was 1.5 mm (SE ± 0.2; 95% CI 1.1, 2.0; $p < 0.001$), and the statistical analysis did not show any differences in the study groups in terms of PPD values both at the beginning of the study and in its 26th week. Moreover, in the case of the parameter assessing inflammation (BoP) no statistically significant differences between the groups were proven, neither at the beginning of the study nor during the final examination after 26 weeks. The fact that the study included smokers may constitute a disruptive factor [10]. The obtained reduction in the clinical parameters confirms the necessity of conducting nonsurgical treatment, especially in terms of reducing periodontal pocket depth and the location of connective tissue attachment, and confirms the gold-standard status of the scaling and root planning procedure in periodontal treatment [11–13]. Mechanical cleaning is necessary in order to improve clinical parameters, and the applied preparations can only support and consolidate the treatment results [14,15].

With regard to the microbiological tests, there are currently no reports confirming effective reduction of periopathogens in periodontal pockets after local application of piperacillin in combination with tazobactam in patients with moderate and advanced periodontitis. No statistically significant differences between the studied groups were found in our own studies. After local application of Gelcide in combination with SRP, a significant increase in *Treponema denticola* and decrease in *Micromonas micros* as well as a considerably higher number of *Capnocytophaga gingivalis* were found in the microbiological test in relation to the control pockets. After application of SRP alone, a decrease in the number of *Fusobacterium nucletaum* and *Micromonas micros* was observed. In the studies conducted by Lauenstein et al. [10], reduction in pathogens (including *Porphyromonas gingivalis, Tanerella forsythia,* or *Agregatibacter actinomycetemcomitans*) after local application of piperacillin with tazobactam, in comparison with application of only mechanical cleaning, was obtained in both groups after 26 weeks. On the other hand, significant reduction of *Fusobacterium nucleatum, P. micra,* and *T. denticola* was obtained in the group in which the treatment was combined with local antibiotic therapy.

Due to bacteria organization within the biofilm present in the periodontal pockets and the presence of periopathogenic bacteria in the tissues of the host, eradication of periopathogens and long-term improvement of clinical and in particular, microbiological parameters, is virtually impossible. In the case of local application of doxycycline, the results of clinical and microbiological tests are inconclusive. Reports on considerable improvement of clinical parameters and reduction in periopathogens after SPR combined with intrapocket administration of antibiotic [16,17] are available. Long-term observations, however, do not indicate considerable benefits [18]. At the same time, some scientists do not prove significant differences in results with regard to the independent use of SRP in periodontal disease treatment [19]. Difficulties in obtaining complete reduction of periopathogens, regardless of the type of therapy, were also presented in the studies by Mobelli et al. [20] after local application of tetracycline fibers [21]. The study included seventeen patients, with whom microbiological analysis of subgingival samples collected mesially–distally from 852 areas at the beginning and after a month of using the preparation was conducted. In the basic study, 46 samples, from 10 positive individuals, showed positive results for *P. gingivalis*; 82 samples, from 5 individuals, were also identified as positive in terms of *A. actinomycetemcomitans*. The presence of *A. actinomycetemcomitans* and *P. gingivalis* was not confirmed in the material collected from 3 patients. Microbiological tests conducted one month after periodontal treatment showed that 89% of areas which initially showed positive results in terms of pathogens were negative, while 16 areas which were initially identified as negative showed positive results. With regard to *A. actinomycetemcomitans*, 77% of areas which were identified as positive in terms of bacteria presence in the first test were later identified as negative, but 5 areas which were initially identified as negative showed positive results. During the examinations of patients whose results of the microbiological test remained positive, another attempt at applying local treatment using tetracycline fibers was made. Despite another attempt at treatment, microbiological tests still showed the presence of *P. gingivalis* in 5 individuals; and in 4 patients, the results for *A. actinomycetemcomitans* remained positive. Those 9 patients were finally subjected to systemic antibiotic treatment (3 × 250 mg metronidazole and 3 × 375 mg amoxicillin/per day for 7 days). Despite all the efforts, after 3 months *P. gingivalis* was again detected in 3 individuals and *A. actinomycetemcomitans* was isolated from 1 area.

Organization of periopathogens within the biofilm, their diversity, and ability to quickly recolonize bacteria still constitutes a great therapeutic challenge. Nonsurgical methods (SRP) are still the basis for treatment of periodontium diseases, and local or general application of antibiotic therapy is still ineffective in terms of periopathogen elimination [22]. In summary, this study showed similar improvement of clinical parameters in patients treated with SRP combined with single administration of local antibiotic (piperacillin/tazobactam) or without it, without showing statistically significant differences in the microbiological tests results. Piperacilin in combination with tazobactam is used primarily for the general treatment of nosocomial infections. Clinical data indicate that the combination of piperacillin/tazobactam is effective in the treatment of moderate to severe polymicrobial infections, including intra-abdominal, skin, and soft-tissue and lower respiratory tract infections. It shows high efficiency in the case of infections caused by *Escherichia coli,* many *Bacteroides* and *Klebsiella species, Staphylococcus aureus,* and *Haemophilus influenzae*, with low resistance after the treatment. The combination of piperacillin and tazobactam in dentistry has been not well documented. There are no medical studies that investigate the effectiveness against pathogens that cause periodontitis. Despite the broad spectrum of gram-negative bacteria, there is no evidence of the effects on periopathogens. It should be emphasized that, in the case of topical application in periodontal pockets, there is currently no available literature. It is required to conduct clinical trials confirming the spectrum of the drug's action on periopathogens, together with correlation of the clinical results. In vitro studies should also be considered to systematize and confirm the efficacy of piperacillin/tazobactam on specific periopathogens.

Further clinical studies on larger groups of patients and long-term observations of the obtained results or changes occurring in them in order to optimize the algorithms of nonsurgical treatment of periodontal diseases and minimize application of general antibiotic therapy is required.

**Author Contributions:** Author A.S.-J. caried SRP, author J.Z. carried drug administration and article review and editing, author M.S. carried study design and draft preparation, author T.K. carried examination and project supervision.

**Funding:** This research received no external funding

**Conflicts of Interest:** The authors declare no conflict of interest.

## References

1. Papapanou, P.N.; Sanz, M.; Buduneli, N.; Dietrich, T.; Feres, M.; Fine, D.H.; Flemmig, T.F.; Garcia, R.; Giannobile, W.V.; Graziani, F.; et al. Periodontitis: Consensus report of workgroup 2 of the 2017 World Workshop on the Classification of Periodontal and Peri-Implant Diseases and Conditions. *J. Clin. Periodontol.* **2018**, *89*, S173–S182. [CrossRef] [PubMed]
2. Shaddox, L.M.; Mullersman, A.F.; Huang, H.; Wallet, S.M.; Langaee, T.; Aukhil, I. Epigenetic regulation of inflammation in localized aggressive periodontitis. *Clin Epigenetics* **2017**, *9*, 94. [CrossRef] [PubMed]
3. Adamczyk, M.; Orzechowska, A.; Napora, M.; Wyszyńska, M.; Górska, R. Częstość występowania chorób przyzębia u mieszkańców Warszawy w wieku 35–44 lat na podstawie wskaźnika CPITN Borgis. *Nowa Stomatol.* **2013**, *2*, 93–97.
4. Górska, R.; Pietruska, M.; Dembowska, E.; Wysokińska-Miszczuk, M.; Włosowicz, M.; Konopka, T. Częstość występowania chorób przyzębia u osób w wieku 35–44 lat w populacji dużych aglomeracji miejskich. *Przegl. Epidemiol.* **2015**, *69*, 537–542, 643–647.
5. Konopka, T.; Dembowska, E.; Pietruska, M.; Dymalski, P.; Górska, R. Periodontal status and selected parameters of oral condition of Poles aged 65 to 74 years. *Przegl. Epidemiol.* **2015**, *69*, 537–542.
6. Głowacka, B.; Konopka, T. Needs for gerodontological treatment in the elderly living in Lower Silesia. *Dent. Med. Probl.* **2019**, *56*, 89–96. [CrossRef]
7. Pei Shian, T.; Pulikkotil, S. Comparison and agreement among variouscase definitions of periodontitis: A secondary data analysis. *J. Periodontol. Implant Dent.* **2014**, *6*, 40–46. [CrossRef]
8. Goodson, J.M.; Haffaje, A.; Socransky, S. Periodontal therapy by local delivery of tetracycline. *J. Clin. Periodontol.* **1979**, *6*, 83–92. [CrossRef]
9. Kinane, D.F.; Stathopoulou, P.G.; Papapanou, P.N. Periodontal diseases. *Nat. Rev. Dis. Primers* **2017**, *3*, 17038. [CrossRef]

10. Lauenstein, M.; Kaufmann, M.; Persson, G.R. Clinical and microbiological results following nonsurgical periodontal therapy with or without local administration of piperacillin/tazobactam. *Clin. Oral Investig.* **2013**, *17*, 1645–1660. [CrossRef]
11. Jervøe-Storm, P.M.; Al Ahdab, H.; Semaan, E.; Fimmers, R.; Jepsen, S. Microbiological outcomes of quadrant versus full-mouth root planning as monitored by real-time PCR. *J. Clin. Periodontol.* **2007**, *34*, 156–163. [CrossRef] [PubMed]
12. Knöfler, G.U.; Purschwitz, R.E.; Jentsch, H.F. Clinical evaluation of partial- and full-mouth scaling in the treatment of chronic periodontitis. *J. Periodontol.* **2007**, *78*, 2135–2142. [CrossRef]
13. Roman-Torres, C.V.G.; Bryington, M.S.; Kussaba, S.T.; Pimentel, A.C.; Jimbo, R.; Cortelli, J.R.; Romito, G.A. Comparison Of Full-Mouth Scaling and Quadrant-Wise Scaling in the Treatment of Adult Chronic Periodontitis. *Braz. Dent. J.* **2018**, *29*, 296–300. [CrossRef] [PubMed]
14. Al Hulami, H.; Babay, N.; Awartani, F.; Anil, S. The effect of locally delivered doxycycline as an adjunctive therapy to scaling and root planing in smokers. *Saudi Dent. J.* **2011**, *23*, 143–148. [CrossRef] [PubMed]
15. Szulc, M.; Zakrzewska, A.; Zborowski, J. Local drug delivery in periodontitis treatment: A review of contemporary literature. *Dent. Med. Probl.* **2018**, *55*, 333–342. [CrossRef]
16. Ahamed, S.; Jalaluddin, M.; Khalid, I.; Moon, N.; Shafi, T.K.; Ali, F.M. The use of controlled release locally delivered 10% doxycycline hyclate gel as an adjunct to scaling and root planing in the treatment of chronic periodontitis: Clinical and microbiological results. *J. Contemp. Dent. Pract.* **2013**, *14*, 1080–1086.
17. Sandhya, Y.P.; Prabhuji, M.L.; Chandra, R.V. Comparative evaluation of the efficacy of 10% doxycycline hyclate in the periodontal treatment of smokers—A clinical and microbiological study. *Oral Health Prev. Dent.* **2011**, *9*, 59–65.
18. Bogren, A.; Teles, R.P.; Torresyap, G.; Haffajee, A.D.; Socransky, S.S.; Wennström, J.L. Locally delivered doxycycline during supportive periodontal therapy: A 3-year study. *J. Periodontol.* **2008**, *79*, 827–835. [CrossRef]
19. Tsalikis, L.; Sakellari, D.; Dagalis, P.; Boura, P.; Konstantinidis, A. Effects of doxycycline on clinical, microbiological and immunological parameters in well-controlled diabetes type-2 patients with periodontal disease: A randomized, controlled clinical trial. *J. Clin. Periodontol.* **2014**, *41*, 972–980. [CrossRef]
20. Mombelli, A.; Schmid, B.; Rutar, A.; Lang, N.P. Local antibiotic therapy guided by microbiological diagnosis. *J. Clin. Periodontol.* **2002**, *29*, 743–749. [CrossRef]
21. Mombelli, A. Microbial colonization of the periodontal pocket and its significance for periodontal therapy. *Periodontology 2000* **2018**, *76*, 85–96. [CrossRef] [PubMed]
22. John, M.T.; Michalowicz, B.S.; Kotsakis, G.A.; Chu, H. Network meta-analysis of studies included in the Clinical Practice Guideline on the nonsurgical treatment of chronic periodontitis. *J. Clin. Periodontol.* **2017**, *44*, 603–611. [CrossRef] [PubMed]

MDPI
St. Alban-Anlage 66
4052 Basel
Switzerland
Tel. +41 61 683 77 34
Fax +41 61 302 89 18
www.mdpi.com

*Applied Sciences* Editorial Office
E-mail: applsci@mdpi.com
www.mdpi.com/journal/applsci

www.ingramcontent.com/pod-product-compliance
Lightning Source LLC
Chambersburg PA
CBHW061446210926
51850CB00008B/1631
* 9 7 8 3 0 3 6 5 1 2 9 9 0 *